The Transition from Lower to Middle Palaeolithic and the Origin of Modern Man

International Symposium to commemorate the 50th anniversary of excavations in the Mount Carmel Caves by D. A. E. Garrod

University of Haifa, 6-14 October 1980

Edited by

Avraham Ronen

Applied Scientific Research Co.,
University of Haifa Ltd.

BAR International Series 151
1982

B.A.R.

B.A.R., 122 Banbury Road, Oxford OX2 7BP, England

B.A.R.-S151, 1982: 'The Transition from Lower to Middle Palaeolithic and the Origin of Modern Man'

ISBN 9780860541905 paperback
ISBN 9781407329741 e-book
DOI https://doi.org/10.30861/9780860541905
A catalogue record for this book is available from the British Library
This book is available at www.barpublishing.com

TO THE MEMORY OF

C.B.M. McBURNEY

HONORARY CHAIRMAN

Prof. BENJAMIN MAZAR

ORGANIZING COMMITTEE

Prof. OFER BAR-YOSEF
Prof. GERHARD BOSINSKI
Mr. ABRAHAM EITAN
Prof. ASOK K. GHOSH
Prof. ARTHUR J. JELINEK
Prof. AVRAHAM RONEN, Coordinator

CONTENTS

PARTICIPANTS

Prof. O. Bar-Yosef	Institute of Archaeology, The Hebrew University, Jerusalem, Israel.
Prof. Dr. G. Bosinski	Institut für Ur- und Frühgeschichte, Universität zu Köln, 5000 Köln 1, West Germany.
Prof. Dr. K. Brunnacker	Geologisches Institut, Universität zu Köln, 5000 Köln 1, West Germany.
Prof. J.D. Clark	Department of Anthropology, University of California, Berkeley, California 94720, U.S.A.
Prof. W.R. Farrand	Department of Geological Sciences, The University of Michigan, Ann Arbor, Michigan 48109, U.S.A.
Prof. Dr. G. Freund	Institut für Ur- und Frühgeschichte, Universität Erlangen – Nürnberg, 8520 Erlangen, West Germany.
Prof. A.K. Ghosh	Department of Anthropology, University of Calcutta, Calcutta 700019, India.
Dr. N. Goren	Institute of Archaeology, The Hebrew University, Jerusalem, Israel.
Prof. A.J. Jelinek	Department of Anthropology, University of Arizona, Tucson, Arizona 87521, U.S.A.
Mr. M. Lamdan	Curator, The M. Stekelis Museum of Prehistory, Haifa, Israel.
Prof. H. Laville	Institut du Quaternaire, Université de Bordeaux I, 33405 Talence, France.
Prof. P.M. Masters	Scripps Institution of Oceanography, La Jolla, California 92093, U.S.A.
Prof. B. Mazar	Institute of Archaeology, The Hebrew University, Jerusalem, Israel.
Dr. K. Paddayya	Department of Archaeology, Deccan College, Poona 6, India.
Dr. M. Piperno	Instituto Italiano di Paleontologia Umana, Rome, Italy.
Dr. R. Ray	Department of Anthropology, University of Calcutta, Calcutta 700019, India.
Dr. L. Reisch	Institut für Ur- und Frühgeschichte, Universität Erlangen – Nürnberg, 8520 Erlangen, West Germany.
Prof. D.A. Roe	Donald Baden-Powell Quaternary Research Centre, University of Oxford, Oxford, England.
Prof. A. Ronen	Department of Eretz-Israel Studies, University of Haifa, Israel.

Dr. A.G. Segre	Instituto Italiano di Paleontologia Umana, Rome, Italy.
Prof. E. Trinkaus	Department of Anthropology, Peabody Museum, Harvard University, Cambridge, Mass. 02138, U.S.A.
Dr. A. Tuffreau	Musée des Antiquités Nationales, 78103 Saint-Germain-En-Laye, France.
Prof. K. Valoch	Institut Anthropos, Moravske Museum, 65937 Brno, Czechoslovakia.
Prof. B. Vandermeersch	Laboratoire de Paléontologie des Vertébrés et de Paléontologie Humaine, Université de Paris VI, 75230 Paris, France.
Mr. E.E. Wreschner	Department of Anthropology and Sociology, University of Haifa, Israel.

AUDITORS

Dr. B. Arensburg	Department of Anatomy and Anthropology, The Sackler Medical School, Tel Aviv University, Tel Aviv, Israel.
Ms. A. Belfer-Cohen	Institute of Archaeology, The Hebrew University, Jerusalem, Israel.
Mr. B. Brimer	Assistant Curator of Prehistoric Period, the Israel Museum, Jerusalem, Israel.
Mr. M. Davies	Director, The M. Stekelis Museum of Prehistory, Haifa, Israel.
Mr. A. Eitan	Director, Department of Antiquities and Museums, Ministry of Education and Culture, Jerusalem, Israel.
Mr. I. Gilead	Department of Bible and Near Eastern Studies, Ben Gurion University, Beer Sheva, Israel.
Dr. P. Goldberg	Institute of Archaeology, The Hebrew University, Jerusalem, Israel.
Prof. M. Goldstein	Department of Anatomy and Anthropology, The Sackler Medical School, Tel Aviv University, Tel Aviv, Israel.
Dr. G. Gvirtzman	Geological Survey of Israel, Jerusalem, Israel.
Dr. M. Inbar	Department of Geography, The University of Haifa, Haifa, Israel.
Prof. D.L. Inman	Scripps Institution of Oceanography, La Jolla, California 92093, U.S.A.
Dr. M. Klein	Department of Geography, the University of Haifa, Haifa, Israel.
Prof. D. Nir	Department of Geography, The Hebrew University, Jerusalem, Israel.
Dr. T. Noy	Curator of Prehistoric Period, The Israel Museum, Jerusalem, Israel.

Ing. Y. Olami	Archaeological Survey of Israel, Jerusalem, Israel.
Prof. L. Picard	Department of Geology, The Hebrew University, Jerusalem, Israel.
Dr. M. Prausnitz	Department of Antiquities and Museums, Ministry of Education and Culture, Jerusalem, Israel.
Dr. Y. Rak	Department of Anatomy and Anthropology, The Sackler Medical School, Tel Aviv University, Tel Aviv, Israel.
Prof. V. Rohrlich	Department of Geotechnology and Mineral Engineering, Technion – Israel Institute of Technology, Haifa, Israel.
Ms. A.M. Tillier	Laboratoire de Paléontologie des Vertébrés et de Paléontologie Humaine, Université Paris VI, 75230 Paris, France.
Ms. M. Weinstein	Department of Eretz-Israel Studies, Laboratory of Prehistory, The University of Haifa, Haifa, Israel.
Prof. D. Yaalon	Department of Geology, The Hebrew University, Jerusalem, Israel.
Mr. J. Zias	Curator of Prehistoric Period, Dept. of Antiquities and Museums, Ministry of Education and Culture, Jerusalem, Israel.

SYMPOSIUM PROGRAM

Sunday, Oct. 5

Arrival

Monday, Oct. 6

Morning		**THE THEME**
	Chairman:	Prof. A.J. Jelinek
	Lectures:	A. Ronen O. Bar-Yosef
Afternoon		**THE NEAR EAST**
	Chairman:	Prof. O. Bar-Yosef
	Lectures:	A.J. Jelinek P.M. Masters W.R. Farrand
Evening		**FESTIVE OPENING SESSION**
	Chairman:	Mr. A. Eitan
	Greetings:	Prof. A. Ronen Mr. G. Avner, President of the University of Haifa Prof. O. Bar-Yosef
	Lecture:	Prof. B. Mazar, Honorary Chairman "Carmel: Holy Mountain in Ancient Times"

Tuesday, Oct 7

MOUNT CARMEL EXCURSION

Sefunim, Oren, Tabun, Skhul, El-Wad, Kebara;
Coastal plain stratigraphy and Middle Palaeolithic red loam;
Rock engravings.

Wednesday, Oct. 8

Morning		**EUROPE**
	Chairman:	Prof. W.R. Farrand
	Lectures:	G. Bosinski G. Freund
Afternoon		**EUROPE**
	Chairman:	Prof. B. Vandermeersch
	Lectures:	D. Roe A. Tuffreau H. Laville
Evening	Visit to the M. Stekelis Museum of Prehistory.	

Thursday, Oct. 9

Morning

S.E. EUROPE AND ASIA

Chairman: Prof. G. Bosinski

Lectures: L. Reisch
R. Ray

RECENT DEVELOPMENT IN ISRAEL

Chairman: Prof. O. Bar-Yosef

Lectures: M. Lamdan
N. Goren
E. Wreschner

Afternoon Visit to the DAGON Museum.
Sightseeing tour of Haifa.

Friday, Oct. 10

Morning

THE ORIGIN OF MODERN MAN

Chairman: Dr. B. Arensburg

Lectures: B. Vandermeersch
E. Trinkaus

CONCLUDING DISCUSSION

Chairman: Prof. A. Ronen

Participants: A.J. Jelinek
G. Bosinski
D. Roe

Afternoon Visit to Akko

Evening Reception

EXCURSIONS

Saturday, Oct. 11

GALILEE

Caves of Hayonim and Qafzeh; Acheulian open-sites of Baram and Yiron; Natufian site Einan; Hula Museum of Prehistory at Kibbutz Mayan Baruch.

Sunday, Oct. 12

JORDAN VALLEY

Nahal Amud Caves; Tiberias; Ubeidiya; Jericho and the Dead Sea.

Monday, Oct. 13

NEGEV

Massada; Mousterian site of Rosh Ein-Mor; Upper Palaeolithic of Boker area; Rock engravings at Ramat Matred.

Tuesday, Oct. 14

JERUSALEM

The Old and New City; Excavations near the Wailing Wall; the Hebrew University campus; the Israel Museum.

Wednesday, Oct. 15

Departure

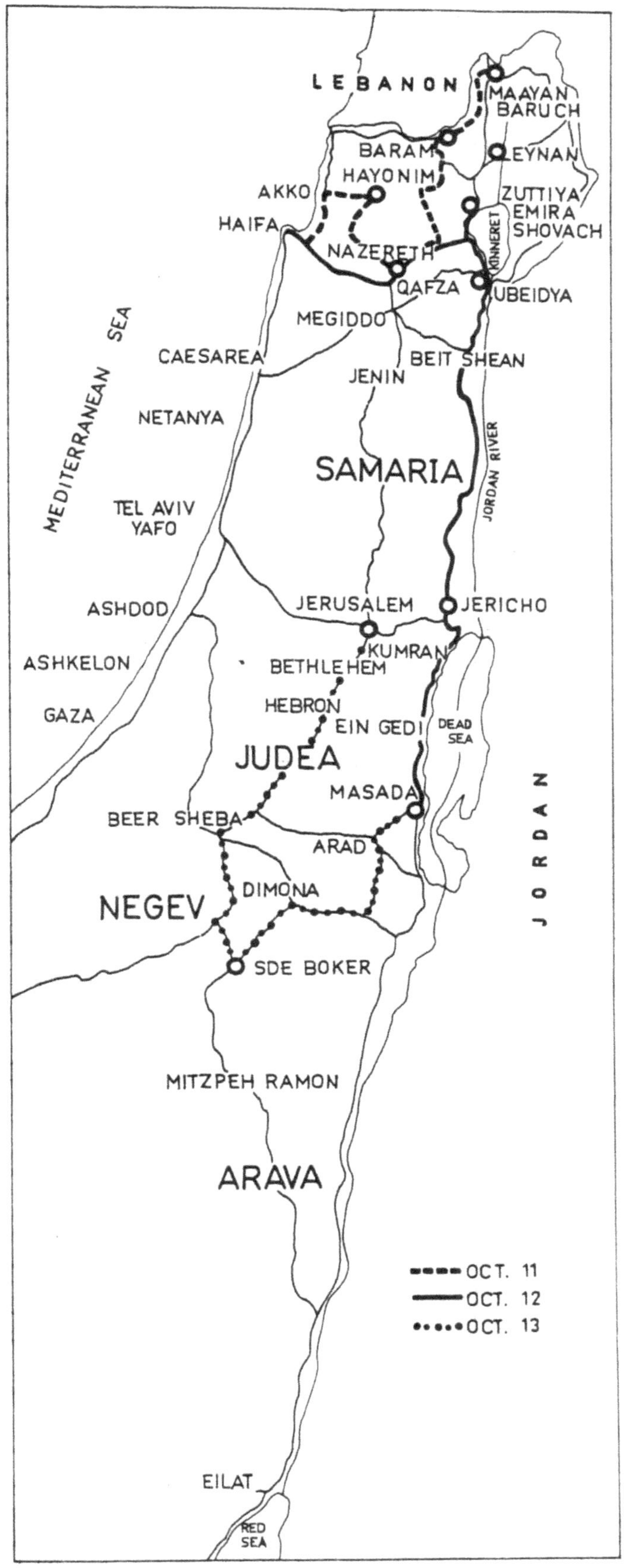

Excursions to sites in Galilee, the Jordan Valley, the Negev and Jerusalem.

PREFACE

Prehistoric investigations have been actively carried out in Israel since their initiation in 1925, date of the pioneer excavations in the caves of Emira and Zuttiya in the Galilee. Activity in this field of research has accelerated since the State of Israel was established in 1948. One of the highlights of prehistoric investigations in Israel and the entire Near East were the excavations in the caves of Mount Carmel by D.A.E. Garrod. The 50th anniversary of these excavations provided a most appropriate occasion for an international meeting, the first to be held in Israel. One of the caves initially explored by Garrod, Tabun, is the subject of new investigations, first by A.J. Jelinek (1967–1971) and then by the present writer (since 1973).

The three classical caves, Tabun, El-Wad and Skhul, provided a long sequence of cultures ranging from the Lower Palaeolithic to historical periods. Thus, a wide range of topics would have been appropriate for the theme of our symposium. The present theme – the transition from Lower to Middle Palaeolithic – was chosen first because of the link between the old and new excavations at Tabun, which has one of the best sequences of Lower and Middle Palaeolithic cave deposits in the Eastern Mediterranean; second, this theme has been far less discussed than, say, the transition between Middle and Upper Palaeolithic or from Hunting/Gathering to food production. Yet, the onset of Middle Palaeolithic is a turning point in human evolution: it saw the appearance of modern man and the first burials ever; hence, the definitive proof of the knowledge of death and of preoccupation with spiritual concepts. Our symposium had the aim of offering a survey of events which may delineate the transition – compare the latest occurrences of the Lower Palaeolithic to the earliest occurrences of the Middle Palaeolithic. It was felt that a cross-Old World survey rather than a regional approach would best serve our purpose.

The first colleague who was approached to serve on the organizing committee was Prof. C.B.M. McBurney, a student of Dorothy Garrod and a distinguished scholar of Mediterranean prehistory. He immediately responded affirmatively, and during the first months of preparations had been of tremendous help. Aware of his illness, his enthusiasm was mixed with fear that he might not make it to the symposium. Unfortunately, his feelings were correct and he died shortly before our gathering. Knowing how important his contribution would have been to our symposium, this volume is dedicated to the memory of Charles McBurney.

The symposium was held at the University of Haifa under the joint auspices of the University and the Israel Exploration Society. Contributions were prepared and circulated in advance. The discussions during conference days were recorded, typed and then sent to the participants for corrections and adjustments. This volume contains the papers offered in the symposium, each followed by the discussion which ensued. Only a minimal editing took place after the authors' final corrections. A few colleagues sent their contributions but could not attend the meeting: Dr. Ghosh, Dr. Paddayya and Dr. Valoch; two papers were requested after the symposium and were written specially for this volume: by Clark on Africa and Piperno and Segre on Italy.

The organization of the symposium and excursions as well as the preparation of this volume required help from numerous colleagues and collaborators. We wish to thank all of them for having accorded us the privilege to listen and talk to each other.

Mr. G. Avner, then President of the University and Rector O. Schild were closely involved with the preparations and extended much help in crucial moments. Prof. A. Grabois, former Dean, and Dean Y. Mansour have recognized the symposium as a Faculty event. All the University authorities having been connected with the project did everything possible to facilitate and smooth things: the Departments of Public Affairs, Finances, Publications and Student Authorities. Participants were lodged in the student dormitories, and we extend our sincere thanks to the staff for having made their stay so pleasant. Mrs. Aliza Aharoni, Symposium Secretary, managed to organize hotels, restaurants and itineraries and also to handle everyone's comfort and small problems with care and with a smile. She made our guests feel at home.

We are especially grateful to Mrs. Eleanor Kennedy Shaw for her incredibly precise memories from the time she took part in Garrod's excavations. She has supplied us with invaluable information on daily life in the camp and has kindly given permission to reproduce several of her photographs. In editing the proceedings of the symposium we were fortunate to have Mr. Paul Hirschhorn to improve the English of parts of this book; Ms. Naomi Limor and Mr. Soli Rechtman assisted with correspondence, Ms. Ofra Lazar assisted with graphics. We are grateful to Dr. A. Hands and the staff of B.A.R. for their support in producing the book. However, it would not be in its present form if not for the excellent typeset of Dr. S. Lindenbaum.

Financial support from the following Institutions enabled the Symposium to become a reality, for which our sincere thanks are due to: The University of Haifa; American Friends of Haifa University; The Israel Exploration Society; The Israel National Academy of Sciences; The Wenner-Gren Foundation for Anthropological Research; The Department of Antiquities and Museums, Ministry of Education and Culture; The Department of Scientific Relations, Ministry of Foreign Affairs; Unico Inc.

The preparation and publication of this volume benefitted by the support and interest granted by Dr. R. Hecht and the Hecht Foundation; Mr. J. Perrot, Director of the Centre de Recherches Préhistoriques Français de Jerusalem; The Israel Prehistoric Society, and Mr. U. Thon, Head of the Applied Scientific Research Co., The University of Haifa.

We are deeply indebted to all of them.

A. Ronen
July 1982

1

INTRODUCTION

CARMEL – A HOLY MOUNTAIN IN ANCIENT TIMES

Benjamin Mazar

It gives me great pleasure to attend this international conference on prehistory, which is taking place here, at the University of Haifa on Mount Carmel. We should thank the sponsors of the conference, the President of the University of Haifa, Mr. Avner, and Prof. Ronen, for linking it with an important milestone in the history of prehistoric research in the Middle East, namely, the 50th anniversary of the beginning of systematic excavations in the Mt. Carmel caves. These excavations were primarily carried out by the late Miss Dorothy Garrod and her associates, on behalf of the British School of Archaeology in Jerusalem. I am happy to accept the kind invitation to give a short lecture on "Carmel: A Holy Mountain in Ancient Times" as a prelude to the scientific deliberations of the conference.

Mt. Carmel, a mountain graced with sheer slopes and a majestic appearance, rises 546 m above sea level, and has since remote antiquity attracted the attention and the veneration both of the inhabitants of this country and of the seafarers sailing along the eastern Mediterranean coast. It is not surprising, therefore, that through the ages Carmel inspired numerous traditions closely connected with the prophet Elijah and his deeds on this holy mountain.

Mt. Carmel is possibly first mentioned in an Egyptian document from the time of Pharaoh Pepy I, in the 24th century B.C.E., which describes a seaborne military expedition landing at the foot of a mountain called "the Antelope Nose". Special interest is attached to the geographical list of Pharaoh Thutmosis III, in the 15th century B.C.E., where the mountain is called the "Sacred Cape". Already at that time, Mt. Carmel seems to have been one of the outstanding sacred mountains of Canaan, reputed to be the abode of the gods, somewhat like Mt. Olympus in Ancient Greece, and inspiring awe in the mortals by their majesty and might. In particular, Phoenician and Greek seafarers regarded Mt. Carmel with veneration, and attached great importance to the holy site at the foot of the mountain. The god worshipped in the 1st Millenium B.C.E. in this place was Ba'al Shamem, the great Phoenician god of the heavens and the seas, the ancient patron of the renowned metropolis of Tyre, who, as a result of Phoenician maritime commerce and far-flung colonization, achieved a cosmopolitan and universal character. As for the Greek seafarers, the equivalent of this god was no other than Zeus, head of the Greek pantheon. It is interesting to note that the 4th century B.C.E. anonymous geographer known as Pseudo-Skylax, in his Periplus, calls the Carmel Óros hierón Diós, that is, "the mountain sacred to Zeus". Tacitus describes Carmel, on the boundary between Tyre and Judea, as the name or epithet of both mountain and god, and the site of the god's cult is mentioned not as a temple, but only as an altar. This is the place where

Vespasian, on his way to Judea, sacrificed to Zeus and received the oracle forecasting his future as emperor of Rome. To this should be added the fragment of a statue found at the foot of Stella Maris, which bears a Greek inscription dedicated to the Heliopolitan Zeus, god of Carmel.

The holiness and the splendor of Mt. Carmel are expressed even more forcefully in biblical literature. Isaiah draws the parallel between "the glory of Lebanon", the "majesty of Carmel and Sharon", and the "glory of Yahweh – the splendor of our God". At times the Carmel is described by the prophet Jeremiah as "rising sheer from the sea", it is thickly wooded and serves as a pasture, and is shrouded in mystery. Owing to its exalted status, Mt. Carmel has served as a place of seclusion and refuge for prophets and men of spirit, among them Elijah and his successor Elisha in the 9th century B.C.E. The Elijah cyclus tells us that the prophet communed with his God, Yahweh, the God of Israel, on the summit of Mt. Carmel, and that he prayed there for the rain that would put an end to the drought afflicting the country at that time.

Chapter 18 of the First Book of Kings describes in detail the dramatic struggle of Elijah against the prophets of Ba'al, which took place in the cultic precinct at the foot of Mt. Carmel, above the Kishon river. This is the gathering place of the tribes, where stood the altars both of Yahweh, the God of Israel, and of the Tyrian Ba'al, Ba'al Shamem, whose priests gathered there to practice their cult. It is not surprising that the altars of Ba'al and Yahweh should exist in close proximity, since Mt. Carmel and its sacred precinct were on the boundary between Israel and the kingdom of Tyre. Moreover, the site must have served as an official meeting ground ever since David and Solomon, his son, concluded a "covenant of brothers" with Hiram, king of Tyre. This covenant was strengthened in Ahab's reign, and the close relations found expression in marriages between the two royal houses, in political and commercial relations and even in the growth of religious syncretism in Israel. The Bible attributes this development to Jezebel, wife of Ahab, king of Israel, and daughter of the king of Tyre. It was under Jezebel's patronage and direction that the devotees of the cult of Ba'al oppressed the "faithful of Israel", until the fateful events which took place in the sacred precinct on Mt. Carmel. According to the biblical account, Elijah eradicated the cult of Ba'al in the sacred place, repaired the altar of Yahweh which had been destroyed by Jezebel's followers, sacrificed on this altar in a miraculous manner, and called on the assembled crowd to declare: "Yahweh, he is the God", that is, the God of Israel is God "par excellence", thus renewing the covenant between the people of Israel and their God, which many generations earlier, Moses had entered into on Mt. Sinai.

The great importance attached to the holy site on Mt. Carmel during the Israelite Monarchy may be gathered from a lyrical passage in the blessing of Moses relating to the tribe of Zebulun, whose territory ranged from Western Galilee to the Mediterranean coast. This passage, preserved in Deuteronomy, reads: "The tribes will come together on the Mountain; there they will offer righteous sacrifices, for they taste the riches of the seas". This is a song of thanksgiving to the God of Israel, offered up in the sacred site on Mt. Carmel by the tribes, accompanied by the offering of sacrifices and expressing gratitude to God for economic prosperity and the development of shipping. It should be dated to the reign of Ahab or his son Yehoram, in the middle of the 9th century B.C.E.

Finally, it should be noted that Mt. Carmel and its sacred precinct are also mentioned in two inscriptions of Shalmaneser III of Assyria, from the year 841 B.C.E., dealing with his campaign against the kingdom of Israel. The march of the Assyrian army ended at Mt. Ba'li rāsi, on the sea coast near the border of the kingdom of Tyre. There Shalmaneser received a heavy tribute from Jehu, the new king of Israel, and his surrender, as well as that of Ba'al'ezer, king of Tyre. Shalmaneser also erected there a stele to commemorate his victory. The name Ba'ali rāsi must be an abbreviation of the full Phoenician name Ba'al -rās -Karmel (the Ba'al of the Cape of Carmel). This event must have brought about a decisive change in the policy of the Israelite Monarchy, which was marked henceforth by vassalage to Assyria and the end of the league with Tyre, which led to the aboliton of the cultic precinct on Mt. Carmel as a national-religious center. We cannot identify the site of this precinct and it is only tentatively that I suggested the site of Tell Qīri, within the area of Kibbutz Hazorea, which is being excavated by Prof. A. Ben-Tor.

The veneration of holy sites on Mt. Carmel has not ceased in the succeeding generations, up to modern times. They have not been forgotten in popular traditions, connected mainly with the Muḥraqa ("Place of the Fire") in the southeastern part of Mt. Carmel, where we are shown the site of Elijah's altar, and with Elijah's Cave, near the coast, in the northern Carmel. It is not surprising that Mt. Carmel has come to be known among Christians as St. Elias, and among Moslems as Mar Eliās.

MT. CARMEL CAVES – THE FIRST EXCAVATIONS

Avraham Ronen

The excavations of the Mt. Carmel caves have contributed extensively to our knowledge of the prehistory of the Levant, defining our concepts in the process. I would like to describe the site, to trace the excavations from the beginning to the appearance of the final report in 1937 and the way concepts were formed and revised over that period.

The Site

Nahal Hamearot, or the Valley of the Caves, drains Mt. Carmel westward to the Mediterranean, entering it near Athlit about 20 km. south of Haifa. The western slope of the Carmel at this point consists of Cenomanian reef forming a hard limestone cliff. Nahal Hamearot cuts through this cliff to emerge into the coastal plain, and it is here, in its south bank, that the Mt. Carmel caves were formed. From west to east, they include Tabun, El-Wad and Skhul (Figs. 1-3).

The first we hear of the caves is late in the 19th century, when they are briefly mentioned by Lawrence Oliphant, an English mystic with a passionate love for the country and the Jewish people (Oliphant 1886:14-15). The next to mention them is the German traveler Graf von Müllinen, who visited Palestine at the turn of the century and wrote a book on the history and geography of Mt. Carmel. He describes the caves in somewhat greater detail than Oliphant, noting that the largest of them has a wall of big limestone blocks across its entrance (Müllinen 1908:190), with a smaller sandstone wall in the cave's interior. Müllinen, curiously enough, uses the Lebanese term *ramleh* rather than the present-day local Arab one, *kurkar*.

A great leap forward in the caves' status came in 1927, when the British Mandatory Government Public Works Department initiated the Haifa Harbour Project. Article 306 of the Special Instructions for the Main and Lee Breakwaters specified: "The location and the quarry is in the Wadi el-Mughara . . . to produce hard stone of good quality . . ." (Anonymous 1929:109).

With the start of blasting operations at the cave of Skhul, the Department of Antiquities assigned Mr. C. Lambert to check the caves.

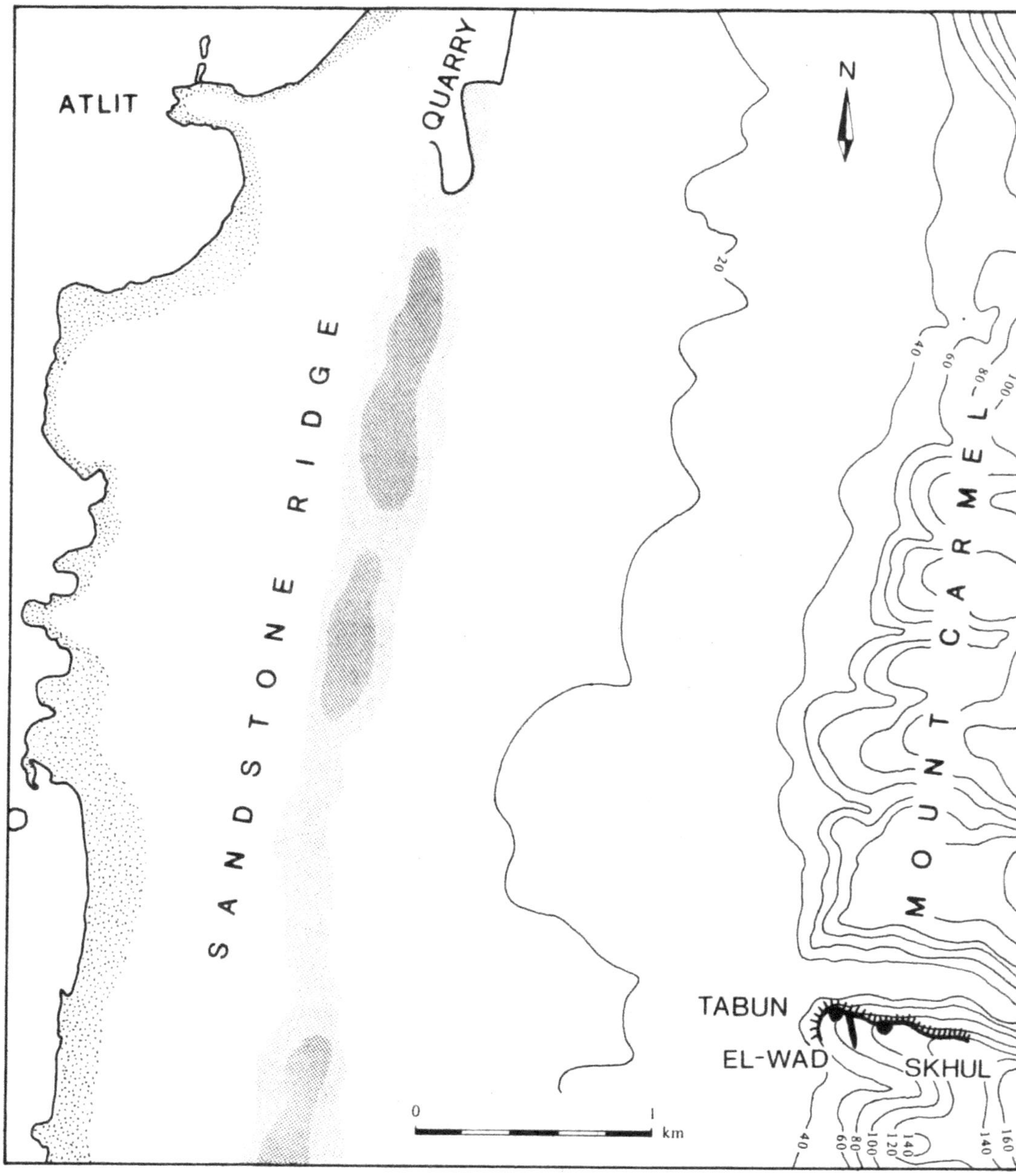

Figure 1 Location of the Mt. Carmel caves.

Preliminary Investigations

Lambert's work, carried out in November 1928, is described in a report by the Director of Antiquities. After a detailed description of the area and the caves, he concludes that "Of all the caves that have been mentioned it is the Mugharat al-Hamama with the wall across its opening; its high domed roof of rock; the remarkable tunnel at the back; the signs of constructed terraces in front of it and the quantities of surface flints outside it, that first excites interest." (Director's report 1928).

Figure 2 The western cliff of Mt. Carmel dissected by the Valley of Caves, looking east (Courtesy Dept. of Antiquities and Museums).

Figure 3 Mt. Carmel caves by 1930. Left: El-Wad, center: Jamal, right: Tabun. Skhul, located further left, is not visible (Courtesy Dept. of Antiquities and Museums).

The wall across the opening included several cyclopic blocks (Fig. 10). Similar walls are known from other caves, either as here across the entrance (Sefunim), or as a retaining wall on terraces (Sefunim, Qafzeh). These occurrences are difficult to date, but a similar feature was recently discovered at the open-air site of Nahal Sefunim (Lamdan, in preparation), and was definitely assigned to the PPNB period. The other features may well date to the same period. Lambert is the only one to apply the name El-Hamama to the cave of El-Wad. He also tells us that it was treated with respect by the local Arabs, who would compete for the privilege of having their flocks winter in the cave in the belief that it would keep the nanny-goats from miscarrying and induce them to bear twins (the name of the nearby cave – Skhul, i.e. kids – may be suggestive of the same belief). The 60 m. long tunnel, on the other hand, was avoided by the shepherds, in the belief it was endless.

Another indication of the cave's special status is the *sabil* grove of prickly pears in front of it. Planted by a pious individual, its fruit is free for the taking (*sabil* is Arabic for 'public fountain').

Lambert dug five trenches in the cave of El-Hamama (El-Wad), three inside and two on the terrace. Below the disturbed upper layer on the terrace he uncovered a layer with stone walls, grinding implements, and two burials – the first ever found at Mt. Carmel, and harbinger of a whole string of burials eventually discovered in this layer, later known as Natufian.

Inside the cave, Lambert found a stony upper layer, the stones increasing in size as he went down. This layer, going to a depth of 1.20 m., consisted of a mixed assemblage of flint artifacts, grinding stones, and Roman shards and glass. Below that, he came on an undisturbed layer with an industry similar to that found on the terrace. A human skull turned up here, as well as an object of art – the first prehistoric art ever discovered in the Near East. It was "a young animal, probably a calf, standing with head thrown back, in an attitude which faintly recalls one of the painted bisons in the cave of Altamira. It represents a development of prehistoric art for which no previous finds either in Palestine or in Syria has in the least prepared us, and is therefore of primary importance" (Director's report, 1928).

Lambert felt that the surprising and unprecedented association of flint artifacts, quern fragments and burials "is likely to be questioned"; but he concluded that "the circumstances of their discovery allow of no other interpretation; they seem to provide a definite link between the Palaeolithic . . . and the Neolithic" (Director's report, 1928).

The two trenches dug in the back of the tunnel showed this area to be disturbed.

Lambert's excavations "were undertaken with the sole object of testing the archaeological value of the site, and were discontinued when that object had been attained."

Not only had the object been attained, it bore practical results too. Father Alexis Mellon, S.J., then Head of the Pontifical Biblical Institute at Jerusalem, wrote after a visit to the cave: "Tout ce que j'ai vu a produit en moi la conviction que nous avons là le site préhistorique le plus important de la Palestine . . . il est donc, à mon avis, nécessaire de pratiquer en cet endroit des fouilles methodiques et complètes." (Cited in Director's report, 1928). Another distinguished visitor was Prof. Albright, Director of the American School of Oriental Research in Jerusalem. Albright felt "no hesitation in saying that the site promises to be more important than any other prehistoric site yet examined in Palestine. . . . I earnestly hope that it will be possible to continue researches so auspiciously begun, which are certain to revolutionize our knowledge of the age when agriculture and cattle breeding were first introduced into Palestine, not less than six thousand years ago" (Director's report, 1928). "For these reasons" concluded the Director of

Antiquities "the site, which includes the rock and valley areas, must be retained among . . . registered antiquity sites, and must be afforded every measure of protection required by the law . . . it must not only be reserved for investigation by those most competent in such matters but must, after investigation, be preserved as a permanent monument."

Consequently, stone quarrying for the Haifa harbour was moved westward, to the sandstone ridge near Athlit[1]. These quarries supplied two million tons of stone during the years of construction, 1928–33. Their exploitation revealed, for the first time, the existence of a red loam with Mousterian implements *in situ* (Garrod and Gardner 1935; Ronen 1977) (Figs. 1, 4).

The Director of Antiquities alerted the British School of Archaeology in Jerusalem and the American School of Prehistoric Research, resulting in a joint undertaking, to investigate the Mt. Carmel caves under the direction of Dorothy A.E. Garrod. The project, supported by a grant from Mr. Robert Mond, got under way in 1929 and lasted six years, out of which 21½ months were spent on actual excavation.

Garrod's Excavations

In 1929, Palestine had just under one million inhabitants, serviced by 912 km. of asphalt-paved roads and about 3,000 motor vehicles (Enc. Hebraica Vol. 6). That year, 1929, saw the first steps on the way to modernization: the idea to exploit the country's major natural resources, the Dead Sea minerals, was born (it was realized in 1931), and Haifa harbour was under construction, to be inaugurated four years later, in 1933.

Information on the prehistory of the country in that year, of 1929, was no more than fragmentary. There had been Turville-Petre's excavations in the Amud Valley (the caves of Zuttiya and Emira) in 1925–26 (Turville-Petre 1927), Garrod's work in the cave of Shukba in 1928, and Neuville's excavations in the Judean Desert, the two latter still unpublished then. Though they could not be linked satisfactorily into a coherent sequence as yet, these ventures showed the potential for further exploration, both for human remains (the Zuttiya skull) as well as for lithic assemblages, either comparable to Europe or original.

The Mt. Carmel expedition camp was well organized, though no doubt of limited comfort for the 37-year-old Garrod (Figs. 5–9). It consisted of one long barrack-type, single-storey building made of mud bricks and covered with a straw roof, which served as headquarters, working space, and residence for the senior staff. The expedition maintained friendly relations both with the nearby Arab villages, which supplied the workers, and with the neighboring Jewish settlements. The doctor at Athlit could be called on in case of need, while the expedition car was occasionally used as an ambulance to carry the sick from Athlit to the hospital in Haifa. Furniture from the main building was stored between seasons with one of the workers. Some members of the staff, including Garrod, spoke Arabic. A special jargon developed in the course of the project: the women who washed flint tools were called "flint-bints" – less derogatory than it sounds, as *bint*

[1] We could not find any documentation or correspondence regarding the sandstone quarry in the Dept. of Antiquities; only photographic documents exist in the Port Authority archives.

Figure 4 The Quarry in the sandstone ridge near Athlit, 1930. Mousterian Red Loam is visible in the fresh cut (Courtesy Haifa Port Authorities).

Figure 5 Expedition's main building, bedrooms on left (Courtesy Mrs. Kennedy Shaw).

Figure 6 The camp, looking north.

Figure 7 Dorothy Garrod (left) and Jacquetta Hawkes.

Figure 8 Expedition's Ford "Elijah", October 1932 (Courtesy Mrs. Kennedy Shaw).

Figure 9 Dorothy Garrod in front of her room, Sunday morning, 1933 (Courtesy Mrs. Kennedy Shaw).

Figure 10 El-Wad and the wall across the entrance. Digging temporarily discontinued because of bats carrying 'Persian flea' (Courtesy Mrs. Kennedy Shaw).

also happens to be the Arabic for girl – while the headquarters building was "Tibn Towers", *tibn* being the Arabic for straw (Mrs. Kennedy Shaw, pers. comm.).

The Mt. Carmel project prevented Garrod from continuing at Shukba as she had planned. El-Wad (Fig. 10) was the first cave to be excavated, as a natural consequence of Lambert's work.

El-Wad

By the end of the first season, the stratigraphy of El-Wad was pretty clear to Garrod. The outer chamber was found to be almost completely disturbed, in a jumble of potsherds and flint implements of various periods. Only a small undisturbed remnant with microlithic industry was encountered, containing a collective burial of ten skeletons (four adolescents and six children) in an extended position. Immediately below one of the skeletons, Garrod found the second work of art discovered in Palestine – a small pebble roughly carved into the shape of a human head. She concluded that the object of art found by Lambert the year before was bound to have come from the same collective burial, which Lambert's trench must have impinged on (Garrod 1929:221). Beneath the microlithic layer Mousterian industry was found embedded in a hard breccia, which was left unexplored in the season of 1929.

The inner chamber of El-Wad presented Garrod with a 2.5-m. thick undisturbed sequence, "more complete than any yet known in Palestine" (Garrod 1929:221). It was composed of:

1. An upper, disturbed layer with finds ranging from Bronze Age to Recent, 0.6 m.
2. Microlithic, 0.1 m.

3–5. Three Upper Palaeolithic layers, 1.9 m. thick.

6. Mousterian, unexplored at that time. Bedrock had not been reached.

Layer 3, the top Upper Palaeolithic, recalled the European Chatelperronian phase to Garrod. Layer 4 resembled the Aurignacian, while the lowest layer, 5, had an original aspect, characterized by "very delicate flint points which cannot at present be classified" (Garrod 1929:222).

Garrod had noticed that the microlithic industries at El-Wad and Shukba were similar, and considered that "as it will be convenient to have a name for this culture, I propose to call it Natufian, after the Wadi en-Natuf at Shukba, where we first found it in place" (Garrod 1929:222). Garrod believed the El-Wad material to be slightly later than that of Shukba at the time. What strikes one is that once proposed, the term Natufian was not employed by Garrod pending her final report in 1937: she continued using the terms 'Mesolithic' or 'Microlithic' for these assemblages in all her interim reports.

In the following season of 1930, Garrod concentrated on the Mousterian. She encountered, below the Upper Palaeolithic, a mixed assemblage with Middle and Upper Palaeolithic components. A worn and rolled Mousterian element was mingled with Upper Palaeolithic items that were not as worn, and in some cases even quite fresh. There was no sediment between the closely packed flint artifacts, a circumstance which Garrod took to indicate a period of erosion that had shifted the smaller but left the heavy pieces in place. Next and last came a 7.5 m.-deep depression with Mousterian industry in a grey phosphatic material.

By the end of that season the following stratification was given by Garrod for the inner chamber (Fig. 11):

Layer A – Bronze Age to Recent, ca. 0.5 m. thick.

Layer C – (There was no B) Capsian, 0.8 m.

the following succession confirmed:-

A. Bronze Age to Recent 50cm.
C. Capsian 80cm. average
D¹. Middle Aurignacian γ. 60cm. average.
D². " " β. 20cm.
E. " " α. 60cm.
F. Layer of erosion. Aurignacian & Mousterian.
G. Mousterian. (present only in the alcove opening on the N.W. of this chamber.)

Terrace.

A sounding made on the terrace in 1929 was enlarged, & the following succession was found:-

A. Bronze Age to Recent. 70cm. — 1.60m.
B. Mesolithic. 30cm. – 90cm.

The Mesolithic layer was here found to be completely undisturbed, and at its base we uncovered a kind of primitive enclosure, made of a single layer of flat slabs of limestone, slightly shaped & carefully fitted together, defining a small area where the rock had been dressed to form a more or less level surface. In this area a well-shaped basin, 35 cm. deep, had been cut in the rock. Immediately to the N.W. of this wall were three burials, each containing a single skeleton, & two other burials were found at the base of the deposit, near the mouth of the cave.

Inner Corridor. Two soundings were made in the inner corridor of the cave, but in each case the deposit was found to be disturbed down to the rock floor.

D. A. E. Garrod
19. VII. 30

Figure 11 A part of Garrod's report on El-Wad dated 19.VII.1930 (Courtesy Dept. of Antiquities and Museums).

Layers D1, D2 and E – Upper Palaeolithic, altogether 1.5 m. thick.
Layer F – Erosion with a mixture of Middle Palaeolithic and Upper Palaeolithic.
Layer G – Mousterian in deep pit.

Garrod noted the absence of sterile deposits between the archaeological layers, an observation that holds true not only for El-Wad but for most of the caves in the Levant.

Following is a summary of Garrod's opinions on the industries of El-Wad, going from bottom to top. The Mousterian of layer G was described by Garrod as fully Levallois, and compared to the Upper Mousterian of Europe, though lacking an independent dating. In layer F, the same type of Mousterian as in G was said to prevail. The nature of the Upper Palaeolithic component mixed with the layer F series was not clearly defined (and still remains unclear today).

The oldest Upper Palaeolithic of layer E was characterized by the tiny points which, following the 1930 season, received the tentative designation of 'Krems points'. In time they became also known as 'Font Yves points' (Garrod 1936:127). Actually, this tool has not been clearly defined. Garrod initially described it as "made on small, narrow blades, and have the tip worked to a very sharp point by nibbling retouch on one or both edges" (Garrod 1937:48). Elsewhere, she more specifically referred to the blanks as "a narrow bladelet sharply pointed" (1953:19). Meanwhile, however, she had classified as El-Wad points "a perfectly typical group, very evenly graduated in length from 29.5 mm to 90 mm." (Garrod 1954:162), thereby subsuming quite large tools under this type. This difference between theory and practice is apparent to anyone who has dealt with the collections: there exists in fact an even graduation in size among a large variety of pointed items, ranging from bladelets to large blades. All are pointed by retouch that varies from steep to semi-steep. Its location varies as well: it may affect one or both sides of the tip alone, the tip and part of the edge or edges, or a complete edge or edges. In addition, these points may be more or less curved. The vague definition made the term applicable to a variety of size, location of retouch or curvature, and may explain why Neuville preferred not to use the definition proposed by Garrod and to include the El-Wad point under the broader category of "pointes à dos retouché" (Neuville 1951:88). Recently the name 'El-Wad point' was agreed upon at the conference on Near Eastern typology, defining only microlithic points (Hours 1974).

Layer D industry was rich in steep scrapers and burins, resembling the European Aurignacian ('Middle Aurignacian' of the early 30's). The Upper Palaeolithic layer C, still considered as Capsian, was characterized by a remarkable abundance of grattoirs of all kinds and of burins, many made of tabular flint. This assemblage, eventually known as the Athlitian, also contained a few Chatelperron points, according to Garrod.

Only few human remains were found in the Upper Palaeolithic of El-Wad: the lower jaws, fragments of a skull, and an astragalus. A paucity of human remains from the Upper Palaeolithic is a general feature throughout the Levant, in marked contrast to their relative abundance in the Middle Palaeolithic.

The Mesolithic of layer B marked a sharp break from the assemblages of the preceding beds. Microliths, especially crescents, made up the major part of the assemblage, in addition to numerous sickle blades 'for cutting straw', in Garrod's words. Garrod ascribed the sickle blades and the gloss on their edge to some primitive form of agriculture in the Mesolithic of El-Wad, prior to the invention of pottery (Garrod 1931a:10). We found by experiments that the Natufian sickle blades, with their straight cutting edge, were suitable for cutting green grass rather than dry straw; on the other hand, the denticulated sickle which appeared in the PPNB is most effective for use on dry growths. We now believe that the two forms of cutting edge denote the shift from wild cereals –

which must be cut green, slightly before it is fully ripe, to prevent the grains' scattering – to domesticated cereals, which can be left till ripe without risking loss of grains.

The industry of layer A was defined by the same label Garrod always employed to describe the surface layers at her sites, 'Bronze Age to Recent'. A PPNB arrowhead was, however, found in the dump of El-Wad (Fig. 12), indicating that the stony layer of El-Wad probably spans the same time range as that of Sefunim cave – from the Early Holocene to 3,000 BC or later (Ronen 1968; in preparation). We believe the formation of stony layers in that period to be a large-scale phenomenon (Ronen 1971).

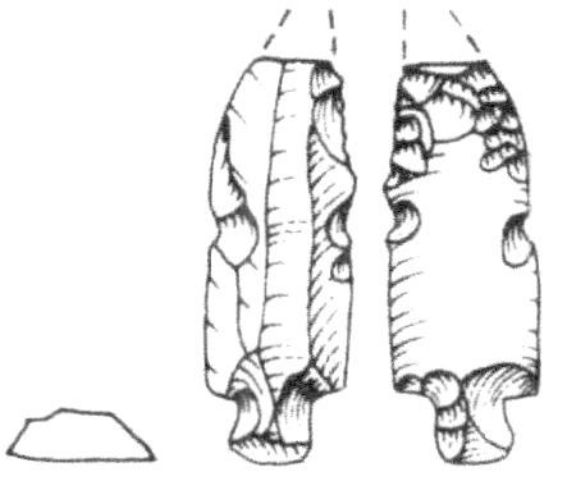

Figure 12 Arrowhead found in front of El-Wad, probably from layer A; natural size (Courtesy Mrs. U. Malbin).

Two layers were distinguished on the terrace:
Layer A – a top disturbed bed, very stony, 0.5 to 2.0 m. thick.
Layer B – Mesolithic, 0.5 to 2.0 m. thick.

Layer B had a number of interesting features. Four basins were cut into the bed-rock in an area leveled in advance, one basin displaying a rim of fine workmanship. The basins were surrounded by a pavement made of carefully fitted stone slabs, behind which a trough had been cut in the bedrock for drainage. Several walls were also found in layer B, including a long one made of large blocks.

The largest Stone Age cemetery known until then was uncovered in the Mesolithic on the terrace of El-Wad. It contained the remains of at least 60 individuals, the majority of which were badly preserved. All were in a more or less flexed position. A few had elaborate decorations made of dentalium shell and occasionally bone beads. The decorations, in all likelihood, had been attached to long disappeared garments (Garrod 1931b: 101; 1932:8–11). Given the absence of any signs of grinding or pounding in the basins described above, and the elaborate constructions in proximity to the large cemetery, Garrod cautiously suggested that the whole was connected with the cult of the dead (Garrod 1931a:10).

Only the terrace of El-Wad was excavated in the 1931 season. At its conclusion, Garrod could subdivide the Mesolithic into an upper part, B1, and a lower part, B2. The difference between them can be summarized as follows:

1. Lunates are smaller in the upper part than in the lower. In the upper division the back is generally made by steep retouch as against a preponderance of bifacial retouch ('Helwan') in the lower division.
2. Micro-burins were abundant in the upper division and extremely rare in the lower.
3. Sickle blades, on the other hand, were rare in the upper part but plentiful in the lower, and –
4. Bone implements as well as objects of art were rare in the upper part, whereas both were relatively numerous in the lower.

Accordingly, Garrod could now assert that the terrace structures – cup marks, pavement and wall – belonged to the lower subdivision, along with the burials H.27 and H.53, prominent for the necklaces found with both. The microlithic of the cave's interior was assigned, on typological grounds, to the lower division of the Natufian. Consequently, a temporal ordering emerged in the burial customs: the oldest contained group burials as well as individual inhumations, all tightly flexed; also, the decorated individuals belonged here. The later burials had only individual inhumations and were less tightly flexed. The skeletons inside the cave were, as already mentioned, in an extended position. Their chronological position relative to the terrace burials could not be precisely determined (Fig. 13).

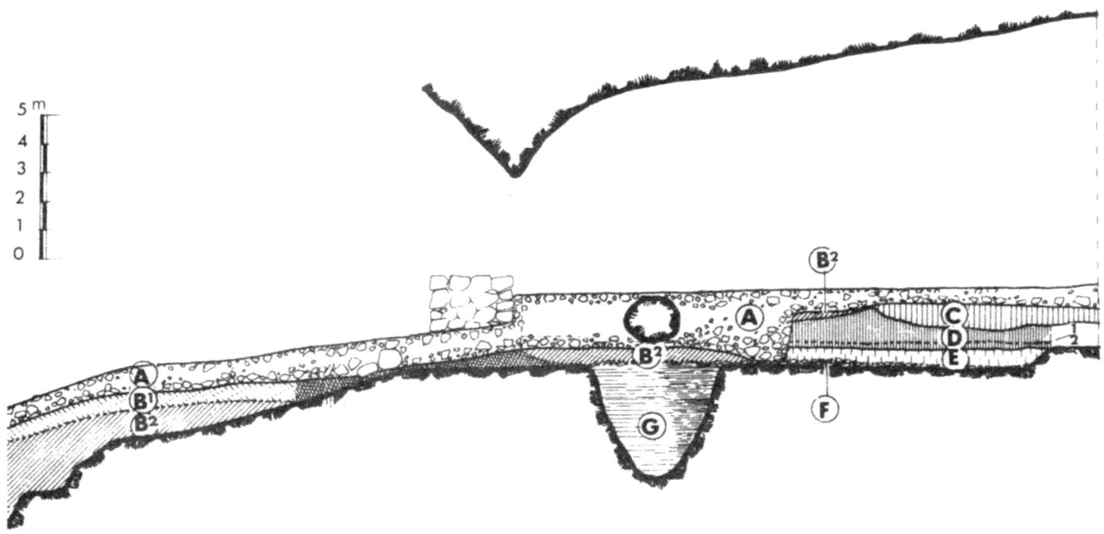

Figure 13 Stratigraphical section of El-Wad (after Garrod and Bate 1937).

Finally, Garrod revised her former assumption that Shukba was older than El-Wad; after the 1931 season, all the signs pointed to Shukba belonging to the upper subdivision of the Natufian. These revised opinions of Garrod were, indeed, basically corroborated by recent research on the evolution of the Natufian (Bar Yosef and Valla 1979).

Skhul

The site consists of a small cavity with a rock shelter, or terrace, in front of it. A small test excavation carried out here by Ms. Mary Kitson-Clark in 1929 showed a Mousterian deposit with well-preserved fauna. Skhul had not been explored in 1930, and was taken over by Theodore McCown in the following year. McCown found the usual mixed upper layer of 0.5 – 0.7 m. thickness, containing among other things what he considered to be Aurignacian remains (McCown 1932:13). If this could be confirmed, it would represent a unique situation in the northern part of Israel, where Aurignacian has never been found on terraces or in the open. As matters stand, therefore, we prefer to think that what was found in layer A of Skhul was some similar material – Epipalaeolithic or even Neolithic which might be mistaken for Aurignacian. Below the disturbed bed there was a Mousterian layer with abundant Levallois elements, some scrapers and a few burins. The faunal remains included bos, rhinoceros and cervidae.

In 1931, McCown unearthed the first skeleton at Skhul, that of a 2½ – 3-year-old child. It was immediately classified as "Homo neanderthalensis" (McCown 1932:15). The following season was more productive: after the huge limestone blocks left over from the test explosion five years ago were removed, nine more skeletons came to light

– grounds enough for Keith and McCown to coin the term *Palaeanthropus palestinus* (Garrod 1933:190).

All the skeletons were embedded in an extremely hard breccia (except for no. 2, found in a soft deposit). In the absence of mechanical equipment, McCown had to remove the skeletons in blocks of the breccia by hand, using only such tools as chisels, drills and wedges. Each block was plastered, then skidded to the camp where it was packed in a wooden case and concrete poured into the empty spaces for the voyage from Haifa to London (McCown 1933, 1934). The next stage was up to A. Keith and his team, who spent several years chiseling the skeletons out of their blocks bone by bone. The results of these efforts were revolutionary: "With a wide experience of fossil man to guide me I have to admit that the human beings now being uncovered by Mr. McCown are the most imposing specimens of fossil humanity I have ever seen . . . The Neandertal Palestinians make, I think, a rather nearer approach to the type of modern humanity than to the Neandertal Europeans." (Keith 1934:19). Skhul remained the largest known Middle Palaeolithic cemetery for a long time after (McCown and Keith 1939), and only recently lost that title to Qafzeh.

About 500 cubic meters of deposit were removed in two seasons or six months of work. Skhul was cleared almost entirely to bedrock, except for an area at its north end. "Archaeologically, the site is exhausted as far as further excavation is concerned." (McCown 1933:9-15).

In the final report, McCown (1937) elaborated his stratigraphical observations. He now distinguished two Mousterian layers, an upper B1 in which the patina was much more varied and abundant than in the lower B2 division (Fig. 14).

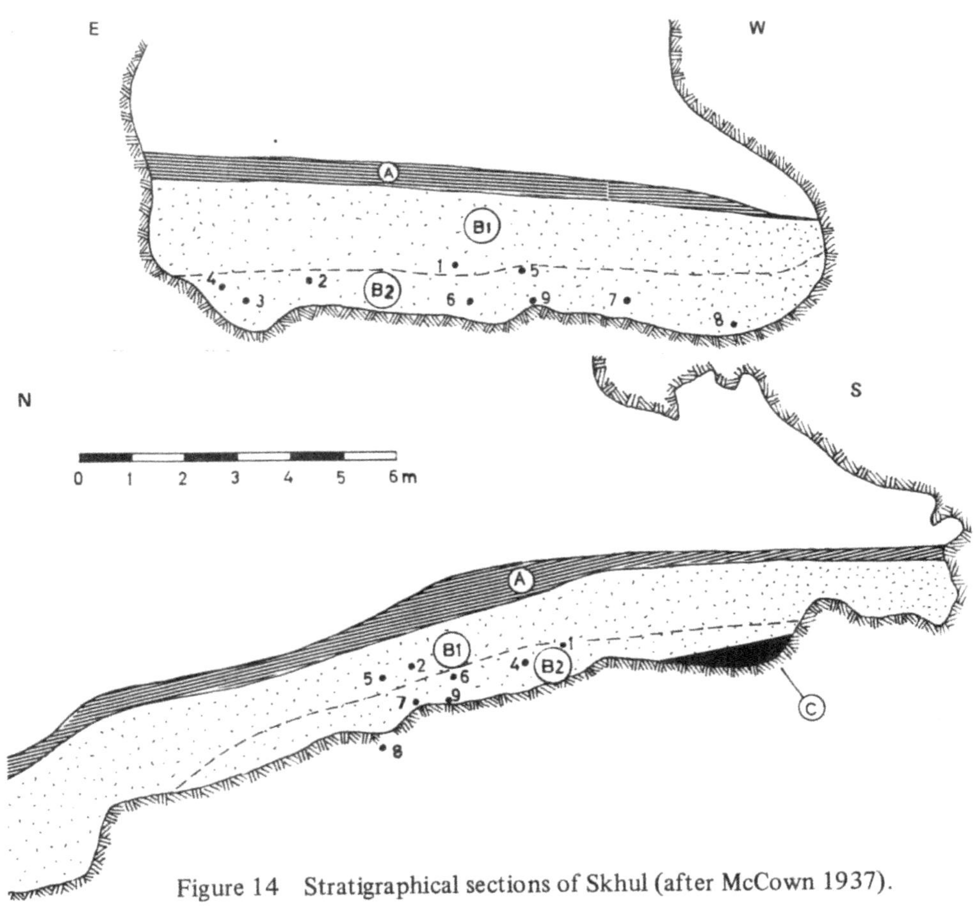

Figure 14 Stratigraphical sections of Skhul (after McCown 1937).

In the final report, we also find the first mention of pre-Mousterian deposits: two depressions in the bedrock were filled with worn and rolled flint implements, without deposits in between, and obviously the result of erosion. Below the worn artifacts were pockets of sandy deposit with unworn flint implements, including one handaxe (three other handaxes were found in the Mousterian). All the pre-Mousterian deposits were called 'Layer C'.

The erosion which had affected the flint concentration at Skhul recalls the similar phenomenon in layer F at El-Wad. The sandy sediment at the base of the Skhul deposits may be the equivalent of the Lower Palaeolithic at Tabun, which likewise consists of sand layers (Ronen 1976). The handaxe at Skhul C supports this possibility, and the odds are that the other handaxes attributed by McCown to layer B2 originate in the lower sand bed as well.

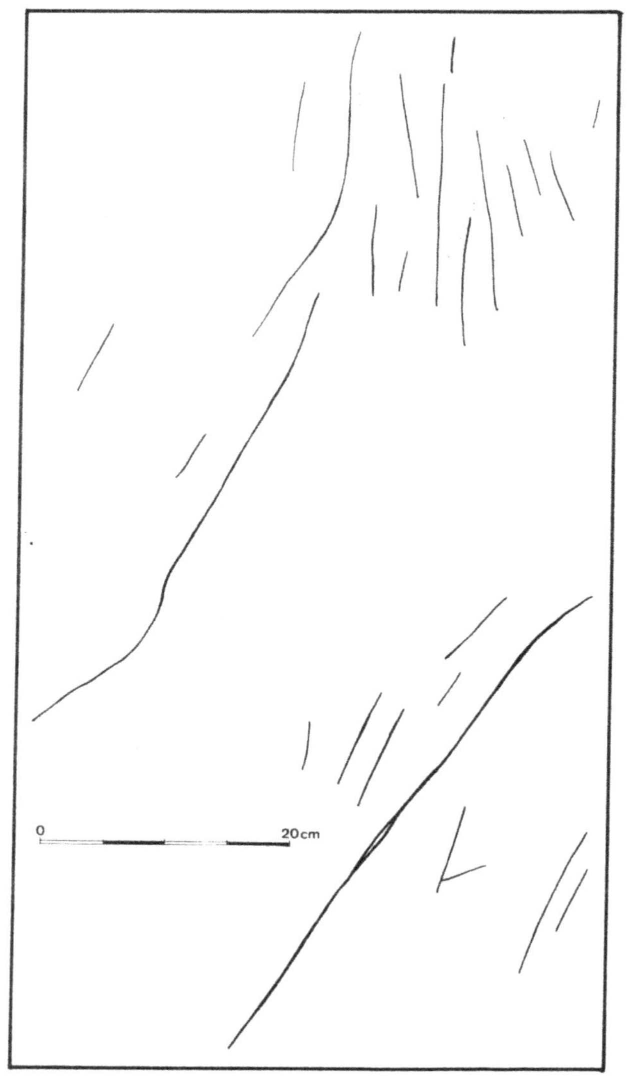

Figure 15 Rock-cut engravings in Skhul (from Ronen and Barton 1981).

Is the erosion phase at Skhul the same as that noticed at El-Wad? Possibly, but it cannot be confirmed as yet. Speculating on the origin of the erosion episode at Skhul, McCown observed that sheets of stalagmite had formed in the Mousterian deposit, and that the sediment had hardened near the walls, which left "no doubt that springs issuing from the fissures in the limestone were intermittently active." For all that, McCown deemed it unlikely "that one of these (springs), or even a combination of several, would create a flow of water sufficient to abrade the flints to the degree existing." If not spring activity, then what caused the erosion bed? "The remaining inference", suggests McCown, "is that the valley bottom was higher at a given moment . . . and the deposit was partially affected for a limited time by stream action" (1937:106). I daresay this explanation has the same power as the spring theory, though we still don't know the precise elevation of the streambed in Mousterian times.

McCown was the first to have noticed rock-cut engravings in the Mt. Carmel area. He briefly commented on the discovery in Skhul of a group of "smooth grooves of varying length, depth and direction. Some were above, others below the original level of deposit." (McCown 1937:92, footnote). He failed to specify which deposit covered the engravings in part, but we assume it was the superficial layer A rather than the Mousterian B. The grooves are certainly man-made (Fig. 15) and, as it turns out, are one of several concentrations of its kind found along the western flanks of Mt. Carmel since (Ronen and Barton 1981). Similar engravings in other circummediterranean countries have been attributed to the Early part of the Holocene. In common with those at the other Mt. Carmel sites, the Skhul engravings were made at quasi-hidden spots and were definitely not intended to be widely seen. Their function is unknown. Suggested explanations vary between markings protecting from the evil eye, territorial signs, or astronomical registrations (Ronen and Barton 1981).

Tabun

The Tabun cave was tested briefly in the first season at Mt. Carmel, in 1929. It seemed to be so small that Garrod considered its excavation would require no more than one season (Garrod 1934c:7). In fact, it turned out to be the most complicated cave of the Mt. Carmel group, and its excavation took longer than any of the others. Work on it was resumed in 1931, and the following layers were distinguished (Fig. 16):

Layer A – disturbed as in the other caves but less rich than elsewhere, ca. 1 m. thick.

Layer B – Upper Mousterian with faunal remains, almost exclusively *Dama mesopotamica*, 3.5 m. thick. The faunal evidence was interpreted as pointing to forest conditions, i.e., to a humid climatic phase. A single 'very crude' handaxe was found in layer B (Garrod 1932:10).

Layer C – 1.7 m. thick, with a Middle Mousterian industry. The complete skeleton of a female was found here, plus a separate lower jaw 0.9 m. below it.

Layer D – Lower Mousterian, 2.3 m. thick. All the Mousterian layers contained Levallois components, but layer D was characterized by a long and narrow Levallois point.

Layer E – Garrod initially called its industry Lower Mousterian. It was 5 m. thick, and contained handaxes as well as the thick scrapers which would subsequently be known as the typical Yabrudian element. Garrod subdivided it into four units, Ea-d, based on the typology of the handaxes. A human femur was found in the upper part of layer E.

Layer F – was termed Upper Acheulian. Its handaxes were more abundant and better manufactured than those of layer E. It was 1.9 m. thick.

The lowest layer, G, had a coarse flake inductry, devoid of handaxes. Garrod termed this assemblage Tayacian. 2 m. maximum thickness.

Garrod first referred to layer E as 'Acheuleo-Mousterian' following the 1932 season, "though I am not entirely satisfied with this label" (Garrod 1934a:87). She placed the layer E assemblages with the "group of sites unsatisfactorily labelled "pre-Mousterian", such as Ehringsdorf, Krapina and Grimaldi." (1934a:88).

In the upper part of layer E, Garrod noticed the phenomenon which later became one of the most controversial issues in the Mt. Carmel data. In her words: "A very interesting feature of Tabun Eb is the presence of a group of implements of Upper Palaeolithic type. These include Chatelperron points, end-scrapers and gravers, and a whole series of narrow blades with a nibbling retouch of the edges. There can be no doubt that these implements are perfectly in place: the great thickness of the overlying deposits rules out any possibility of there being a later intrusion. The whole technique of their manufacture is in marked contrast to that of the typical implements of this layer, and I consider them to be due to contact with a very early Aurignacian rather than a development *in situ* of the Acheuleo-Mousterian industry." (Garrod 1934c:9). The term pre-Aurignacian was hinted here, a full 17 years before it became formalized (Rust 1950).

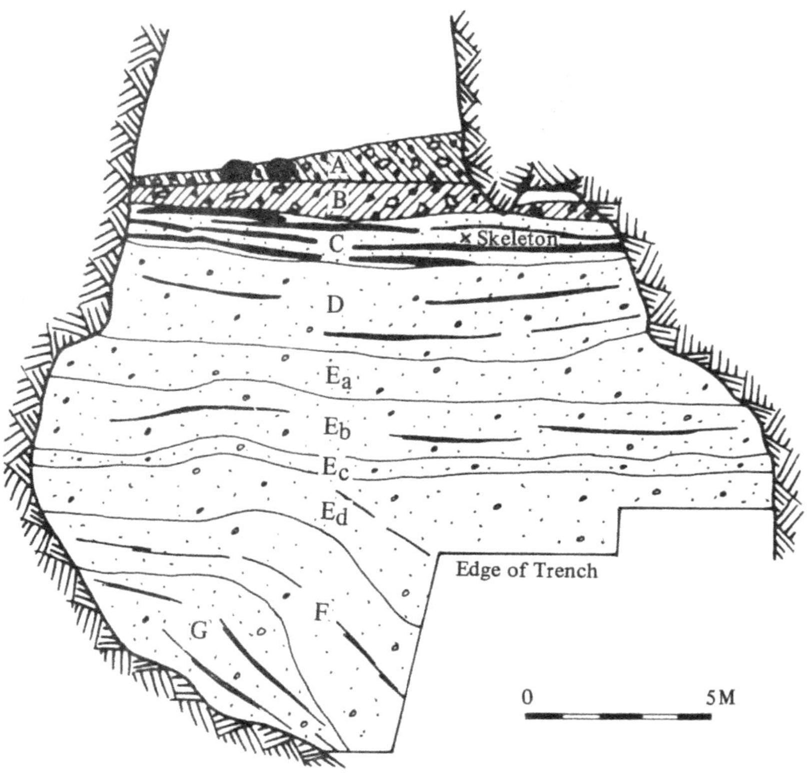

Figure 16 Stratigraphical section of Tabun (after Garrod and Bate 1937).

In 1933, Garrod suggested a chronological framework for the Tabun sequence based on two assumptions: that the industries of Israel and Europe were similar for one thing, and that this similarity indicated contemporaneity for another (Garrod 1934a: 89).

Layer		
B	–	Würm Glacial
C D	–	Riss-Würm Interglacial

E	–	Retreat of Riss
F G	–	Maximum extension of Riss Glacial

The seventh and last season at Tabun lasted from April 2 to August 25, 1934, and Garrod submitted her final field report to the Department of Antiquities on September 31 of that year (Garrod 1934b). What appears to have impressed her in particular is the first discovery of faunal remains in layer E. These contained hippopotamus, rhinoceros and elephants. The two elephant tusks found at Tabun were the second such find in Palestine, the other being the elephant from the fauna-bearing beds at Bethlehem, explored by Miss Bate in the same year (Bate 1941). Since then, elephant remains have been found in numerous Acheulian sites in Israel.

The debates which took place after the publication of 'The Stone Age of Mount Carmel' (Garrod and Bate 1937) will not be dealt with here. Rust (1950) introduced new names for the cultural units which Garrod had tried to define by means of traditional nomenclature; Bordes (1962) questioned the chronology proposed by Garrod and Rust alike, especially the position assigned to the pre-Aurignacian (Garrod 1956). Higgs (1961) challenged the assumed contemporaneity between the Tabun and Skhul human remains. These objections and others are likely to come up in the discussion which will follow on the Lower and Middle Palaeolithic in the course of the Symposium. I would like to conclude with some brief remarks on another well-known concept related to the Mt. Carmel project, namely Garrod's 'Transitional Phase' from the Middle to the Upper Palaeolithic.

As mentioned above, Garrod initially considered the oldest Upper Palaeolithic in Mt. Carmel, layer F of El-Wad, to be a simple mixture of two components due to a phase of erosion, and accordingly, divided the assemblage on typological grounds to a Middle Palaeolithic (F2) and an Upper Palaeolithic (F1) (Garrod and Bate 1937:116, 120). This interpretation was consistent with the then-known basal industry of the Upper Palaeolithic of Palestine – the Qafzeh and et-Tabban series of Neuville (1934; 1951:73-74), characterized by numerous retouched and backed blades and by rare Aurignacian elements. Garrod took the industry of Emira cave (Turville-Petre 1927) to be mixed as well.

Ultimately, Garrod learned from publications about the 'Transitional' character of the early Upper Palaeolithic discovered at the Lebanese sites of Abu Halka (Haller 1943) and Ksar Akil (Ewing 1947). These assemblages were said to have Middle and Upper Palaeolithic groups intimately mingled, leading Garrod to the opinion that a similar type of 'Transitional Phase' had existed in Israel, a fact that had been "misunderstood for more than twenty years" (Garrod 1951:129). This new idea of hers was founded on a simple quantitative basis: "The evidence of the four sites, Abu Halka, Ksar Akil, El-Wad and Emira, outweighs for me the apparent contradiction of et-Tabban and Qafzeh" (Garrod 1951:123).

That being the case, the El-Wad F assemblage was now regarded as an undisturbed 'Transitional' industry. El-Wad G, previously thought of as Mousterian, became part of this Transitional phase too. The Emira cave assemblage was likewise redefined as an undisturbed transitional industry, though Garrod was somewhat troubled by the presence of El-Wad points in this series (1951:128). What Garrod came to regard as the principal characteristic of the initial Upper Palaeolithic phase was the intimate mixture of Middle and Upper Palaeolithic elements, each represented by its distinct technical and typological characteristics in conjunction with the presence of Emira points. Henceforth, the term 'Transitional industry' was implicitly applied to two distinct occurrences in Israel:

that of Qafzeh–Tabban and that of El-Wad–Emira.

It was only afterwards that Garrod got a chance to examine the Abu-Halka material for herself. She amply confirmed Haller's observations (Garrod 1955), which in reality indicated substantial differences from the El-Wad and Emira assemblages: in Lebanon, Upper Palaeolithic tools were manufactured on Levallois blanks, while typical Middle Palaeolithic tools were rare or absent; chanfreins were numerous, as were backed and retouched blades. All these features were unlike the Israeli material except for the last which, though unlike the El-Wad–Emira 'Transitional', was well in line with the 'Transitional' of Qafzeh.

The handling of the Abu Halka material could have led Garrod to reconsider her view on the 'Transitional Phase' in Israel. Instead, she only suggested that it was a "local facies" (Garrod 1955), but did not revert to her original opinion of a mixed assemblage.

Garrod painted in broad brush-strokes. She was content to understand and demonstrate the large traits of cultural evolution in the Mt. Carmel caves. She was reluctant to resort to new terminology as a device to better understanding, attempting rather to define new occurrences giving existing terms a subtle new shade of meaning. In the rare cases where she did propose new terms, cultural or typological, she did so almost apologetically, and the definitions were too general to be of a precise use. Her reports show a careful observer uninterested in details, and what emerges from them is the general picture that has been elaborated and, on the whole, substantiated over the past 50 years.

References

Anonymous 1929 *Haifa Harbour. General Considerations and Specification* Office of the Crown Agents for the Colonies, London.

Bar Yosef, O. and F.R. Valla 1979 L'evolution du Natufien, nouvelles suggestions *Paléorient* 5: 145-152.

Bate, D.M.A. 1941 The Bone-bearing Beds of Bethlehem *Nature* 147: 783.

Bordes, F. 1962 Sur la chronologie du Paléolithique au Moyen Orient *Quaternaria* 5: 57-69.

Director of Antiquities 1928 A preliminary Note describing the investigations made at the Antiquity Site at Wadi El-Mughara in November, 1928.

Ewing, J.F. 1947 Preliminary Note on the Excavation at the Palaeolithic Site of Ksar Akil, Republic of Lebanon *Antiquity* 21: 186-196.

Garrod, D. 1929 Excavations in the Mugharet El-Wad, Near Athlit. April–June, 1929 *Palest. Explor. Fund Quarter.* 62: 220-222.

Garrod, D.A.E. 1930 Report on the Excavations at the Mugharet El-Wad. April–June, 1930. Field Report dated 19.VI.1930.

Garrod, D.A.E. 1931a Excavations in the Caves of the Wady El-Mughara, 1929–1930 *Bull. Americ. School Prehist. Res.* 7: 5-11.

Garrod, D.A.E. 1931b Excavations at the Mugharet El-Wad, 1930 *Palest. Explor. Fund Quarter.* 64: 99-103.

Garrod, D.A.E. 1932 Excavations in the Wady El-Mughara, 1931 *Bull. Americ. School Prehist. Res.* 8:6-11.

Garrod, D.A.E. 1933 Magharet Es-Sukhul *Quart. Dept. Antiq. Palest.* II: 189-190.

Garrod, D.A.E. 1934a Excavations at the Wady El-Mughara, 1932-3 *Palest. Explor. Fund Quart.* 67: 85-89.

Garrod, D.A.E. 1934b Wady Mughara. 7th Season; 2nd April - 25th August, 1934. Field Report No. 30/6.

Garrod, D.A.E. 1934c Excavations at the Wady El-Mughara (Palestine) 1932-33 *Bull. Americ. School Prehist. Res.* 10: 7-11.

Garrod, D.A.E. 1936 A Summary of Seven Seasons' Work at the Wady El-Mughara *Bull. Americ. School Prehist. Res.* 12: 125-129.

Garrod, D.A.E. 1951 A Transitional industry from the base of the Upper Palaeolithic in Palestine and Syria *J. Roy. Anthr. Inst.* 81: 121-130.

Garrod, D.A.E. 1955 The Mugharet el-Emireh in Lower Galilee: type station of the Emiran industry *J. Roy. Anthr. Inst.* 85: 1-22.

Garrod, D.A.E. 1956 Acheuleo-Jabrudien et "Pre-Aurignacian" de la grotte de Taboun (Mont Carmel) – Etude stratigraphique et chronologique *Quaternaria* 3: 39-59.

Garrod, D.A.E. and E.W. Gardner 1935 Pleistocene Coastal Deposits in Palestine *Nature* 135: 908.

Garrod, D.A.E. and D.M.A. Bate 1937 *The Stone Age of Mount Carmel I* Oxford.

Haller, J. 1943 Notes de préhistoire Phénicienne: L'Abri de Abou Halka *Bull. Musée de Beyrouth* 6: 1-20.

Higgs, E.S. 1961 Some Pleistocene Faunas of the Mediterranean Coastal Areas *Proc. Preh. Soc.* 27: 144-154.

Hours, F. 1974 Rémarques sur l'utilisation des listes-types pour l'étude du Paléolithique supérieur et de l'Epipaléolithique du Levant *Paléorient* 2: 3-18.

Keith, A. 1934 Note by Sir Arthur Keith *Bull. Americ. School Prehist. Res.* 10: 18-19.

Lamdan, M. Nahal Sefunim Rock Shelter. *In:* Ronen, A. *The Cave of Sefunim, Mount Carmel* (in preparation).

McCown, T.D. 1931 The Mugharet es-School (sic!). Excavations of 1931. Field Report 6/7/31 No. 30/6.

McCown, T.D. 1932 A Note on the Excavation and the Human Remains from the Mugharet Es-Sukhul (Cave of the Kids), Season of 1931 *Bull. Americ. School Prehist. Res.* 8: 12-15.

McCown, T.D. 1933 Fossil Men of the Mugharet Es-Sukhul, Near Athlit, Palestine, Season of 1932 *Bull. Americ. School Prehist. Res.* 9: 9-15.

McCown, T.D. 1934 Excavations at the Wady Al-Mughara (Palestine) 1932-33 *Bull. Americ. School Prehist. Res.* 10: 13-18.

McCown, T.D. 1937 Mugharet Es-Skhul. Description and Excavations. *In:* Garrod, D.A.E. and D.M.A. Bate *The Stone Age of Mount Carmel I* Oxford, pp. 91-111.

McCown, T.D. and A. Keith 1939 *The Stone Age of Mount Carmel II, The Fossil Human Remains from the Levalloiso-Mousterian* Oxford.

Müllinen, Graf von 1908 Beiträge für Kenntnis des Karmels *ZDPV* 31: 190ff.

Neuville, R. 1934 Le préhistorique de Palestine *Révue Biblique* 43: 237-259.

Neuville, R. 1951 Le Paléolithique et le Mésolithique du Désert de Judée *Arch. Inst. Paléont. Hum.* Paris.

Oliphant, L. 1886 Excavations on Carmel *Palest. Explor. Fund Quart.* 19: 7-15.

Ronen, A. 1968 Excavations at the Cave of Sefunim (Iraq el-Barud), Mt. Carmel. Preliminary report *Quartär* 19: 275-288.

Ronen, A. 1971 Post-Pleistocene Stony Layers in East Mediterranean Sites *Quartär* 22: 73-93.

Ronen, A. 1976 The Skhul burials: an archaeological review *IX Congres UISPP,* Nice, Colloque XII: 27-40.

Ronen, A. 1977 Mousterian Sites in Red Loam in the Coastal Plain of Mount Carmel *Eretz-Israel* 13: 183-190.

Ronen, A. *The Cave of Sefunim, Mount Carmel* (in preparation).

Ronen, A. and G.M. Barton 1981 Rock engravings on western Mount Carmel, Israel *Quartär* 31/32: 121-137.

Rust, A. 1950 *Die Höhlenfunde von Jabrud (Syrien)* Neumünster.

Turville-Petre, F. 1927 *Researches in Prehistoric Galilee 1925–1926* British Sch. Arch. in Jerusalem, London.

SOME REMARKS ON THE NATURE OF TRANSITIONS IN PREHISTORY

Ofer Bar-Yosef

This paper is not a systematic survey of what is generally called "transitions" or "transitional industries" in the literature of Old World prehistory. It is, rather, an eclectic, personal overview written from the standpoint of someone who is currently engaged in the study of the latest hunters-gatherers in the Levant and the various ways through which they became farmers and pastoralists. This societal and economic change was once described as "The Neolithic Revolution", a term that stems from the same approach that divides the prehistoric record into a developmental, ranked sequence (Childe 1951).

It is superfluous to repeat here the main steps through which eminent scholars have tried during the 19th and early 20th centuries to establish an order in the available archaeological record. This was done by several authors in the last two decades (Daniel 1964: Roe 1970; Lamberg-Karlowsky and Sabloff 1979, to mention only a few examples). However, the title of this symposium necessitates a few brief remarks.

Terms such as "Lower Palaeolithic", "Middle Palaeolithic", etc. were formed in order to have a better and more accurate classification of isolated artifacts and lithic assemblages in sequential time-units, using a relative chronology based on geological observations. The most frequent artifact-types in the earlier collections led to the usage of synonyms such as "Pebble Tool and Handaxe Cultures", "Flake Tool Cultures" etc. (Oakley 1964). In such a scheme, or even in a more detailed one, the definition of "transitional industries" is essential. It helps to fit in those assemblages which share common tool-types with earlier and later periods. Or, in other words, it means that there are no specific traits which designate these assemblages and they merely have "mixed" features. This nature of "transitional industries" is similar to the "Intermediate" phases in the historical record. It combines the connotations of both cultural degeneration and the up growth of new cultural elements.

Long ago prehistorians recognized that there are several kinds of "transitions". Some are time-trajectory while others are spatial. Each "transition" has a generally accepted social interpretation based on historical analogies. For example, a continuous lithic sequence with minute and regular changes is seen as evidence for local development, inferring evolution within the same population. Radical changes are interpreted as major shifts in population, evidence for immigration, or even extermination of the local people by newcomers. Gradual changes in spatial dimensions are considered to be the result of diffusion while abrupt changes may mark the border between two cultural/ethnic territories. All these are very well known and might even be correct.

During the last decade many of the traditional explanatory mechanisms were seriously questioned. In the search for covering laws in archaeology, many scholars have tried and are still trying to uncover the causes for and the rules of human behavior. Many social units were and are still under reexamination as to their changing patterns of behavior. Among these units, hunters-gatherers are the most attractive for study for Palaeolithic archaeologists, as they are considered the last survivals of pre-agricultural societies. The symposium "Man the Hunter" (Chicago 1966) marked the onset of these systematic efforts and the rise of ethnoarchaeology as an evolving discipline. These in turn provide better insights as to the potentials of socio-economic interpretations of the archaeological residues. Detailed faunal studies, lithic micro-wear analysis, geoarchaeology and micromorphology of site's sediments, etc. developed at more or less the same time stressing even further the large variety of approaches available in the elucidation of cultural processes during the Palaeolithic period.

Today it seems that there is general accord in viewing the evolution of mankind as a series of major steps. These, in a schematic framework are as follows:

a. Hunters-gatherers (originally scavengers and gatherers) evolving from bands to tribes.

b. Farmers and pastoralists (from tribes to chiefdoms and early states) leading and transforming into urban communities.

c. Industrial societies (from states to super-states).

The transition from one stage to another was often termed as "The Agricultural Revolution" and the "Industrial Revolution". (Several scholars would add the "Urban Revolution" in between the two.)

A similar status is sometimes given to the emergence of Man as the "Tool Maker", but a lot of scepticism was expressed concerning this issue based on the increasing knowledge of primates behavior. But there is no doubt that from a certain point in the geological time-scale, early hominids are found along with archaeological remains (artifacts, broken bones, etc.) which are classified as "sites" and understood as products of a human cultural system.

In spite of the deficiencies in our data of the "Lower Palaeolithic" times (Jelinek 1977) one can follow a gradual change in time. This change is mainly determined by the increasing stone-artifact complexity (Isaac 1972), as these are better preserved than other cultural elements. Due to its length in time, the "Lower Palaeolithic" is often considered to have been in homeostasis status, but close scrutiny shows that this is a false impression. For the sake of this short paper, it might be enough to examine its latest portion.

Archaeologically, this part of the sequence is quite similar in several characteristics to its next time-unit, the "Middle Palaeolithic". For example, the appearance of Levallois technique in Middle and certainly Late Acheulian assemblages was recognized long ago (Bordes 1952). The similarity in site locations, site sizes, loosely defined territories, etc. as well as the technical abilities in hunting (as inferred from the bones of the hunted game), all point to a continuum. This continuum was accompanied by small innovations, technical improvements (perhaps even increasing complexity in social structure) etc., but in the author's view, there is no clear cultural break which today deserves the continuous usage of the terms "Lower" and "Middle Palaeolithic" in the construction of a viable chronological-cultural framework.

It is slightly more difficult to argue the same for the "Upper Palaeolithic" because for generations we have been overwhelmed by the West European remains which include the parietal art, mobile art objects, magnificent bone tools, etc. One would be tempted to see the novelties of the West European Upper Palaeolithic as the outcome of a major "revolution" – the arrival of Modern Man. Unfortunately, accumulating evidence destroys this paradigm. Early *Homo sapiens* remains turned up in Mousterian levels (like in Qafzeh) and Neanderthal remains were recently found in a Chatelperronian layer. The latter phenomenon confirms the suggestion of Bordes to view this industry (also known as "Lower Perigordian") as part of the uninterrupted continuum beginning with the Mousterian of Acheulian Tradition. Similar observations can be mentioned concerning other cultural aspects such as the use of red ochre, the presence of dwelling remains, built up installations, etc.

Without a detailed and elaborated argument it seems that the later part of the Palaeolithic sequence represents only an accelerated technological improvement which was perhaps accompanied with more complex social systems. (In view of the scarcity of studies concerning palaeodemographic interpretations of Upper Palaeolithic sites it will be difficult to ascertain this point.) It is only then, at the close of the Pleistocene and early Holocene millenia that the "agricultural revolution" happened in several nuclear areas.

In view of these very generalized remarks, let us turn now to the actual subject of the symposium – "The Transition from the Lower to the Middle Palaeolithic" in the Levant – in order to exemplify some of the arguments mentioned previously. A few queries concerning this issue might provide the main guidelines for the following discussion:

1. Using terms such as "Lower" and "Middle Palaeolithic", into what part of the sequence should the "Acheuleo-Yabrudian" (or "Acheulian of Yabrudian facies") be classified?
2. Are there overall marked differences in settlement patterns, procurement strategies, intra-site organization, etc. between the Late Acheulian, "Acheuleo-Yabrudian" and the Mousterian in the Levant? If there are such differences, can they be explained as the results of shifting ecological conditions alone?
3. Is there a lineal evolution from *Homo erectus* into the early types of Modern Man or the Neanderthals in the Levant? Are there detectable changes in the hominid record at the onset of the Late Pleistocene? And finally, how can this local segment of human evolution be related to the mentioned archaeological units?

The relevance of these questions is better understood on the background of the available archaeological evidence.

The Late Acheulian industries are known from the collections of many findspots and sites all over the Levant including its arid zones. Basic typological and metrical studies were made by the late Gilead, who proposed a subdivision of these industries into several groups (Gilead 1970; Bar-Yosef 1975). Unfortunately, the detailed chronology of these is as yet not very well established. Recent U/Th dates indicate that some sites might be as old as 250–350,000 years (Schwarcz *et al.* 1980).

It is generally accepted that the Late Acheulian is followed chronologically by a specific industrial complex. It includes industries named "Yabrudian" "Acheuleo-Yabrudian" (or "Acheulian of Yabrudian facies") as well as "Amudian" and "Pre-Aurignacian" (Rust 1950; Bordes 1955, 1977; Garrod 1970). These are now unified, following the proposition of A. Jelinek (1981, this volume) under the term "Mugharan Tradition".

The territorial distribution of the Mugharan Tradition sites is limited to the central part of the Levant, form the El-kowm depression to the coastal sites in Lebanon and Mt. Carmel. The intensive surveys in the Negev, either in its northern or central parts, have failed to uncover any sites containing this industry (Valla *et al.* 1979; Marks 1977;

Goring-Morris, pers. comm.). Does this distribution reflect the arid conditions of the Riss-Würm Interglacial (stage 5e in the $O^{16/18}$ sequence), or is it merely a cultural phenomenon?

The varying proportions between bifaces and scrapers in the different assemblages, now attributed to the Mugharan Tradition, make it a worthless task to try and classify them as either a part of the "Lower" or "Middle Palaeolithic". If the mere presence of bifaces is the only criterion, then it is sufficient to recall the Mousterian of Acheulian Tradition which is known in France to be of the Middle Palaeolithic sequence (as well as of Würmian age). But if it is the special forms of bifaces that should be taken into account, then one may argue for assigning the Mugharan Tradition to the Lower Palaeolithic, in view of the frequency of similar bifaces in the Late Acheulian.

The Levantine Mousterian complex is not under discussion in this symposium. Its sequence as arranged on the basis of excavations at Tabun (Garrod and Bate 1937; Copeland 1975; Jelinek 1981, this volume), is not free from queries concerning the chronological framework. However, in a generalized assessment it seems that the settlement pattern, intra-site organization, intensity of lithic production, etc., the Mousterian shares common features with its preceding archaeological unit. In four sites (Bezez, Yabrud I, Zuttiyeh and Tabun) the two occur stratified one on top of the other. The only detectable difference is in the overall geographic distribution: The Mousterian is well represented in the arid zones of the Levant. This phenomenon is interpreted as the result of better environmental conditions which prevailed there during the early part of the Last (Würmian) Pluvial (Horowitz 1979; Marks 1977).

Finally, the available hominid remains might be arranged in such a way that will satisfy various schools in human palaeontology. The fragmentary skull of Zuttiyeh is considered as a late *Homo erectus*. There is therefore a considerable hiatus between its age and the successive remains, perhaps in the order of 50–80,000 years. Then, it is either the woman of Tabun or the Qafzeh population who come first, and later the human remains from Skhul, Amud and Kebara appear. Until the intra-Mousterian chronology is established satisfactorily, the chronological placement of each hominid will be controversial.

Summing up the fragmentary Levantine sequence, it seems to the author that no exceedingly important socio-economic changes can be inferred from the available data that necessitate the differentiation between the "Lower" and the "Middle Palaeolithic". Therefore it might be advisable to abandon these terms, especially in their old and strict meanings and use them only as a "slang" word or "jargon" while carrying out oral discussions. The abandonment of these out-of-date terms will also enforce the tendency to think and work in a more adequate approach viewing the archaeological remains within a framework of cultural systems. In the case of the Levant it means that large-scale careful research of Late Acheulian occurrences (through systematic excavations and detailed surveys) is very much needed. This will provide a proper basis for comparisons with the changing patterns of human behavior as demonstrated by the Mousterian record.

References

Bar-Yosef, O. 1975 Archaeological Occurrences in the Middle Pleistocene of Israel. *In:* K.W. Butzer and G.Ll. Isaac (eds.) *After the Australopithecines: Stratigraphy, Ecology and Culture Change in the Middle Pleistocene* Aldine, Chicago, pp. 571-604.

Bordes, F. 1952 Technique Levallois et "Levalloisien ancien" *L'Anthropologie* 56: 554-556.

Bordes, F. 1955 Le Paléolithique Inférieur et Moyen de Jabrud (Syrie) et la Question du Pré-Aurignacien *L'Anthropologie* 59: 486-507.

Bordes, F. 1977 Que sont les Pré-Aurignacien et le Iabrudien? *In:* B. Arensburg and O. Bar-Yosef (eds.) *Eretz-Israel* 13 (M. Stekelis Memorial Volume), pp. 49-55.

Childe, G. 1951 *Social Evolution* C.A. Watts, London.

Copeland, L. 1975 The Middle and the Upper Palaeolithic of Lebanon and Syria in the Light of Recent Research. *In:* F. Wendorf and A.E. Marks (eds.) *Problems in Prehistory: North Africa and the Levant* SMU Press, Dallas, pp. 317-350.

Daniel, G. 1964 *The Idea of Prehistory* Penguin Books, Baltimore.

Garrod, D.A.E. 1970 Pre-Aurignacian and Amudian: A Comparative Study of the Earliest Blade Industries in the Near East *Fundamenta* A2: 224-229.

Garrod, D.A.E. and D.M.A. Bate 1937 *The Stone Age of Mount Carmel* Vol. 1, Oxford University Press.

Gilead, D. 1970 Handaxe industries in Israel and the Near East *World Archaeology* 2: 2-11.

Horowitz, A. 1979 *The Quaternary of Israel* Academic Press, New York.

Isaac, G.Ll. 1972 Chronology and the Tempo of Cultural Change during the Pleistocene *In:* W.W. Bishop and J.A. Miller (eds.) *Calibration of Hominid Evolution* Scottish Academic Press, Edinburgh, pp. 381-430.

Jelinek, A.J. 1977 The Lower Palaeolithic: Current Evidence and Interpretations *Annual Review of Anthropology* 6: 11-32.

Jelinek, A.J. 1981 The Middle Palaeolithic in the Southern Levant. *In:* P. Sanlaville, and J. Cauvin (eds.) *Prehistoire du Levant* Colloque CNRS 596, Lyon, pp. 265-280.

Lamberg-Karlowsky, K. and J.A. Sabloff 1979 *Ancient Civilizations: The Near East and Mesoamerica* Benjamin/Cummings Publishing Co., Menlo Park.

Marks, A.E. 1977 Introduction: A Preliminary Overview of Central Negev Prehistory. *In:* A.E. Marks (ed.) *Prehistory and Palaeoenvironments in the Central Negev, Israel* Vol. II: The Avdat/Aqev Area, part 2 and the Har Harif, SMU Press, Dallas, pp. 3-34.

Oakley, K.P. 1964 *Man the Tool Maker* University of Chicago Press, Chicago.

Roe, D. 1970 *Prehistory. An introduction.* University of California Press, Berkeley and Los Angeles.

Rust, A. 1950 *Die Hohlenfunde von Jabrud (Siryen)* Karl Wachholz, Neumunster.

Schwarcz, H., P. Goldberg and B. Blackwell 1980 Uranium Series Dating of Archaeological Sites in Israel *Israel Journal of Earth Sciences* 29: 157-165.

Valla, F.R., I. Gilead and O. Bar-Yosef 1979 Prospection préhistorique dans le Néguev Septentrional *Paléorient* 5: 221-231.

RED OCHRE, THE TRANSITION BETWEEN LOWER AND MIDDLE PALAEOLITHIC AND THE ORIGIN OF MODERN MAN

Ernst E. Wreschner

As the central theme of our wider subject should deal with the origin of modern man I would like to elaborate on one aspect of what Albert Damon called 'the humanization of hominidization'. It concerns the choice of color in the evolution of human behavior. By the very nature of color phenomena in archaeological associations they are the red pigments of mineral origin, usually referred to as red ochre, which occupy our attention here. I assume that the red color practices have a special significance for an evaluation of human behavior in the transition between Lower and Middle Palaeolithic. On the other hand they pose serious problems by the difficulties of the material that has to serve as evidence.

It has to be noted that color choice in culture is a subject of various fields of studies and the literature dealing with its various aspects is impressive. To mention some, such as ethnography (Turner 1967), color preference (Eysenck 1941; Pickford 1973), linguistics (Jacobson 1941; Berlin and Kay 1970), psychology (Murray and Daebler 1957), symbology (Leach 1976; Sahlins 1977) and neurobiology (Padgham and Saunders 1975; Zollinger 1979). You may notice that prehistory is, till the beginning of the eighties, not among the contributors. It is only very recently that it joins the debate by drawing attention to red color choice in past human behavior (Wreschner 1980).

Prehistorians and anthropologists usually agree that red ochre is a prominent feature in later prehistoric contexts, of *Homo sapiens sapiens* origin, probably because they can discern, since the Upper Palaeolithic, a widespread and variegated use of red color pigments in symbolic expressions. However the persistence of similar practices during the succeeding culture stages, even till the present, demands a closer look on finds and find circumstances of red pigments in Middle Palaeolithic and, because of existing relevant claims, in Lower Palaeolithic culture stages. Needless to say that it is prehistory which has to record these data in their chronological and cultural contexts. This can best be accomplished by their synchronic documentation in a diachronic model. Thus it conveys their stratigraphical placement in time and culture, their sequential occurrence, and whenever and wherever possible, their association with human types. This model allows as well for the integration of relevant ethnographic data as similar evidence from the uppermost time level and the related specific cultural details.

Before we attempt an evaluation let us first deal with some problems that concern ochre finds in archaeological contexts and in specific archaeological associations. A first attempt to compile an index of ochre finds on record resulted in a culturally affiliated classification of 251 sites, including burials. I suppose that there exist more which are not yet listed in this survey that shows the following distribution: Lower Palaeolithic – 3; Mousterian – 15; Upper Palaeolithic (Perigordian, Aurignacian, Solutrian) – 79; Late Palaeolithic (Magdalenian) – 42; Late Upper Palaeolithic and Epi-Palaeolithic outside Europe (Late Palaeolithic, Kebaran, Mushabian, Natufian, Ibero-Maurusian) – 20. The remainder comprise Mesolithic, Neolithic, Palaeo-Indian, Late Archaic, European Copper and Bronze Age and Siberian and African Iron Age sites.

Besides the regrettable fact that reports often lack the desirable detailed information on ochre finds and find circumstances, there exists a serious handicap for a solid evaluation of ochre finds. This is the total absence of information and knowledge on properties and behavior of powdered ochres in relation to archaeological remains, to bones, stone tools, grinding utensils, shells and other objects in soils in general and in ochre-stained soils in particular. Finds of end-scrapers with ochre traces suggest that hides could have been dyed. In burials, dyed garments disintegrated and there might be a possibility that their ochre content became fixed on, or stained, the skeleton. Hence the feasible conclusion on sprinkling of the body when in fact it might have been secondary.

It should be noted that the term "Red Ochre" as a collective determinator for red pigments of mineral provenience is unsatisfactory. We need therefore precise mineral classification and determination by analytical methods as well as color classification by a recognized color chart. Munsell would do the job. But what we need foremost are experimental studies by reconstructing those ochre phenomena which are encountered in archaeological associations. Thus red color traces on flake edges and blades, supposed to have served for ochre scraping (Bar-Yosef and Phillips 1977: 81) might be studied by microwear analyses. I further suggest to try establishing validity for intentional rubbing of ochre pieces by microwear research. Keely has recently drawn attention to these possibilities (1980: 172).

Though, with some measure of justification, Butzer can remark that "Many claims for ochre are not based on competent analyses or contextual evaluation but on unreasoned archaeological field conclusions" (*in:* Wreschner 1980), I can not agree, however, with Butzer's conclusion that "as a result, perhaps a majority of the claims for ochre in particular archaeological associations are either inconclusive or simply wrong" (*ibid.*). It could surely benefit our investigations if Butzer would have noted the sites he has in mind. I can only guess from his line of argumentation in his comment that his criticism concerns foremost the claims for the Lower Palaeolithic, but involves Middle Palaeolithic sites as well.

The three claims for finds of red color pigments in Lower Palaeolithic contexts are for Olduvai BKII (Leakey 1958: 1099-1103), Ambrona (Howell 1966: 129) and Terra Amata (Lumley 1966, 1975). The most impressive of these three is Terra Amata. Lumley claims 73 pieces of ochre with different color shades from yellow-brown to dark red (pers. comm.). The different colors point to firing of feasibly limonite or goethite. The question is if this had been done by man or if it could be accidental. Analyses of the material are not yet published. One can think of an accidental firing by a sweeping brush fire which could turn yellow-brown ferrohydroxides into reddish ferrioxides, provided that the mineral pieces were then on open ground, possibly even contemporary with the occupation. An investigation of this hypothesis is feasible. At Ambrona Howell reported an ochre slab which was meanwhile identified as a reddish siltstone which had been intentionally brought to the site. Both finds are associated with an Acheulian industry, assigned to *Homo erectus*. Chronologically these discoveries are unique. So are the cultural achievements of the Terra Amata people as claimed by Lumley and critically reviewed by Villa (1978).

I must stress that these Lower Palaeolithic red color claims are not followed by similar phenomena in all the succeeding stages of the Acheulian anywhere in Europe. Sporadic finds of ochre pieces are claimed for a few Sub-Saharan Acheulian sites by Beaumont (pers. comm.). We have to reckon therefore with a pronounced find gap of red color pigments between Terra Amata/Ambrona and Middle Palaeolithic (Mousterian) assemblages with abundant presence of lumps of red ochre at Abri Les Tares, La Ferrassie Layer C, Montgaudier Abri Lartet, L'Ermitage, Le Pech-de-L'Aze (France), Castillo (Spain), Ilskaya and Volgograd (Russia), and Qafzeh and Geula Caves (Israel).

The absence of ochre in Acheulian contexts, except perhaps Southern Africa, confronts us with the problem that about 7,000 generations of Acheulian populations have left no evidence for red color behavior since Ambrona and Terra Amata. Does that reflect in fact their indifference to the red color? Or could it be that the presence of ochre was not recognized during excavations by so many workers in the field?

Considering this situation and the evidence for ochre collecting in earlier Mousterian assemblages, the first clues for red pigment use by Neanderthals is of prime significance for an evaluation of evolutionary trends in behavior and cognitive capacities. They are obviously Neanderthals which created various forms of human burials, attached objects to the body – among them pieces of red ochre – and began to deposit animals and animal remains in a ritual form, as the discoveries at Nahr Ibrahim and Molodova testify. Though we shall never know the specific meaning attached to these activities, but those people left the unmistakable message for symbolic behavior, for cognitive expressions of beliefs and for their ability to create referential symbolic forms of a social and communicative nature. The basic similarity of attaching objects to the dead and the use of red ochre in Neanderthal mortuary customs and in those of *Homo sapiens sapiens* is striking. Both point to beliefs concerned with death, with the departed and the remaining, we may say with kin. The animal depositions suggest beliefs related to fertility and procreation, a theme which *Homo sapiens sapiens* developed in many variations. Whether we have here indications for contact and diffusion or independent inventions is impossible to discern.

The ochre practices in the Middle Palaeolithic and their persistence patterns in all later culture stages certify to their validity as a fossil indicator for mental and cognitive capacities of their practitioners. Jerison pointed out that "the human behavior – tool-making, social behavior, learning – are all peculiarly human because their execution involves major cognitive components" (1975: 52). Though tool-making has a much greater antiquity and distribution than ochre practices we can nontheless see in the latter a limited but significant parallel to the former. In both we can observe the manifestations of cognitive capabilities not of *Homo sapiens sapiens* only, but also of *Homo sapiens neanderthal*. The question poses itself if it is possible to extend this line of thought further into the Lower Palaeolithic. Before I am trying to elaborate on this possibility, I would like to draw attention to an interesting aspect in ochre behavior. When we take the Lower Palaeolithic finds and all the evidence from the Middle Palaeolithic, including in burials, we find that ochre appears in all pre-*Homo sapiens sapiens* contexts in the form of lumps, but nowhere as powder. It seems that natural pieces of ironhydroxides had been collected and fired to obtain red ochre in the form of pencils. There is no evidence available for use of powdered ochre, no sprinkling of the dead and no bedding of corpses on ochre. Grinding utensils are absent. In sharp contrast to the preceding stages, it is only since the Upper Palaeolithic that the powdered form of red ochre or ochre-fat or other semi-liquid mixtures become applied in burials, to figurines, amulets, spearthrowers, probably for dyeing hides, and of course in art. Grinding utensils and palettes become frequent and paint horns for carrying pigments make their appearance. The ochre pit mining installations at Lovas, Hungary, dated 32,000 BC testify to the efforts to meet the demand for red color pigments by Upper Palaeolithic hunters.

If we can acknowledge that the red color had such an attraction for cognitive creative behavior of *Homo sapiens sapiens* and *Homo sapiens neanderthalensis*, why do we hesitate to think this feasible for types which are still classified as *Homo erectus*? The capacities of Lower Palaeolithic hominids to preconceive tool forms, to kindle fire, to develop forms of social organization are obviously accepted. Curiously their ability to perceive and mediate environment by specific cultural means which some people like to call non-utilitarian, is doubted or outrightly dismissed. Mediating environment calls for the endowment of things and phenomena with meaning. This is what is meant by the humanization of hominidization.

The case of the red color pigments shows that red things could be endowed with meaning, and scholars have thought this feasible because of their blood-like color (Leach 1976). Edwards and Clinnick (1980: 381-383) have drawn attention to other "non-utilitarian" activities of Lower Palaeolithic man, besides ochre.

In the light of the substantiated Middle Palaeolithic evidence, but also of the still existing Acheulian gap, I venture to pose what is surely a speculative question: Can the Lower Palaeolithic red color phenomena perhaps indicate sporadic developments of a cognitive nature, of isolated achievements of some hominids that remained in their time an episode in the evolution of human behavior and which, by causes unknown, were not passed on? This speculation reflects not only the difficulties of the material but begs for the investigation of all behavioral indications, among them red color phenomena, in Lower Palaeolithic contexts. Hopefully yet hidden material might come to light provided that workers in the field are aware of these possibilities.

References

Bar-Yosef, O. and J.L. Phillips 1977 *Prehistoric Investigations in Gebel Maghara, Northern Sinai* Qedem 7, Institute of Archaeology, the Hebrew University, Jerusalem.

Berlin, B. and P. Kay 1970 *Basic Color Terms* University of California Press, Berkeley.

Edwards, St. W. and R.W. Clinnick 1980 Keeping the Lower Palaeolithic in perspective *Man* 15: 381-383.

Eysenck, H.J. 1941 A Critical and Experimental Study of Color Preference *Amer. J. Psychology* 54: 385-394.

Howell, F.C. 1966 Observations on the Earlier Phases of the European Lower Palaeolithic *American Anthropologist* 68(2): 129.

Jacobson, R. 1941 *Kindersprache, Aphasie und allgemeine Lautgesetze* Uppsala.

Jerison, H.J. 1975 Fossil Evidence of the Evolution of the Human Brain *Annual Review of Anthropology* 4: 27-58.

Keely, L.H. 1980 *Experimental Determination of Stone Tool Uses. A Microwear Analysis* University of Chicago Press.

Leach, E. 1976 Color Symbolism. *In: Culture and Communication* Cambridge University Press.

Leakey, L.S.B. 1958 Recent Discoveries at Olduvai Gorge, Tanganyika *Nature* 19: 1099-1103.

Lumley, H. de 1966 Les Feuilles de Terra Amata à Nice: premiers resultats. *Bull. Mus. Anthrop. Monaco* 13: 29-51.

Lumley, H. de 1975 Cultural Evolution in France in its paleoecological setting during the Middle Pleistocene. *In:* K. Butzer and Glyn Isaac (eds.) *After the Australopithecines* Mouton, The Hague.

Muray, D.C. and H.L. Daebler 1957 Colors and Mood-Tones *J. Applied Psychology* 41: 279-283.

Padgham, C.A. and J.E. Saunders 1975 *The Perception of Light and Color* Academic Press, New York.

Pickford, R.W. 1973 *Psychology and Visual Aesthetics* Hutchinson Educational, London.

Sahlins, M. 1977 Color and Cultures. *In:* J.L. Dolgin, D.S. Kemnitzer and D.M. Schneider (eds.) *Symbolic Anthropology* Columbia University Press, New York.

Turner, V. 1967 Color Classification in Ndembu Rital: A Problem in Primitive Classification. *In: Forest of Symbols* Cornell University Press, Ithaca.

Villa Paola 1978 *The Stone Artifact Assemblages from Terra Amata* Ph.D. Thesis, University of California, Berkeley.

Wreschner, E.E. 1980 Red Ochre and Human Evolution. A Case for Discussion *Current Anthropology* 21: 631-644.

Zollinger, H. 1979 Correlations between the neurobiology of color vision and the psycholinguistics of color naming *Experientia* 35: 1-8.

2

THE LEAVANT

AN AMINO ACID RACEMIZATION CHRONOLOGY FOR TABUN

Patricia M. Masters

The first attempt at producing an amino acid racemization chronology for the Neanderthal sites of the Near East was reported at the UISPP Conference in Nice (Bada and Helfman 1976). The dates derived from the D/L aspartic acid measurements were calibrated using an Upper Paleolithic bone sample from the cave at Sefunim. Due to the imprecision of calibrating by cultural association only, the date lists (Appendix A) for the caves of Tabun, Skhul, and Qafzeh were described as preliminary. Since the publication of these dates I have continued working on a more rigorous calibration procedure for these sites. The results of these efforts are herein reported.

Theory and Practice of Amino Acid Racemization Dating

The amino acids comprising the proteins of plants and animals are all L-amino acids. D-Amino acids, the stereoisomers of the proteinaceous L-amino acids, can be synthesized readily in chemical systems but are never incorporated into the proteins of living biological organisms. After death, the structural proteins of bone, shell, coral, wood, etc. are preserved in favorable depositional environments. Over thousands of years the spontaneous chemical reaction racemization converts L-amino acids into D-amino acids. Thus, fossilized materials of biological origin can be dated by measuring their contents of D-amino acids (Masters and Bada 1978). The half-life for the racemization of aspartic acid (one of the twenty common amino acids found in proteins) at 20°C is 15,000 years. The reaction reaches an equilibrium with a 50:50 mixture of the two stereoisomers, or when the D/L ratio equals 1, and this is the dating limit for any particular amino acid.

Racemization dating has several inherent advantages over other dating techniques. It can produce dates directly on bone. It requires merely 5–10 grams of fossil bone which permits dating of valuable hominid materials. Its time range extends to roughly 500,000 years in mid-latitude environments. The method is thus uniquely suitable for dating the Middle Paleolithic hominids of the Near East, provided a reliable rate constant can be determined.

In Situ Calibration of the Aspartic Acid Racemization Rate Constant

The accuracy of a racemization date depends not only on the state of preservation of the bone sample to be analyzed but also on the calibration of the racemization rate constant. Unlike radiometric dating methods having constant radionuclide decay rates, the rate of the racemization reaction varies with the temperature of the depositional environment. In applying the racemization method, two calibrating procedures are possible. First, a bone that has been dated by some independent technique (usually radiocarbon dating) is analyzed for its D/L aspartic acid content. The age of the bone (t) and the D/L ratio are substituted into the equation

$$\ln\left[\frac{1 + D/L}{1 - D/L}\right] - 0.14 = 2kt \qquad (1)$$

and it is solved for k, the rate constant for the aspartic acid racemization reaction. This rate constant is proportional to the average temperature of the depositional environment over the burial time t. Thus, temperature fluctuations are accounted for using this method of calibration.

In 1963 Vogel and Waterbolk published a date of 39,700 ± 800 BP (GrN -2534) for Tabun B on charcoal taken at a depth of 1 meter below surface. During the field seasons directed by Professor Jelinek, remnants of Layers B and C were excavated from Garrod's dump. Sacks of bone assigned to B (labeled TXB) and C (TXC) on the basis of soil type and animal species present were called to my attention at the Tabun field laboratory by A. Ronen during the 1979 field season, and I selected samples for racemization analysis. The TXB ratio, along with the Groningen date, can be substituted into equation (1) to calibrate the glacial rate constant at Tabun.:

$$k = 1.02 \times 10^{-5}\ \text{year}^{-1}$$

Calibration of the Rate Constant by Ground Temperatures

A second means of calibrating the rate constant is by evaluating the ground temperature of the locality under consideration and calculating a rate constant from a second locality having a known temperature and known rate constant. This calculation is carried out using the Arrhenius equation which relates the rate of a reaction (k) to the reaction temperature (T, in degrees Kelvin),

$$\ln\left[\frac{k_2}{k_1}\right] = \frac{E_a}{2.303R}\left[\frac{T_2 - T_1}{T_1 T_2}\right] \qquad (2)$$

where R is the gas constant (1.987 cal $\text{deg}^{-1}\ \text{mole}^{-1}$) and E_a is the activation energy of the racemization reaction (33.4 kcal mole^{-1}) in bone.

Because measurable ground temperatures are representative of the Holocene only, a further calculation must be made to convert the post-glacial rate constant into a glacial rate constant, i.e., one representative of cooler temperatures averaged over both Pleistocene and Holocene times. At two different coastal localities, a limestone cave on the island of Majorca in the western Mediterranean (Schroeder and Bada 1973) and open-air sites in southern California (Bada *et al.* 1979), the mathematical relationship of post-glacial to glacial rate constants is known to be

$$k\left(\frac{<10{,}000\ \text{years}}{>15{,}000\ \text{years}}\right) \cong 1.38 \tag{3}$$

Rate constants which average temperatures over 17–20,000 years have been shown to produce accurate dates over the last ~ 120,000 years (Bada and Deems 1975). Employing equations (2) and (3) in conjunction with present day ground temperatures should yield a rate constant applicable to bone specimens ranging in age up to the last interglacial. It is, of course, desirable to calibrate by both methods whenever possible as a cross check on the accuracy of the rate constant.

Ground Temperature Measurements

In 1976 the Amino Acid Racemization Dating Laboratory at Scripps Institution of Oceanography undertook a study of mean annual ground temperatures at sites around the world where racemization rate constants had been determined from radiocarbon-dated bones. Standard climatological data on average annual temperatures were inadequate for several reasons. Stations where meteorological information was recorded frequently were distant from the archaeological sites of interest. Records for ground temperatures are rare. Air temperatures offer at best an estimate of actual ground temperatures, which are affected by the amount of direct sunlight, the albedo of surface soils, the extent of ground cover, depth, etc. In order to obtain ground temperature measurements at specific localities and at appropriate depths, we needed a reliable, inexpensive recording thermometer. We decided to use the Pallmann method of chemical temperature integration.

As described by Lee (1969), the Pallmann method is a thermometric technique based on the hydrolysis of sucrose. This disaccharide is composed of one molecule of glucose and one molecule of fructose. In an acidified solution, sucrose hydrolyses to fructose and glucose. The rate of reaction is determined by the pH of the solution and by the temperature. With a standard solution of known pH, temperature can be calculated from the equation

$$T_e = \frac{5856.6}{20.1998 - \text{pH} - \log[\frac{\log(c_o/c)}{t}]}, \tag{4}$$

where T_e is the effective temperature in degrees Kelvin, t is the time expressed in days, c_o is the initial sucrose concentration, and c the sucrose concentration at the end of the exposure time. The original method published by Pallmann *et al.* (1940) relied on polarimetric analyses of the sucrose solutions. Our laboratory, however, has applied a gas chromatographic technique to directly measure sucrose, glucose, and fructose contents (Buccigrossi *et al.* in preparation). The term c_o/c is evaluated from the equation:

$$c_o/c = 1 + 0.95[0.485\,(H_f/H_s) + 0.510\,(H_g/H_s)], \tag{5}$$

where H_x represents the chromatographic peak heights of the various sugars.

The sucrose solution was prepared at the University of Haifa by mixing equal volumes of one solution containing 1500 grams sucrose in 1 liter distilled water with a second solution composed of 404 ml 0.2 M sodium citrate and 596 ml 0.2 N HCl. To inhibit the growth of microorganisms, approximately 10 ml formaldehyde was added per liter of final solution. HCl was used to adjust the pH to 3.42. Aliquots of 10 ml were sealed in pyrex tubes and then frozen. Immediately before implantation, the tubes were thawed and packed inside 15 cm lengths of 2.5 cm diameter polyvinyl chloride tubing to prevent breakage.

With the help of Prof. Ronen, two vials were buried in Tabun, two in Sefunim, and three in Qafzeh. The implantation localities at Tabun and Qafzeh are marked in Figures 1 and 2. Four of the original seven vials were recovered and analyzed. Exposure times and calculated temperatures are summarized in Table 1. Included in the table is the calibration locality of Del Mar, California, where the ground temperature was measured with a second batch of sucrose solution, pH 2.95, prepared in La Jolla.

Table 1

Ground temperatures calculated by the Pallmann sucrose method

Provenience*	pH	Date In	Exposure Time (days)	Temperature
Tabun #1	3.42	8 Oct 76	503	21.46
Tabun #2	3.42	”	”	21.44
Qafzeh #1	3.42	1 Dec 76	375	22.13
Qafzeh #2	3.42	”	”	20.42
Del Mar, CA	2.95	11 Mar 77	365	21.35

Israeli localities are mapped in Figure 1 (Tabun) and 2 (Qafzeh). The locality in southern California was in an open-air coastal Indian midden at a depth of 76 cm.

Temperature Adjusted Rate Constants for Tabun, Skhul and Qafzeh

In sites along the coast of southern California, and on the island of Majorca, radiocarbon dates on bone collagen span the Holocene and the end of the Pleistocene. These bones have been analyzed for D/L aspartic acid ratios as well, and the rate constants calculated. Meteorological records indicate very similar present-day climates at these sites. The data are summarized in Table 2. The Pallmann ground temperature (T_1) at Del Mar and the Holocene rate constant from SDM 16709 (k_1) are the reference parameters. Using the ground temperature at Tabun as T_2, k_2 can be calculated from equation (2). This k_2 represents the Holocene rate constant at Tabun. Adjusting k_2 by the factor 1.38

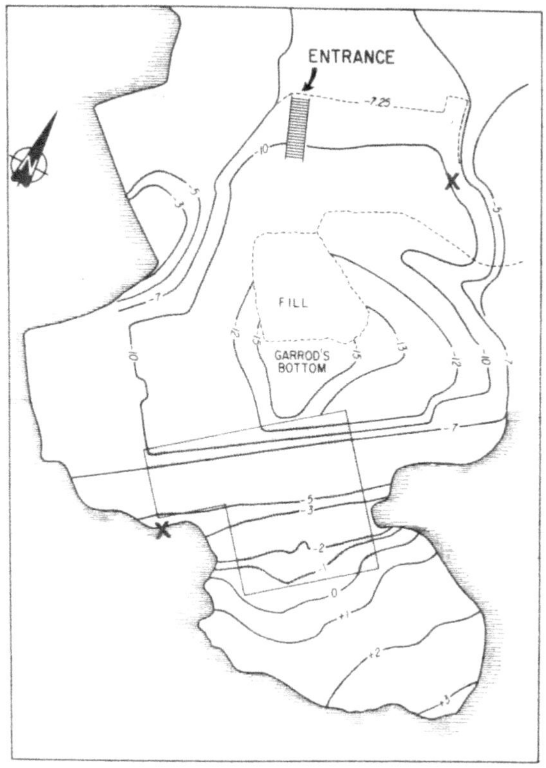

Figure 1 A contour map of Tabun cave showing an outline of the excavated units and burial locations (X) of the Pallmann temperature vials.

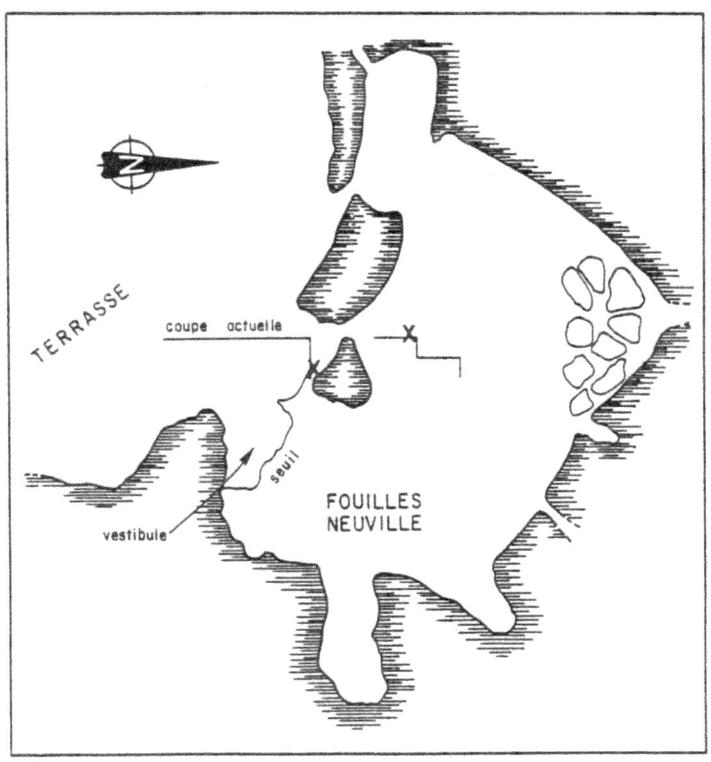

Figure 2 A sketch map of Qafzeh from Vandermeersch (1971) with the Pallmann temperature vial localities marked (X).

from equation (3), yields a rate constant averaged over both Pleistocene and Holocene temperature exposures:

$$k = 1.06 \times 10^{-5} \text{ year}^{-1}$$

The agreement between the *in situ* calibration and the temperature adjusted calibration is very encouraging. The *in situ* calibration can in turn be used to calculate a Holocene ground temperature for Tabun, again utilizing the Del Mar reference parameters of k and T in equation (2). The post-glacial temperature corresponding to $k = 1.02 \times 10^{-5}$ yr^{-1} is 21.02°C. This is 0.44° less than the Pallmann temperature measured in the Tabun deposits. The concordance between the temperatures is good considering the uncertainties in the radiocarbon sample provenience, in the provenience of the bone fragment used for racemization analysis, in the conversion factor of 1.38, and in the Pallmann determinations. Routine field precision of the Pallmann method when two or more batches of sucrose solution are used is ± 0.1°C (Lee 1969).

Table 2

Racemization rate constants in Holocene and Pleistocene bones from coastal locations

Location	Sample	Radiocarbon Age (yrs BP)	D/L Asp	k_{asp} ($\times 10^{-5} yr^{-1}$)	Temp.[a]
La Jolla, CA[b] (USA)	SDM 16709	8,360±75 (Pta-1725)	0.189	1.45	~16°
Laguna Beach, CA[c]	Laguna Woman	17,150±1470 (UCLA-1233A)	0.25	1.08	~16°
Muleta Cave[d] Majorca, Spain	faunal bone	8,570±350 (UCLA-1704C)	0.214	1.72	~17°
"	faunal bone	16,850±200 (UCLA-1704D)	0.273	1.25	~17°
"	faunal bone	18,980±200	0.293	1.22	~17°

[a] Mean annual air temperatures at present.
[b] Ike *et al.* 1979.
[c] Bada *et al.* 1974.
[d] Schroeder and Bada 1973.

Results

Employing these two rate constants, a revised date list for Tabun and Skhul is reported in Table 3. The Tabun calibration is applied to Skhul also because of the sites' close proximity in the Wadi Mughara (Nahal Me'arot). The list includes previously analyzed specimens and new results on TXB, TXC, lower C, Skhul V, Skhul VII, and fragments of associated bone from Tabun I.

Because radiocarbon dates are lacking at Qafzeh, the only means of calibration is to adjust the SDM 16709 rate constant by the ground temperature differences between Del Mar and Qafzeh. The two temperatures measured at Qafzeh are fairly divergent. Although the two vials recovered were not identified by implantation locality, it seems reasonable to assume that the higher temperature of 22.13°C represents the south-facing terrace locality while the cooler temperature of 20.42° comes from the interior of the cave. Two separate calculations were made using equations (2) and (3), and the resultant racemization ages are presented in Table 4.

Table 3

Aspartic acid racemization ages for Tabun and Skhul based on *in situ* and temperature-adjusted calibrations

Sample	D/L Asp	Age $k=1.02\times10^{-5}\,yr^{-1}$	$k=1.06\times10^{-5}\,yr^{-1}$
Tabun TXB	.441	–	38,000
TXC?	.544	53,000	51,000
I[a]	.548	53,000	51,000
lower C			
square 6, bed 19 (burned)	.627	65,000	63,000
square 5, bed 22	.658	71,000	68,000
upper E			
square 14, bed 49	.742	87,000	83,000
Skhul 2-79[b]	.401	35,000	33,000
V[c]	.448	40,000	39,000
VI[c]	.554	54,000	52,000
VII[c,d]	.496	46,000	45,000
IX	.560	55,000	53,000

a Unidentified bone associated with the Tabun woman skeletal remains; obtained from T. Molleson, British Museum. All other Tabun samples are faunal material.

b Faunal bone from terrace remnants recovered with the aid of A. Ronen.

c These bones were refluxed in diethyl ether for approximately one week in order to remove preservatives.

d Second sample submitted by W.W. Howells. First sample yielded a D/L aspartic acid ratio of 0.447.

Table 4

Aspartic acid racemization ages for Qafzeh based on Pallmann ground temperature measurements

Sample	D/L Asp	Age (years)* $k=1.12\times10^{-5}\,yr^{-1}$	$k=9.71\times10^{-6}\,yr^{-1}$
Couche 9 -faunal	.536	47,000	54,000
XVII-Hominid 3	.470	39,000	45,000
XVII-Hominid 6	.478	40,000	46,000
XXI-faunal	.625	59,000	68,000
XXII-faunal	.679	68,000	78,000

* The rate constants are derived according to procedures described in the text from the ground temperatures measured at Qafzeh of 22.13°C ($k=1.12\times10^{-5}\,yr^{-1}$) and 20.42°C ($k=9.71\times10^{-6}\,yr^{-1}$).

The accuracy of the Pallmann ground temperature measurements will obviously have some bearing on the reliability of the rate constants deduced from them by means of the Arrhenius equation. However, the absolute values of T_1 and T_2 actually have little effect on the rate constants or on ΔT, as shown in Appendix B. If the ground temperature measurements are performed using an internally consistent method, the ΔT is probably accurate, and the "reality" of the measured temperatures is not critical. Calculated rate constants are highly sensitive to ΔT, as demonstrated by using mean annual air temperatures differing by $3^{o}C$ between Haifa and Del Mar. This calculation results in a 1.5-fold increase in the rate constant and consequently a 20% reduction in the racemization age.

Intuitively, the Pallmann temperatures appear to be too high. An explanation may be found in the fact that the sucrose hydrolysis rate is an exponential function of temperature, i.e., hydrolysis goes faster at higher temperatures. T_e, the exponential or effective temperature, typically overestimates T_a, the arithmetic mean temperature, in laboratory experiments. T_e rather than T_a may more closely approximate the temperature to which the rate of the racemization reaction is responding because this k is also an exponential function of temperature. But how true this is depends on the activation energy (E_a) for both reactions. The E_a for aspartic acid racemization in bones is known, but we have not yet determined E_a for sucrose hydrolysis.

Discussion

The amino acid racemization dates derived for Tabun show some interesting correlations with the chronology suggested by Farrand (1979). Beginning with the lowest level sampled in this series, Jelinek's Bed 49 corresponds with Garrod's Ea and upper Eb (Garrod and Bate 1937). This level contains the Amudian industry described by Garrod and predates the Levalloiso-Mousterian levels at Tabun. According to Farrand (*op. cit.*), the sandy eolian deposits of upper E resulted from a period of regressing sea level approximately 90,000 years ago, which marked the end of the last interglacial. The racemization age of 87,000 years is in close agreement with the geological interpretation of this level.

The finer, silty eolian deposits of layer D correspond with a much lower sea level, and tree pollens (Horowitz *in* Jelinek *et al.* 1973) also indicate a moist climate coinciding with the first advance of the last glacial cycle. This continental climate would correspond with the oxygen isotope stage 4 (Jelinek *p.c.*) which is thought to span the period of 75,000–64,000 years BP (Shackleton and Opdyke 1973). Unfortunately, the bone I excavated from this layer last October did not contain sufficient amino acids for a date despite structural evidence of good preservation.

Following the unconformity, an amino acid date for lower C, Bed 22 of Jelinek, is 71,000 years. The amount of time represented by the unconformity is unknown. If, as Jelinek (*p.c.*) suggests, layer C is contemporaneous with oxygen isotope stage 3, then it should have been laid down in the period of 64,000–32,000 years BP (*op. cit.*). Sediment analysis (Farrand 1979) and the lower Levalloiso-Mousterian industry (Garrod and Bate 1937) points to the earlier half of this time range, but the amino acid date for Bed 22 seems rather old. Due to the amount of charcoal present in layer C, one cannot rule out the possibility that the Bed 22 faunal specimen was heated, thus producing an erroneously elevated date. Subsequent analysis of a charred bone fragment from lower C (Bed 19) gives a D/L ratio (0.627) comparable to the Bed 22 specimen, indicating that it has probably been heated. The dates for TXC and Tabun I of 53,000 years BP agree very well with Farrand's age assignment and with the inferred oxygen isotope chronology. However, both should be viewed with some caution because TXC does not represent *in situ* material (Jelinek *p.c.*) and the Tabun I date is on unidentifiable bone fragments "associated with" the hominid remains (T. Molleson *p.c.*). If we accept the association with Tabun I, the amino acid date supports Garrod's interpretation of the provenience of the skeletal remains (layer C).

All of the amino acid dates discussed above are calibrated by the TXB bone and its assumed contemporaneity with the charcoal dated to 39,700 ± 800 years BP by the Groningen radiocarbon laboratory. Should a more controlled means of calibrating the aspartic acid racemization rate constant at Tabun become available, it may be necessary to revise these dates in the future. But the concordance between the Pallmann ground temperature calibration and the *in situ* TXB calibration indicates that the date list reported here is certainly more accurate than the series of dates reported in 1976 (Bada and Helfman). Additionally, two of the three racemization dated layers correlate well with chronological information independently derived from geological and cultural analyses.

The suite of racemization dates for Skhul range from 35,000–55,000 years. According to Garrod and Bate (*op. cit.*), all of the burials came from the Levalloiso-Mousterian horizon, and on faunal evidence these deposits were thought to be younger than Tabun C. Higgs (1961) has suggested, also on faunal grounds, that Skhul was occupied as much as 10,000 years later than Tabun B. In reviewing these studies, Jelinek has found the faunal arguments for a 10,000 year gap between Skhul and Tabun B unconvincing (Jelinek *et al.* 1973). The amino acid dates for Skhul V and the faunal fragment 2–79 tend to support Garrod and Bate and Jelinek. Skhul VI and IX, however, appear to be contemporaneous with Tabun C as dated by TXC and Tabun I.

The Skhul hominids represent a mixture of morphological characteristics (Howell 1957). Skhul V and IX are anatomically modern for the most part, while VI and VII resemble Tabun I particularly in post-cranial traits. Amino acid dates for VI and IX as well as Tabun I may indicate the presence of a highly variable population at Mount Carmel in the range of 55,000 years ago but do not support a trend toward "sapiensization" (*ibid*).

Two other factors should be considered in an interpretation of the Skhul dates. I have applied the Tabun TXB calibrated rate constant to Skhul because of the proximity of the two sites. However, the Skhul burials all came from the exposed terrace deposits, so the Tabun rate constant may be underestimating the actual racemization rate at Skhul due to warmer average ground temperatures in the terrace. If this were the situation, the amino acid dates for Skhul are too old.

Second, all the bone samples from Skhul (excluding 2–79) had been treated with preservatives. Refluxing in diethyl ether may remove the preservative, but depending on the type of preservative used, it is difficult to predict the effect on the subsequent amino acid analysis. The first specimen of Skhul VII yielded a D/L ratio of 0.429 (Bada and Helfman 1976 and Appendix A), and the second refluxed sample gave D/L = 0.496. Thus, greater fluctuations in D/L ratios resulting from conservational treatment could be contributing to the wide range of amino acid dates at Skhul.

The revised racemization dates for Qafzeh reflect the same inconsistencies that were reported initially (*ibid*). Recalibrating the rate constant at Qafzeh by Pallmann ground temperature measurements has not resolved the discrepancy between the dates for layer 9 and level XVII. In fact, the rate constants derived from the two Pallmann temperatures worsen the situation. The Upper Palaeolithic layer 9 bone comes from the interior of the cave where the preferred rate constant is $9.71 \times 10^{-6}\ yr^{-1}$. The calculated age for layer 9 then becomes 54,000 years while the terrace rate constant of $1.12 \times 10^{-5}\ yr^{-1}$ yields ages of about 40,000 years for the XVII hominids. The dates for XVII look rather young in comparison with Farrand's sediment chronology (1979). On the other hand, levels XXI (59,000) and XXII (68,000) agree quite well with his geological interpretation. Analysis of additional samples from layer 9 would be advisable, and contamination due to preservatives may also be a problem with the XVII hominids.

Summary

While further refinements of the amino acid racemization chronology for these three sites are yet possible, significant advances have been accomplished since the first series of dates were published in 1976. The reference calibrations recently achieved for coastal southern California in combination with ground temperature measurements there and in Israel have placed the Tabun calibration on a much sounder footing. The concordance between the TXB *in situ* calibration and the temperature adjusted rate constant lends considerable credibility to the amino acid dates at Tabun. Geological evidence for the ages of layers C and E and the amino acid dates for these layers are mutually supportive. At Skhul and Qafzeh, where the calibrations are less exact, and where contamination by preservational treatment is a problem, the dates reported in this paper should still be considered preliminary dates.

Acknowledgements

I thank the following people for providing bone specimens for analysis: W.W. Howells (Skhul), A. Jelinek (Tabun), A. Ronen (Tabun and Sefunim), T. Molleson (Tabun I), and B. Vandermeersch (Qafzeh). A. Ronen, M. Weinstein, and S. Weiner were generous with their time and effort in helping with the temperature vial experiment. Analyses of the temperature vials were carried out by D. Buccigrossi, and D. Darling assisted with amino acid analyses. I am grateful to the L.S.B. Leakey Foundation for supporting this research.

APPENDIX A

Table 1. Preliminary aspartic acid racemization ages for Tabun and Skhul, 1976

Site	Provenience	D/L Aspartic Acid	Age (years)
Sefunim	Layer 8-9 Upper Paleolithic	0.472	27,000[a] $k_{asp} = 1.64 \times 10^{-5}\ yr^{-1}$
Tabun	Lower Layer C Square 5, Bed 22, T69	0.658	44,000
	Upper Layer E Square 14, Bed 49	0.742	54,000(?)
Skhul	Hominid VI[b]	0.540	33,000
	Hominid VII	0.429 (?), 0.539[c]	24,000 (?), 33,000
	Hominid IX	0.526	31,000

[a] Age estimated by A. Ronen.
[b] Sample refluxed in diethyl ether for 6 days in order to remove preservatives.
[c] This D/L ratio was obtained by E. Hoopes and E.T. Peltzer in our laboratory, using gas chromatographic techniques. Gas chromatographic analysis of the Sefunim bone yielded D/L aspartic acid = 0.486, in agreement with the value determined using the amino acid analyzer technique.

Table 2. Preliminary aspartic acid racemization ages for Qafzeh, 1976

Provenience	Sample	D/L Aspartic Acid Ratio	Age (years) Sefunim k_{asp}	Age (years) Adjusted k_{asp}*
Couche 9	faunal	0.536	32,000	39,000
XVII	Hominid 3	0.470	27,000	33,000
XVII	Hominid 6	0.478	27,000	33,000
XXI	faunal	0.625	40,000	49,000
XXII	faunal	0.679	46,000	57,000

* Sefunim $k_{asp} = 1.64 \times 10^{-5}$ at $\sim 18^{\circ}C$ was adjusted to the mean annual air temperature of $17.4^{\circ}C$ at Jerusalem by use of equation (2). This adjusted k_{asp} value is $1.34 \times 10^{-5}\ yr^{-1}$.

APPENDIX B

Dependence of racemization rate constants and ages on values for T_1, T_2 and ΔT in the Arrhenius equation[a]

Del Mar T_1(°C)	Haifa T_2(°C)	ΔT(°C)	k_2[b] ($\times 10^{-5}\ yr^{-1}$)	Tabun I Age (yrs)
16[c]	19[c]	3	1.36	40,000
20	23	3	1.35	40,000
21.35[d]	21.45[d]	0.1	1.06	51,000
16.0	16.1	0.1	1.06	51,000

[a] The value for the k_1 in these calculations is $1.45 \times 10^{-5}\ yr^{-1}$.
[b] Calculated rate constants for Tabun.
[c] Present-day mean annual air temperatures.
[d] Ground temperatures measured by Pallmann sucrose method.

References

Bada, J.L. and L. Deems 1975 Accuracy of dates beyond the ^{14}C dating limit using the aspartic acid racemization reaction *Nature* 255: 218-219.

Bada, J.L. and P.M. Helfman 1976 Application of amino acid racemization in paleoanthropology and archaeology *Union Internationale des Sciences Prehistoriques et Protohistoriques*, IXe Congres, Nice, Colloque I, pp. 39-62.

Bada, J.L., R.A. Schroeder and G.F. Carter 1974 New evidence for the antiquity of man in North America deduced from aspartic acid racemization *Science* 184: 791-793.

Farrand, W.R. 1979 Chronology and palaeoenvironment of Levantine prehistoric sites as seen from sediment studies *J. Archaeol. Science* 6: 369-392.

Garrod, D.A.E. and D.M.A. Bate 1937 *The Stone Age of Mount Carmel* London: Oxford, vol. 1.

Higgs, E.S. 1961 Some Pleistocene faunas of the Mediterranean coastal areas *Proceedings of the Prehistoric Society Series 2,* 27: 144-154.

Howell, F.C. 1957 The evolutionary significance of variation and varieties of 'Neanderthal' man *Quarterly Review of Biology* 32: 330-347.

Ike, D., J.L. Bada, P.M. Masters, G. Kennedy and J.C. Vogel 1979 Aspartic acid racemization and radiocarbon dating of an early milling stone horizon burial in California *Amer. Antiq.* 44: 524-530.

Jelinek, A.J., W.R. Farrand, G. Haas, A. Horowitz and P. Goldberg 1973 New excavations at the Tabun cave, Mount Carmel, Israel, 1967–1972: a preliminary report. *Paleorient* 1: 151-183.

Lee, R. 1969 Chemical temperature integration *J. Applied Meteorol.* 8: 423-430.

Masters, P.M. and J.L. Bada 1978 Amino acid racemization dating of bone and shell. *In:* G.F. Carter (ed.) *Archaeological Chemistry II,* Advances in Chemistry Series No. 171, Washington, D.C.: American Chemical Society, pp. 117-138.

Pallmann, H., E. Eichenberger and A. Hasler 1940 Eine Neue Methode der Temperaturmessung bei okologischen oder bodenkundlichen Untersuchungen *Ber. Schweiz. Botan. Ges.* 50: 337-362.

Schroeder, R.A. and J.L. Bada 1973 Glacial-postglacial temperature differences deduced from aspartic acid racemization in fossil bones *Science* 182: 479-482.

Shackleton, N.J. and N.D. Opdyke 1973 Oxygen isotope and palaeomagnetic stratigraphy of equational Pacific core V28-238: oxygen isotope temperatures and ice volumes on a 10^5 year and 10^6 year scale *J. Quat. Res.* 3: 39-55.

Vandermeersch, B. 1971 Recentes decouvertes de squellettes humains a Qafzeh (Israel): essai d'interpretation. *In:* F. Bordes (ed.) *Origine de l'homme moderne.* Paris: Unesco, pp. 49-54.

Vogel, J.C. and H.T. Waterbolk 1963 Groningen Radiocarbon Dates IV *Radiocarbon* 5: 163-202.

Discussion

D.A. Roe: This is not really a question, but rather two comments on different aspects; and I could say how very nice it is to have such a complicated matter explained so very clearly to us all.

One thing is a passing comment. I do not know if people observed on the screen the name of Klasies River mouth. I thought I would just mention, in case anybody does not know, that this is a site which is really very crucial to the whole theme of this conference, because it is one of those in South Africa which have contributed to the increasing volume of evidence for the very much older age of the South African Middle Stone Age than had previously been thought. There are Middle Stone Age levels there, dating, I think, from more than 120,000 years ago. I think we shall not be talking about South Africa at all on any of the other papers, shall we. So surely this is a point that one could just make in passing.

The other point I wished to make was about the radio carbon dating method (and here I speak with some diffidence because I am not at all qualified to talk about this), but I do not know whether people know that at Oxford, in the Research Laboratory for Archaeology and the History of Art, they are now building a new machine based on an accelerator principle. It is enormously expensive, and I do not pretend to understand it at all, but it is going to produce carbon dates in the order of 70,000–80,000 plus and perhaps after refinements, even older than that; and the kind of size of sample that it will use to do this will be only a few milligrams. I remember asking a question of one of those involved – would it perhaps be possible to date a small quantity of pollen grains? He thought for a while, and said yes.

So maybe the radio carbon method is not yet completely dead, and how very nice it would be to have an independent chronometric method that would be able to check the kind of dates that we have been hearing about this afternoon.

E. Trinkaus: I have two related questions. First, in one of your tables you gave the pH for the sediments in which you calculated ground temperature. To what extent will variance in pH, in addition to variance in temperature, affect the rate of racemization? My second question is, in a sense, related to the first. We were discussing earlier that a lot of the specimens from Skhul were treated with preservative. Might that possibly explain some of the incredible range in dates for what are supposed to be virtually contemporaneous specimens?

P.M. Masters: Starting with your preservation comments first, that was my major concern. We have seen in our laboratory that if you cannot remove the preservatives, particularly the plastic preservatives, it will increase the D/L ratio and that seems to be exactly what is going on in this situation. Secondly, pH is, indeed, one of the chemical factors that affects the rate of racemization. However, to be fossilized and survive in the deposit, the bone won't have been exposed to a low pH. The rate of reaction is not affected at pH as high as about ten, which is not that commonly found in a fossil environment. So we have not found pH to be a major problem. Usually it is contamination.

A.J. Jelinek: I have one other question that concerns the use of temperature data derived from Tabun for the calculation of dates at Skhul. The Skhul material all sat out on a shelf in front of the shelter and I think that the soil temperature would probably, on the average, have been somewhat higher there than it would have been at Tabun. It is possible that the racemization rate at Skhul was somewhat more rapid than at Tabun because the former deposit is not in shadow during the day, as I recall, whereas Tabun is for a significant part of the day.

P.M. Masters: That is something that occurs to me as well, and when we planted the vials at Tabun in 1976 we had automatically assumed that the same temperature would apply to Skhul, being so close. Thinking it over later, I decided it is time to bring back some more vials, something I will finally be able to do this year.

W.R. Farrand: Art Jelinek and I were just arguing about the temperature relationship and the sun relationship at Skhul because it is on a north facing slope, and I am not sure how much of it would have been protected from the sun. That is something we have to see on the spot. That, I think, is something interesting to look at.

The other thing I wanted to point out concerns the dates for Qafzeh: you did not mention, first of all, the inversion of dates which, I think, is a problem we have to worry about. Layer 9 is Upper Palaeolithic, and the other beds with Roman numerals are all Middle Palaeolithic. Stratigraphically, they are older, and you may trace the stratigraphy physically. So that we know that there is inversion in the dates there.

I want to ask you a question about the seeming problem of the hominids. Both hominids giving younger dates and the fauna from slightly older layers giving much older dates. Man versus animals: is it the preservation problem of Trinkaus?

P.M. Masters: We do have preservation problems, and it is my guess that there would be preservatives applied to human material more likely than to faunal material. Another factor involves the probable cooking or heating of faunal bones, and that has been one reason why I have been reluctant to work with faunal materials. Also, if I recall from Prof. Vandermeersch's reports, I think that layer 9 is loaded with charcoal, is that correct? Yes; and, even if the bones themselves were not cooked but lying near a hearth for significant periods of time, my guess is that is why layer 9 at Quafzeh looks as high as it does.

THE MIDDLE PALAEOLITHIC IN THE SOUTHERN LEVANT, WITH COMMENTS ON THE APPEARANCE OF MODERN *HOMO SAPIENS*

Arthur J. Jelinek

Introduction

This discussion will focus on the Middle Palaeolithic cultures of the Palestinian region adjacent to the coast of the eastern Mediterranean and on the derivation of these cultures from their Lower Palaeolithic antecedents. The scheme presented here for the development of these cultures has certain implications for the appearance of modern man in this region which are treated in the conclusions of the paper.

The Southern Levant, as referred to in this discussion, can be seen as distinct from a Central Levant, including the Lebanon and adjacent Syria, and a Northern Levant, including northwestern Syria and the Hatay.

The history of Middle Palaeolithic research in this area has been reviewed recently by Perrot (1968: 287–290) and Hours *et al.* (1973: 254–256). From these discussions it is possible to distinguish several phases of research by pre-historians in the Southern Levant. An initial period of preliminary exploration and poorly controlled excavation is exemplified in the work of Turville-Petre (1927), whose primary contribution lies in having demonstrated the presence of significant Middle Palaeolithic industries and fossil man at the sites of Emireh and Zuttieh in the Wadi Amud. A second period, characterized by extensive excavation with improved techniques for recording context, but a generally selective retention of artifacts, is represented by the work of Garrod at Shukbah, Tabun, El Wad and Skhul (Garrod and Bate 1937; Garrod 1942); Neuville at Qumm Qatafa, Abou Sif, Erq El Ahmar, Et Tabban, and Qumm Naqous in the Wadi Khareitoun, and at Qafzeh with Stekelis (Neuville 1951); Stekelis at Kebara and Abou Uzba (Stekelis and Haas 1952; Shick and Stekelis 1977); and Wreschner (1967) at Geula. These workers were active primarily in the decades prior to and following World War II, which interrupted excavations in this area. A third period of research, beginning in the early 1960s and still in effect, has been characterized by more exacting techniques for recording provenience and geological context, and an increasing interest in preserving all exposed artifactual materials and associated paleoenvironmental evidence. Workers in this period include Watanabe at the Amud Cave (1964; in Suzuki and Takai 1970: 77–114), S.R. Binford at Shubbabiq [Shovakh] (1966), Vandermeersch at Qafzeh (1966, 1967, 1969), Ronen at Sefunim [Iraq el Barud] (1968) and Tirat-Carmel (1974), Marks and his students at Rosh Ein Mor (Marks and Crew 1971; Crew 1976) and Nahal Aqev (Munday 1977), Jelinek at Tabun (Jelinek *et al.* 1973; Jelinek 1975), Bar Yosef at Zuttiyeh

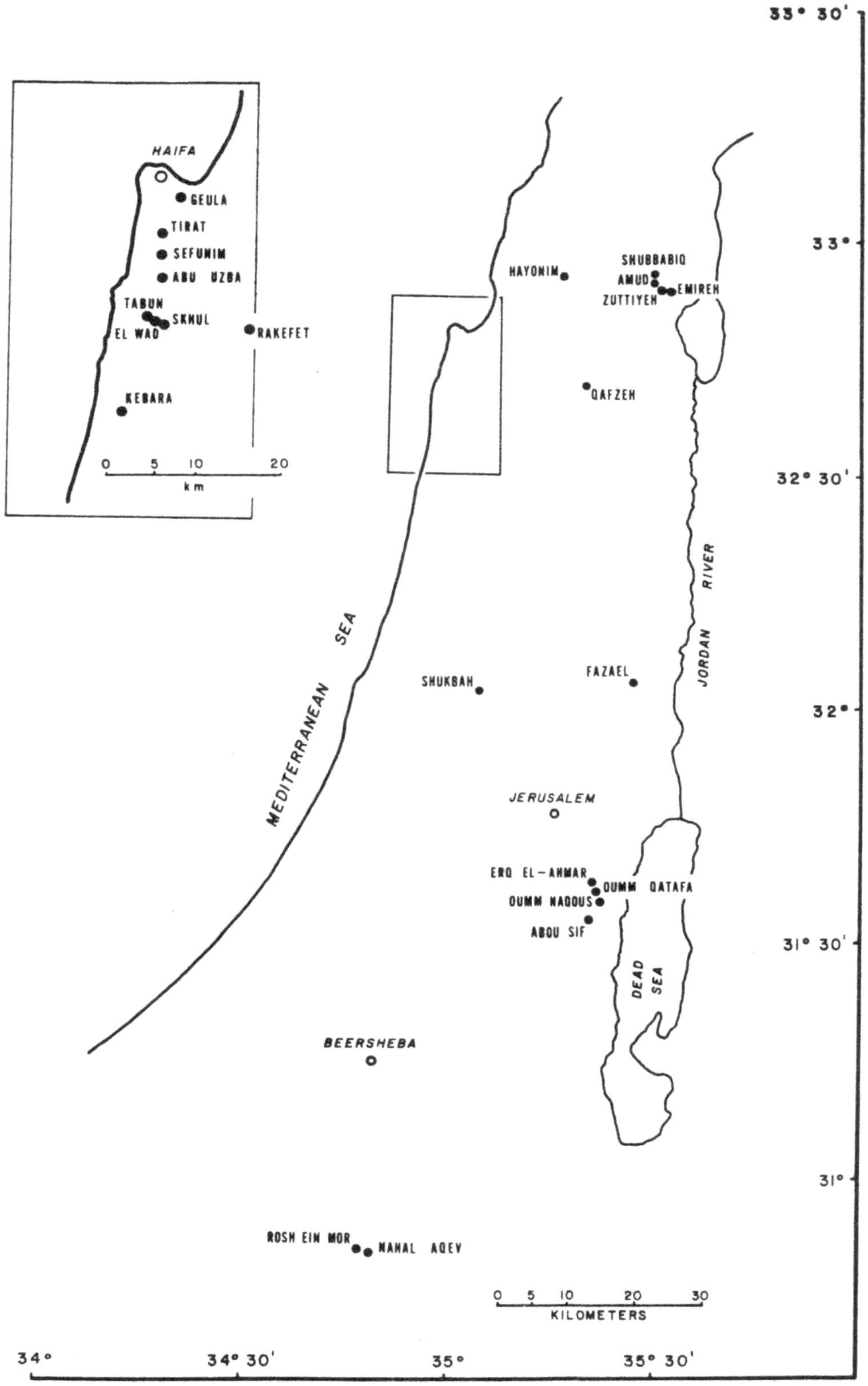

Figure 1 Major Middle Palaeolithic sites in the Southern Levant.

(Gisis and Bar Yosef 1974), Hayonim, and Fazael (Bar Yosef *et al.* 1974) and Higgs and Noy at Rakefet.

As a result of this research we now have a relatively large number of collections from about two dozen excavated sites. However, due to the different excavation techniques in the different periods of research, differing methods of presenting data, and the lack of full publication of many sites, most of these collections can now be compared at only a very general level.

We will, for the moment, put aside the question of distinguishing between the late Lower Palaeolithic and early Middle Palaeolithic and will first survey the industrial sequence from early in the last interglacial/interpluvial stage up through about the middle of the last glacial/pluvial stage. There is good general agreement between two recent syntheses which treat this cultural succession (Copeland 1975; Ronen 1979). Within the interglacial/interpluvial period both authors recognize: (1) industries that are usually termed "Late (or Upper) Acheulian", and are characterized by high relative frequencies of bifaces and a generalized inventory of other retouched tools; (2) industries that are variously termed "Yabrudian", "Acheuleo-Yabrudian", or "Yabrudian of Acheulian Facies", and are characterized by high percentages of *déjeté* and transverse scrapers that frequently exhibit Quina or demi-Quina retouch, abundant simple scrapers and few or no bifaces; (3) industries that are termed "Amudian" or "Pre-Aurignacian", and are characterized by an abundance of Upper Palaeolithic elements (backed points and blades, burins, and end scrapers) and relatively few side-scrapers or bifaces. Within the first half of the last glacial/pluvial period two basic kinds of industries are recognized: (1) industries termed "Abou-Sif" or "Tabun D" type Mousterian, and characterized by an abundance of Levallois points, many of which are elongated, an abundance of prismatic blades, and relatively high frequencies of Upper Palaeolithic elements in some occurrences; and (2) industries termed "Tabun C–B" type Mousterian, and characterized by abundant broad Levallois flakes derived from tortoise-core techniques, with varying quantities Levallois points, prismatic blades, and side scrapers and relatively few Upper Palaeolithic elements. Copeland distinguishes two chronological phases in these last industries; an earlier with few Levallois points, and a later with increasing numbers of relatively short Levallois points and more blades (*op. cit.*: 330–335). She thus segregates three kinds of Mousterian in a chronological succession, with the Abou-Sif or Tabun D type following directly the Yabrudian-Acheulian industries of the last interglacial/interpluvial as a "Phase 1" Mousterian, followed by the two variants of the Tabun C–B type as "Phase 2" (Tabun C) and "Phase 3" (Tabun B). Ronen, in contrast, would interpret the differences observed within a single culture (*op. cit.*: 303). No consistent chronological relationship is distinguished by these authors for the interglacial/interpluvial industries other than the placement of the Amudian/Pre-Aurignacian near the end of that time period.

The Middle Palaeolithic at Tabun

With this general level of agreement, it is possible to examine the two major groups of industries distinguished above in the light of new evidence produced by the recent excavations at the Tabun Cave on Mount Carmel; a site whose long period of occupation and varied cultural sequence have much to contribute to our understanding of man's development in this time period. Dorothy Garrod's excavations at Tabun (Garrod and Bate 1937) first revealed the essential elements of the basic sequence outlined by Copeland and Ronen. No other site of comparable depth has yet been discovered in the Southern Levant. It is thus particularly appropriate that a re-examination of this section should serve as a point of departure for a re-appraisal of those industries.

Methodology

Our excavations at Tabun were confined to a section about ten meters deep in the high profile that formed the south wall of Garrod's excavation (Jelinek *et al.* 1973). They extended from the base of her Layer B down through Layers C, D, and E into what was probably Ed in her sequence. The lower portion of our excavation was offset 3 m to the west to avoid the unstable section of the profile above the travertine spring deposits. Thus we did not sample Garrod's Layer F. However the western portion of the lowest levels of our excavation encountered deposits that appear structurally to pre-date the collapse of Layers F and G into the solution cavity underlying the front chamber. Following the excavation, the preliminary stratigraphy employed in the field was refined and expanded to make maximum use of observable artifact concentrations within each geological context and to isolate sequences of deposition within otherwise uniform geological strata. This revision was made possible by the detailed recording of the natural stratigraphy in drawings and photographs at 1 m intervals in N-S and E-W profiles in the excavation, and by exact provenience recording, including position within a three-dimensional grid and individual artifact attitude, for the 44,000 lithic specimens (greater than 2.5 cm. in maximum dimension) that were recovered. This revised stratigraphy has enabled us to isolate approximately 130 separate contexts in the geologic sequence that are represented by 99 or more lithic artifacts (see Figure 2). Most of the additions to the preliminary sequence of beds have been made in the deposits that are equivalent in elevation to Garrod's Layer D. Here our excavation intersected a major collapse of sediments in the inner chamber of the cave. The complexity of deposits resulting from an extended period of filling and subsidence is reflected in the proliferation of distinct geological deposits in this portion of the section. Within the revised stratigraphy we have distinguished a series of "Major Stratigraphic Units", isolated from one another by clear disconformities that in most cases have truncated the underlying sediments. We assume that most of the periods of non-deposition reflected by the surfaces between the Major Units are due to the karstic subsidence and collapse of nearby sediments into underlying solution cavities. A total of fourteen such units have been identified, eight of which (II–IX) lie within the deposits adjacent to Garrod's Layer D. Figure 2 shows the relationship between Garrod's stratigraphy and our major units. There is some question of exact equivalence in the lower levels; we have no clear equivalent of her Layer Ec (a thin layer characterized by very high frequencies of Micromquian bifaces) and she did not recognize the distinctive nature of our Unit XIV, which apparently precedes the first major karstic collapse. Our full bed sequence will be published in the final report. The equivalents in the revised stratigraphy for those units in the preliminary (field) stratigraphy that have been mentioned in previous publications (Jelinek *et al.* 1973; Jelinek 1975), are as follows: Beds 1–18 (Unit I) have only minor changes in artifact assignment; Bed 19 has

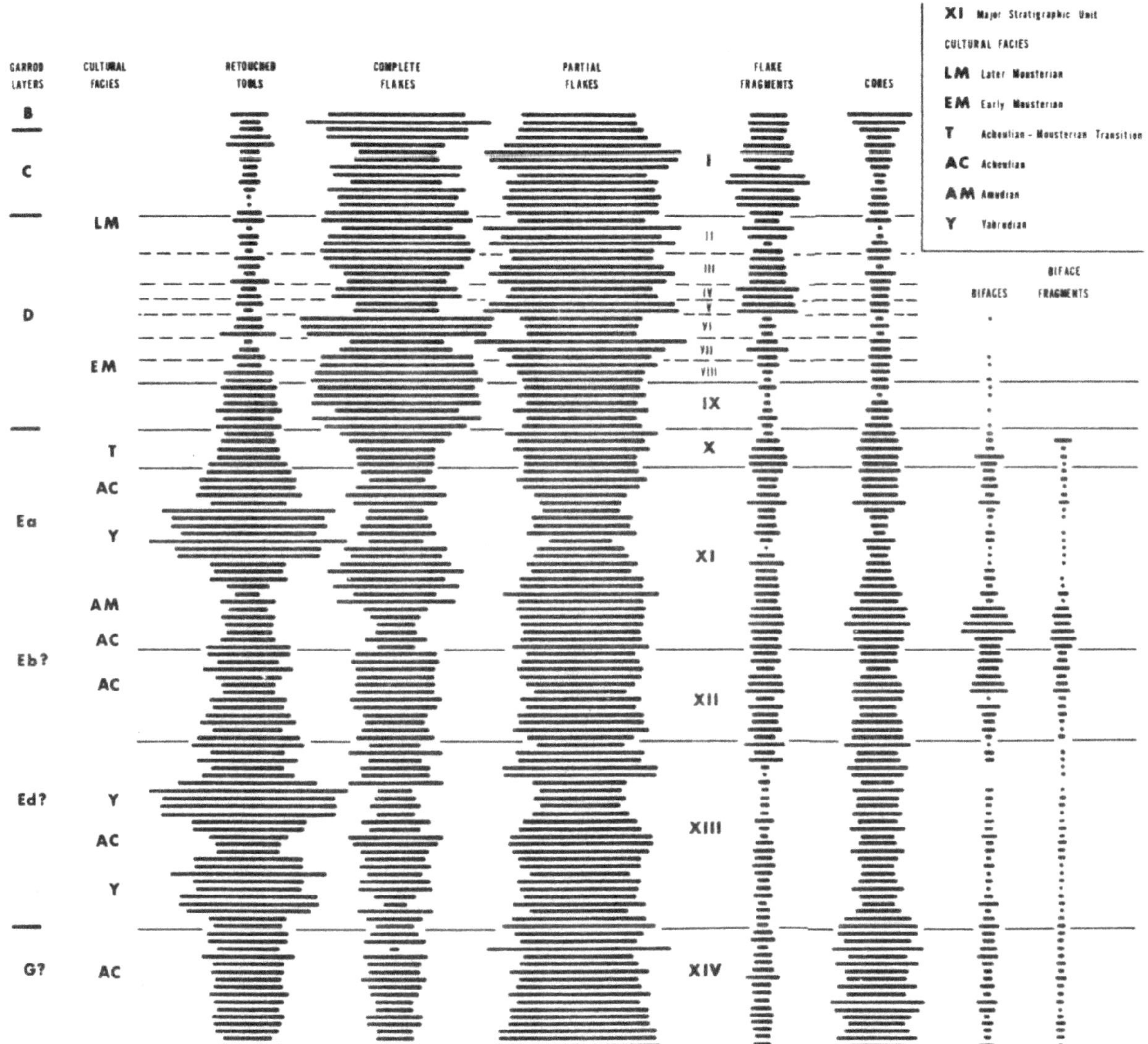

Figure 2 Tabun Cave: Relative frequencies of major lithic categories through time.

had substantial numbers of artifacts changed to Bed 20 (Unit I); Beds 48A and 48B are now Beds 75S and 75I (Unit XI); Bed 80 is now Bed 90 (Unit XIV). The beds in the revised stratigraphy that are included in each Major Stratigraphic Unit are shown in Table I.

Table I

Major Unit	Beds
I	1 – 26
II	27 – 32
III	33 – 36
IV	37 – 40
V	41 – 42
VI	43 – 49
VII	50 – 58
VIII	59 – 61
IX	62 – 69
X	70 – 72
XI	73 – 77
XII	78 – 80
XIII	81 – 85
XIV	90A – 90J

Numerous beds below Unit I have been further subdivided (e.g., Bed 75 above) into smaller units as size and artifact concentrations allowed.

The Acheulian and the Yabrudian

Returning to an examination of the previously described cultural sequence as it is represented in these deposits, and beginning with the earliest levels, we can see the following progression. Unit XIV contains a remarkably uniform sequence through the entire portion that we excavated. In terms of basic categories of lithic artifacts (Figure 2), it is distinguished from later levels by consistently higher frequencies of cores (c. 15–20%), moderate frequencies of retouched tools (c. 20%), relatively low frequencies of complete flakes (c. 15%), relatively high frequencies of partial (broken) flakes (c. 35%) and a low, but consistent frequency of bifaces (c. 3–5%) and biface fragments (c. 2%). The typology of the retouched tools is quite heterogeneous in each layer (Jelinek 1975: 308, 309, 311), but homogenous throughout. The industry clearly conforms to the accepted pattern of the Acheulian and, considering the presence of quite thin and regular forms among the bifaces and a probable mid-to-early last interglacial date (Farrand 1979: 373 –375), it can be assigned to the Late or "Upper" Acheulian.

In contrast to these early levels, the overlying deposits of Units XI, XII, and XIII contain industries that exhibit considerable variation. Due to the collapse of sediments into the solution cavity underlying the front chamber of the cave, there is a time gap of significant, but unknown magnitude between the uppermost levels of Unit XIV and the lowermost levels of Unit XIII. Similarities between the industry found at the bottom of Unit XIII and that of Unit XIV certainly stem in part from the fact that artifacts weathering from the exposed surface of Unit XIV contributed to the content of the initial deposition of Unit XIII. This kind of depositional "heritage" can be expected wherever a long-term stabilized surface occurs in deposits such as those at Tabun (particularly considering the sandy nature of the lower wind-deposited sediments), and it frequently poses a problem when we address the question of continuity from one industry to another. Within the deposits of Units XI, XII, and XIII we can distinguish clear examples of both characteristic Acheulian and Yabrudian industries (including "Acheuleo-Yabrudian" and Yabrudian *sensu strictu* without bifaces). In the data summarized in Figure 2, the clearest typologically Yabrudian levels can also be distinguished by extremely high frequencies of retouched tools (c. 40% and up to > 50%), few or no bifaces, and relatively low frequencies of complete flakes. Three separate episodes of this nature are visible in the sequence; two in Unit XIII and one in Unit XI. The Acheulian, in contrast, is characterized by relatively high frequencies of bifaces (up to 15%), biface fragments, and cores and quite low frequencies of retouched tools (c. 15–20%). Because of the considerable number of samples available and the relatively high degree of continuity in the sequence (the included disconformities do not appear to represent prolonged periods of non-deposition), it has been possible to use the Tabun sequence to examine the kinds of typological relationships that have been used to distinguish between Yabrudian and Acheulian industries in the past. The results of this examination are, in some respects, surprising and may have important implications for the interpretation of similar industries elsewhere. The most distinctive element mentioned in distinguishing Yabrudian industries in the past has been the asymmetrical convergent (*déjeté* or *Winkelkratzer*) scraper (Rust 1950: 14–16). Bordes also mentions transverse and simple convex scrapers as particularly abundant (1955: 488) although he sees as the dominant characteristic of Layer 25 at Yabrud I the *"dominance des racloirs de tous types" (op. cit.)*. It appeared to me, by inspection, that the layers at Tabun that contained relatively high frequencies of *déjeté* scrapers also contained relatively high frequencies of transverse scrapers and that limaces and scrapers with thinned backing (*racloirs a dos aminci*) were virtually confined to these layers. In order to determine whether, in our large numbers of samples, either *déjeté* scrapers (Type 21) or the more comprehensive set of scrapers (Types 8, 21–24, and 27) could serve independently as a defining character of a distinct Yabrudian industry, I tested their frequencies with respect to all scrapers in a regression analysis. The results of this test are presented in Table II.

Table II

Correlation Coefficients Based on Absolute Frequencies of Scraper Types

Relationship	Cases	*r*	*p*
A. Type 21 with all *other* scrapers (Types 8-20, 22-29)	82	.8168	<.001
B. Types 8, 21-24, 27 with all *other* scrapers	86	.9183	<.001
C. Type 21 with Types 8, 22-24, 27	77	.7913	<.001
D. Type 21 with *all* scrapers (Types 8-29)	82	.8617	<.001
E. Types 8, 21-24, 27 with *all* scrapers (Types 8-29)	86	.9588	<.001

These tests show that the frequencies of all of these categories of scrapers are very highly correlated both when we compare mutually exclusive typological categories and when we compare each of the assumed defining sets with the whole population of scrapers. What these results suggest very strongly is that there is a single set of scrapers present in all of our samples and that the kinds of scrapers used to define the Yabrudian are present throughout the sequence of Yabrudian and Acheulian layers in consistent ratios with respect to all other kinds of scrapers. Thus, it appears that typologically the Yabrudian at Tabun can be characterized simply on the basis of the high frequency of all scrapers (as Bordes suggests for Layer 25 at Yabrud), rather than just *déjeté* scrapers or *déjeté* and transverse scrapers. It is possible that the *déjeté* scraper caught the eye of Rust and Bordes as a designating variable because, as a relatively rare form in Western European industries (Bordes 1955: 498), it appeared to represent a distinctive element when it was present in large numbers. This was, in fact, valid reasoning, but it may have obscured the distinctive nature of all of the scraper assemblages as opposed to just the Yabrudian.

An additional feature that has been held to be distinctive of the Yabrudian is the frequency of Quina retouch. While Bordes (1955: 505) refers to the Yabrudian as a "*faciès oriental du type Quina*", he does not present the Quina index for the collections at Yabrud, nor does he specifically mention Quina retouch in his discussion. Some demi-Quina retouch was reported by Solecki (1970: 204) on tools from layers thought to follow the Pre-Aurignacian at Yabrud, but no Quina retouch was noted below these levels. Skinner (1965), however, reports very high Quina indices for Rust's Layers 14, 22, and 25, and Copeland (1975: 326), in a discussion of Yabrud asks, "can there be Yabrudian without Quina retouch?". I examined the three strongest manifestations of Acheulian (based on biface frequencies) at Tabun, in Beds 72, 76, and 79, as compared to the three strongest occurrences of Yabrudian (based on scraper frequencies), in Beds 75, 82, and 83. The highest Quina indices occurred in the Yabrudian of Bed 75S (20.9) and the Acheulian of Bed 79 (20.0). Combined samples of all of the scrapers from the three Acheulian occurrences and all of the scrapers from the three Yabrudian occurrences were compared with the following result.

Table III

Retouch on Scraper Edges from Yabrudian and Acheulian Contexts

Context	Retouch Type Frequency (%)			Total
	Normal	Demi-Quina	Quina	n
Acheulian	88.1	7.7	4.2	402
Yabrudian	88.4	7.9	3.7	1938

All retouched scraper edges were used in this comparison; i.e., each edge of a multiple-edged tool was counted separately. The frequency of each type of retouch in the two populations is virtually identical ($\chi^2 = .3026$, $p = > .80$), which means that at Tabun Quina retouch does not seem to be more prevalent when scrapers are more abundant (in higher frequency) or handaxes are fewer. This too suggests that we are dealing with a single, internally consistent, scraper industry.

Thus it appears that the two major industries of the last interglacial period at Tabun can be defined simply on the basis of: (1) high ratios of scrapers and low ratios of bifaces (Yabrudian), and (2) high ratios of bifaces and low ratios of scrapers (Acheulian). We will then, for the moment, restrict our examination to these two general categories of defining variables. A linear correlation between these categories yields no significant relationship ($r = .0915$ with 85 cases), however an examination of the distribution of the

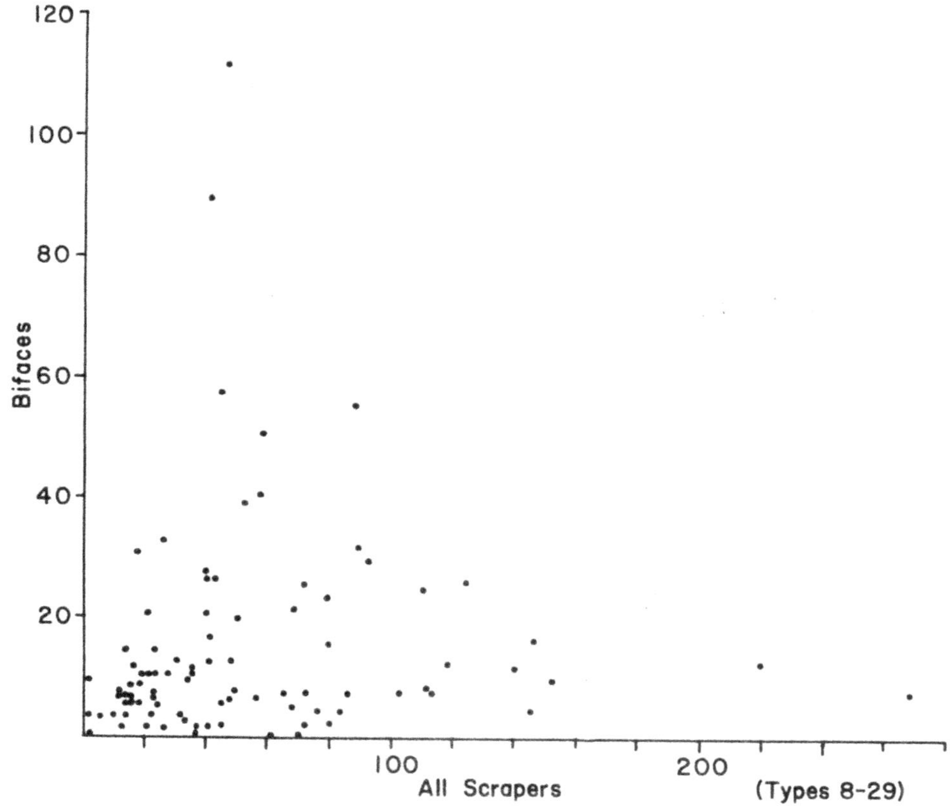

Figure 3 Tabun: Units X-XIV: Absolute frequencies of bifaces and scrapers.

relationship of absolute frequencies (Figure 3) shows a strong tendency for the larger samples to be distributed along either the abscissa or the ordinate axis of the scattergram. This suggests that samples that include over 20 bifaces will very seldom include more than 100 scrapers and, conversely, samples that include over 100 scrapers will very seldom include more than 20 bifaces. These two axes, then, include clear examples of what can be called distinctive Acheulian and Yabrudian industries. The major question that confronts us here arises from the large number of samples that contain less than that minimum number in each artifact category mentioned above. Do these samples represent a mixture of distinctive Yabrudian and Acheulian industries, or do the Yabrudian and Acheulian represent extremes of a single industry which were manifested during periods of more intensive utilization of the cave? One method of clarifying this issue is to examine the relationship between bifaces and scrapers (the defining variables of the two industries) through time. Figure 4 shows the relationship of these variables, expressed as the percentage of scrapers (Types 8–29) in the total of scrapers and bifaces in each sample in our collections. These samples are scaled in a rough approximation of their relationships through time. The striking feature of this diagram is a suggestion of continuity in the industries through Units XIII, XII, XI, and X. This sequence shows that following the period of collapse and stabilization of sediments that terminated Unit XIV there was a long period dominated by the production of numerous scrapers and few bifaces. This pattern dominates Unit XIII and shows only one brief interruption during which a somewhat higher ratio of handaxes is represented. Unit XII shows an increase in the relative frequency of bifaces from earlier to later contexts, and Unit XI begins with even higher frequencies of bifaces. This period of emphasis on biface manufacture is followed by a very rapid decline in biface frequency and a corresponding increase in emphasis on scraper manufacture up to levels similar to those in Unit XIII. Then follows another

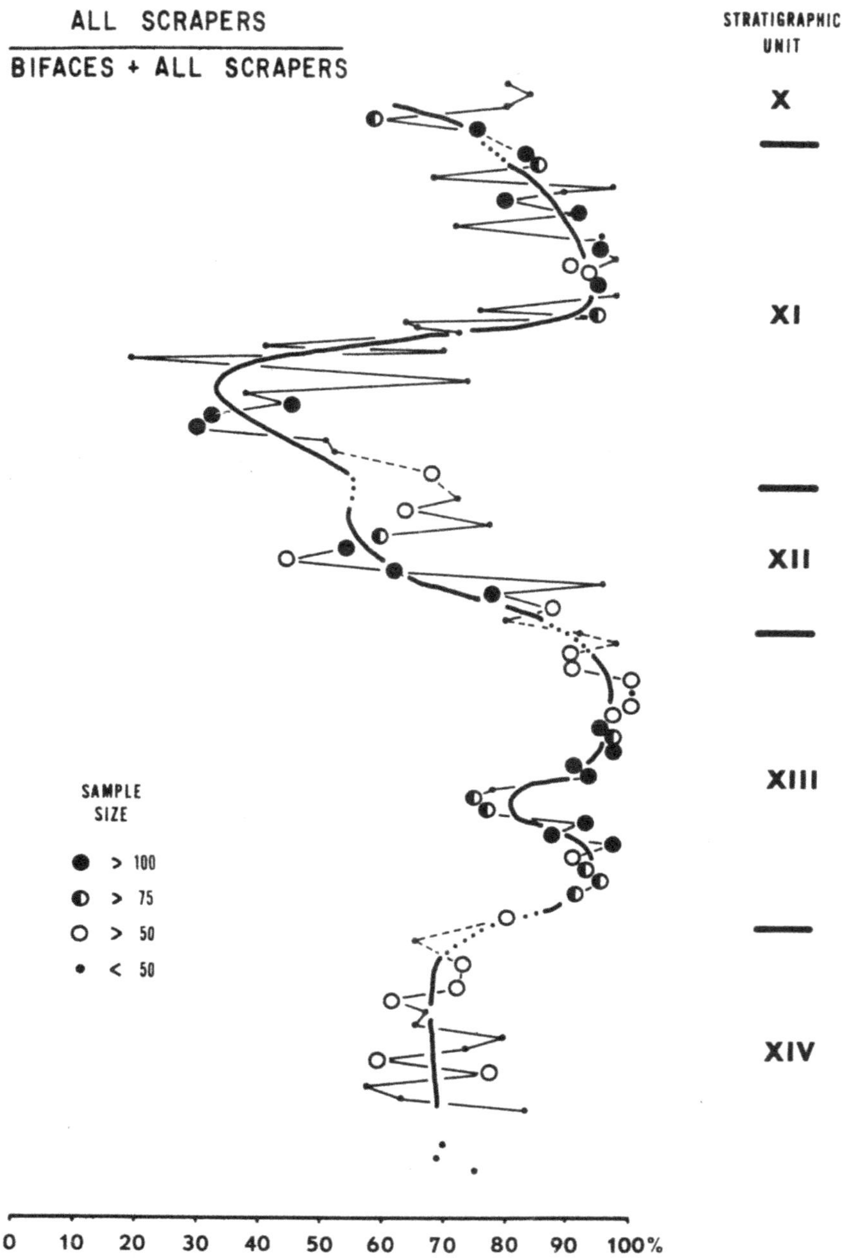

Figure 4 Ratio of scrapers to bifaces.

gradual rise in biface frequency that continues into Unit X, after which bifaces gradually disappear from the artifact inventory.

The Mugharan Tradition

Before discussing the interpretation of the information conveyed by Figure 4, I would like to point out that this is the first time that we have been able to examine such a large number of stratigraphically-controlled samples of these industries. In fact, I have been unable to discover any comparable series of samples in the literature currently available on the Lower and Middle Palaeolithic. This opportunity to study these industries as a long succession of relatively brief occupations in a single restricted setting might thus be expected to provide insights into their variability that have not been possible with shorter and less well controlled series.

The continuity suggested by this succession seems to indicate that what we are observing here are gradual changes in a single industry. It appears that over a long period of time, from the end of the deposition of Unit XIII until well into the deposition of Unit XI, there was a gradual shift in emphasis in the tools which were left in the cave toward relatively more bifaces. This gradual transition follows a long period during which very few bifaces and a great many scrapers were left in the cave and is followed, after a relatively brief transition, by another similar episode. The decline in relative frequency of scrapers toward the end of this sequence suggests the beginning of a trend toward more bifaces similar to earlier ones beginning at the end of Unit XIII. The industry of Unit XIV, which precedes this sequence, is remarkable for its uniformity and its central position with regard to the extremes of these frequencies. The suggestion there is one of a less specialized industry, with all of the elements that were later exaggerated in the two polar extremes of the industries from Units XIII to X. This impression is reinforced by a scraper assemblage in the later beds of Unit XIV that is similar in the frequencies of the "Yabrudian" types to those of the later industries. The general uniformity of the scraper assemblage through the whole sequence, as discussed above, is also very strong evidence for a single cultural tradition. Considering all of this evidence, I would propose that we are dealing with a single, but highly variable, industry, within which two extreme facies can be distinguished. These facies correspond to industries that have been previously designated as Acheulian and Yabrudian. I would suggest that the complex of industries as a whole be designated as the "Mugharan Tradition", and that within this tradition we can distinguish several variants or facies, including a "Yabrudian Facies" and an "Acheulian Facies". These specialized facies contrast with a continuing generalized "Mugharan" industry that resembles the Late Acheulian of Unit XIV. I have not entered into the baptism of this new cultural tradition lightly. When the continuity in this sequence became apparent it was clear that some unifying term would be necessary to encompass the range of variability that we could now demonstrate. I rejected the use of a "Yabrud (or "Yabrudian") Tradition" because of its previous usage in designating only industries with high frequencies of scrapers and the somewhat confusing usage of Skinner (1965: 204–205) in designating one type of his "Group C" Mousterian. I likewise was forced to reject the use of "Acheulian Tradition" for an industry which in one extreme form is characterized by an absence of bifaces. The fact is, that there is no term in the current literature that can, on the basis of accepted usage, be applied to this industrial tradition. I suggest that the name "Mugharan" is appropriate in at least two respects. First, it is derived from the recognized archaeological locality of the Tabun Cave (where this continuity can be demonstrated), the Wadi el-Mughara in the report of Garrod and Bate (1937). Second, the incorporation of "cave" in the meaning of *mughara* in Arabic is appropriate to the archaeological occurrence of this tradition, which is, with one possible exception, confined, in our knowledge, to cave and shelter sites. The term "Tabunian", which might otherwise be an appropriate designation, was applied some years ago by Howell (1959: 15) to the basal industries at Tabun (Layer G) and Ouum Qatafa (Layers E, F, and G) which had previously been designated as "Tayacian". While it now appears that these industries are poor samples of an Acheulian culture (e.g., Jelinek 1975: 312), the re-use of this term for something quite different would be confusing. I would suggest that the Mugharan Tradition be considered the earliest known manifestation of the Middle Palaeolithic in the Southern Levant. This suggestion is based on the composition of the Yabrudian Facies which, typologically, is strikingly similar to the Quina Mousterian of the Middle Palaeolithic of Western Europe, and represents the earliest appearance of the Mugharan Tradition at Tabun.

If we accept the interpretation of a unified Mugharan Tradition, we are left with at least two pressing questions relevant to this material. (1). What is the place of the Amudian in this sequence? (2). Are there any likely explanations of this industrial variability that can be tested with evidence at hand?

The Amudian

The Amudian from our excavations at Tabun was described in an earlier publication (Jelinek 1975: 310–313) on the basis of a small sample drawn from Bed 48B of the provisional stratigraphy (now Bed 75I). Since that time a more thorough analysis of the artifacts and the statigraphy has led to some significant changes in the interpretation of this distinctive industry. One important consideration has to do with the distribution of characteristic Amudian tools (backed knives, end scrapers, and burins) and the associated prismatic blades that frequently served as blanks for tool manufacture. It is now clear that there is a single major concentration of this material in the area of the site that we excavated. This concentration is very heavy at the center of the excavation, which was also topographically the lowest elevation of the surface at the time of the occupation. Amudian elements begin to occur in the upper levels of Bed 76S and reach their highest concentration in several sub-beds within Bed 75I in the revised stratigraphy. These beds rise to the east and west of the main concentration to elevations of well over a meter above the central deposit. It may be that similar fluctuations in elevation led Garrod to interpret the results from her horizontal excavation spits as indicating three separate horizons of Amudian at Tabun (Garrod 1956: 46–47), or it may be that three separate concentrations did in fact exist in the deposits that filled the outer chamber; we may never be able to resolve this question. At any rate, our findings now indicate that a relatively short but intensive occupation or succession of occupations left the characteristic Amudian artifacts at Tabun. When we look beyond the distinctive typological features of the Amudian to the more general aspects of the industry (Figure 2) we can see that the Amudian layers are signaled by a sharp rise in the frequency of complete flakes, relatively few retouched tools and bifaces, and decreasing frequencies of cores. The single most distinctive feature of the Amudian within these general categories is the hitherto unprecedented high frequency of complete flakes. In fact, the high relative frequency of complete flakes alone is sufficient to separate the Amudian levels from the Yabrudian and Acheulian facies. The presence of considerable numbers of prismatic blades in the Amudian is evidence of an emphasis on core preparation that is also unique up to this point in our sequence. In Figure 4, the Amudian is represented by the small samples of tools beginning with the highest ratio of bifaces to scrapers (18.2% scrapers) in the lower portion of Unit XI, and directly follows the most extreme of the Acheulian facies in the sequence (in Bed 76S and 76I). It directly precedes the final manifestation of a Yabrudian facies in the sequence, which in turn is followed by a succession of layers in Unit X which show decreasing frequencies of Mugharan Tradition tools, and rapidly increasing frequencies of complete flakes. These complete flakes include increasing frequencies of Levallois flakes and points and prismatic blades from the top of Unit XI well up into Unit IX. I would like to suggest that there may be similar sequences of causal events leading up to the appearance of the Amudian industry in Unit XI and the Transitional industries of Unit X and that these events may ultimately be linked to an explanation of the variability in the Mugharan Tradition.

Interpretation of the Mugharan Sequence

In order to explore this situation further it is necessary to briefly review the evidence for the chronological placement of the sequence at Tabun and then expand the perspective of our discussion beyond the immediate area of the Southern Levant. The best evidence of chronological placement of the lower deposits at Tabun is derived from the analysis of sediments carried out by Farrand and Goldberg (in Jelinek *et al.* 1973: 155–163). The sedimentary sequence suggests that during the deposition of the earlier deposits (Garrod's Layer G, F, and lower Layer E) sands carried by wind action from nearby beaches contributed substantially to the filling of the cave. This has been taken as evidence of a high stand of the Mediterranean near the mouth of the cave, which Farrand equates with beach deposits at an elevation of 39m at the mouth of the Wadi Mughara (Farrand 1979: 373). That this high sea stand had retreated by the beginning of the deposition of Layer E is indicated by the collapse of sediments in the outer chamber of the cave during the deposition of Layers G and F (*op. cit.*: 375). The increasing ratio of silt to sand in the upper portions of Layer E (Units XII, XI, and X) suggest that by this time the sea was some distance from the mouth of the wadi and was continuing to retreat across the coastal plain. In this respect, it is interesting to examine the geological interpretation of the Amudian deposits at the Abri Zumoffen in the Central Levant, about 75 km north of Tabun. Here the Amudian is incorporated into beach sediments (as the "Beach Industry") and directly overlies beach sediments (Copeland 1978: 42–46). This situation clearly suggests a retreat of the sea within a short time before the Amudian occupation. The Amudian in our sequence occurs well down in Unit XI, during which a sea retreat is assumed to be under way. A slight rise in sand relative to silt in the upper levels of Unit XI centering at about –6 m at the middle of our section (Farrand and

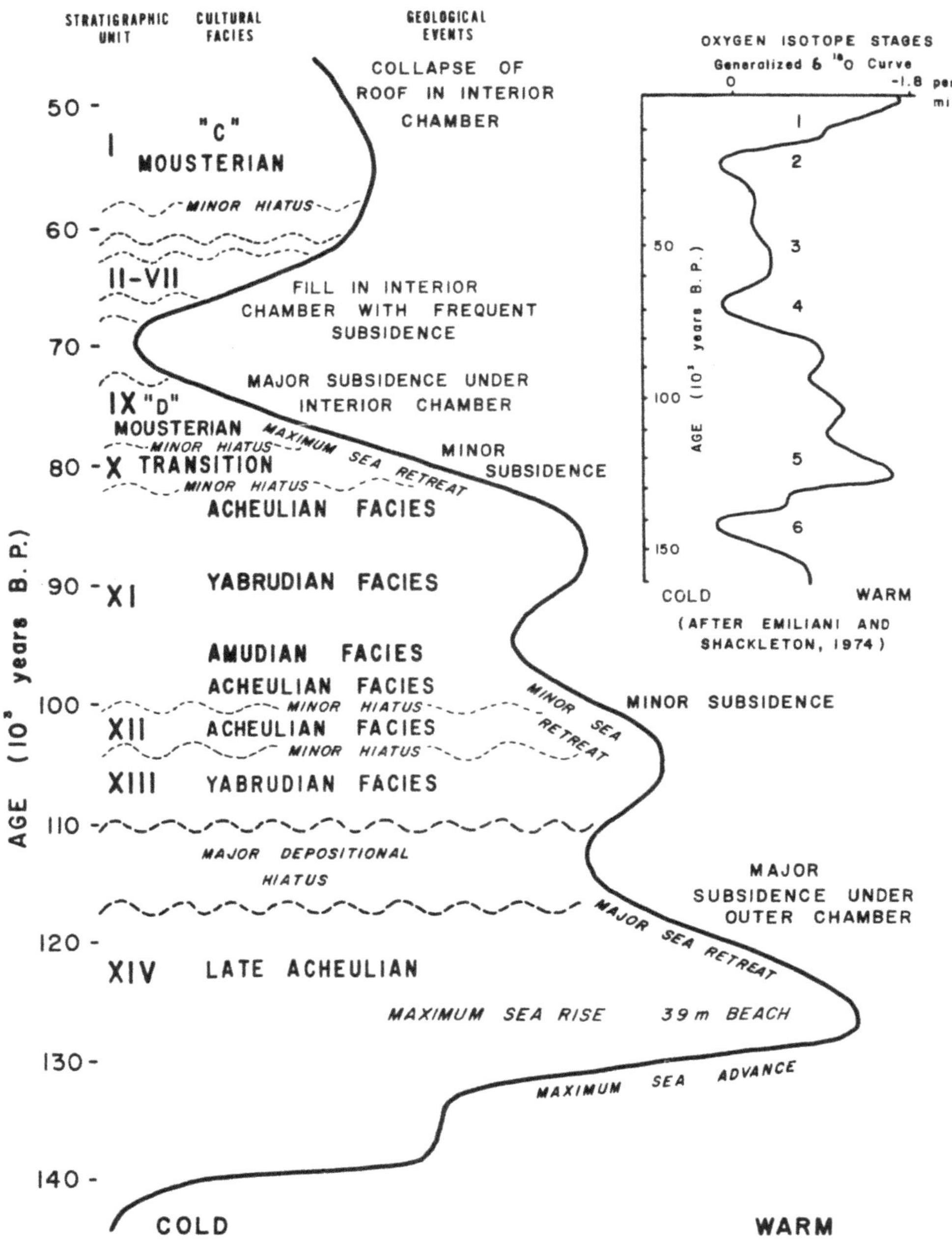

Figure 5 Correlation of the Tabun sequence with palaeoenvironmental conditions.

Goldberg in Jelinek *et al.* 1973: Fig. 3, 159) may indicate a short period of stability or marine resurgence contemporary with the Yabrudian Facies of Unit XI. The sedimentological evidence suggests a minimal contribution from dune sands and a maximum of windblown silt by the time of the deposition of Layer D (beginning with our Unit IX), indicating that the sea was considerably removed from the vicinity of the cave at that time. Beginning with the deposition of Unit IX we have evidence of a major subsidence of deposits in the inner chamber of the cave. It is likely that this collapse of sediments is linked to a further marked lowering of the groundwater table due to a significant further retreat of the sea. Palynological evidence (Horowitz 1979: 253) suggests that by this time a "pluvial" climate was present at Tabun, characterized by cooler and more moist conditions than at present, which supported a relatively dense arboreal flora.

In an attempt to place these events in a more comprehensive framework, I have examined current evidence of global paleoclimatic change based on world-wide changes in oxygen isotope ratios. This evidence, primarily derived from deep sea cores, is now widely accepted as an indicator of past climatic change and seems to be confirmed by independent evidence from the Late and Middle Pleistocene (e.g. Porter *et al.* 1977). Figure 5 presents a correlation of the geomorphological evidence from Tabun with the paleoenvironmental conditions implied by deep-sea oxygen isotope studies. The isotope sequence is derived from the work of Emiliani and Shakleton (1974), within which Stages 3 to 5 are pertinent to our study. Stage 3 is taken as equivalent to the period of Tabun "C" Mousterian on the basis of recent radiocarbon evidence (see Farrand 1979: 376) which indicates an age equal to or in excess of 51,000 years for Layer C (Unit I) in our excavation. The cold peak of Stage 4 corresponds well with the occurrence of Tabun "D" Mousterian (Unit IX), especially when the link of retreating sea and collapse in the interior chamber are considered. The last cold oscillation of Stage 5 could well correspond with the Amudian if the evidence of the beach sequence at Zumoffen is considered. The correlation of the Late Acheulian of Unit XIV with the highest sea stand of the last interglacial has already been noted. This event is clearly indicated by the very marked warm oscillation at the beginning of Stage 5. This correlation is also reinforced by the subsidence of sediments in the outer chamber following the deposition of Unit XIV, which probably followed and was caused by the major sea retreat that can be inferred from the isotope curve.

Altogether, all our evidence points to gradually cooling conditions following an initial brief warm phase at the beginning of the last interglacial/interpluvial about 125,000 years ago. There were two major interruptions in this cooling trend that, according to our evidence, correspond to the most intensive manifestations of the Yabrudian Facies of the Mugharan Tradition. It is interesting to note that the two major occurrences of the obvious Acheulian Facies are here correlated with periods of cooling conditions and retreating sea level (as is the Late Acheulian of Unit XIV). Another element of interest in this sequence concerns the similarities in situation and in technology between the Amudian and the Early ("D") Mousterian. Both fall near or at the end of a trend toward cooler conditions and both make use of techniques of flake production based on the preparation of cores by linear extractions for the production of prismatic blades in significant quantities. The outstanding feature within the general lithic inventory (Figure 2) that both of these industries share is the high frequency of complete flakes. There is little question that these industries differ in their typological inventories and that the techniques used for core preparation in the Mousterian of Unit IX are more sophisticated than those in the Amudian in Unit XI, but the basic similarities in situation and form are undeniable. I would suggest, on the basis of the evidence of the gradual change from an

Acheulian Facies to Early Mousterian now visible from upper Unit XI through lower Unit IX and the present indications of a gradual increase in characteristic Amudian elements from the Acheulian Facies of upper Bed 76 to the Amudian of lower Bed 75 that here, as in the case of the continuity that we can see between the Yabrudian and the Acheulian Facies, there is also continuity. The case is admittedly less secure for the Acheulian-Amudian continuity because the depth of beds involved is smaller. Our best hope of clarifying this relationship is with detailed vertical back-plotting of all of the characteristic Amudian elements. This has not yet been done. In the case of the Acheulian-Mousterian relationship the situation is clearer. Here we have good evidence for a gradual and consistent increase in Levallois technique and a gradual increase in the frequency of characteristic Early Mousterian points and blades that spans at least seven superimposed depositional units through a depth of over 1 m of deposits. There seems little likelihood that this transition resulted from the mixing of two distinct cultural units either during deposition or excavation. Therefore, the underlying similarities in the circumstances of appearance of both industries and the general similarities in some of their chief characteristics lead me to suggest that the Amudian may be a third facies of the Mugharan Tradition in the Southern Levant and that the Early Mousterian in the same area is directly derived from this earlier tradition. I believe that we can show at present the strong possibility that the Yabrudian Facies at Tabun are associated with the more extreme interpluvial/interglacial conditions in the Southern Levant beginning about 110,000 years ago, and that the Acheulian Facies are consistently associated with more moderate conditions. Probably the Amudian, and certainly the Early Mousterian, can be associated with more markedly pluvial/glacial conditions. The explanations for this variability are not yet obvious in the nature of the artifacts and there are no associated materials that we can use to clarify the shifts in resource exploitation that these patterns suggest. Such clarification might, in part, be possible through edge-wear studies of the artifacts, although the compact and consolidated nature of the sediments at Tabun necessitated rigorous cleaning procedures that may have obscured some of these traces. It is hoped that such studies can be attempted in the future. Apart from this, only future excavation in sites with good faunal preservation and other evidence of subsistence is likely to confirm or deny these postulated relationships.

The only other excavated site in the Southern Levant known to contain artifacts of the Mugharan Tradition is the Zuttiyeh Cave, where virtually all of the deposits were removed in a poorly controlled excavation by Turville-Petre (1927). Recent excavations in the remnants of implementiferous breccia on the walls and floor of the cave (Gisis and Bar-Yosef 1974) confirm the presence of bifaces and heavy scrapers at the site, but it seems unlikely that much additional information can be obtained there. There is only a single report of the discovery of Yabrudian Facies artifacts in an open-air context in the Southern Levant (actually in the entire Levant) at Azraq, far in the interior from the Jordan Valley (Zeuner *et al.* 1957: 23–25). The fact that these artifacts were derived from well-digging, when taken in conjunction with the absence of surface occurrences elsewhere, despite the relative abundance of Acheulian surface material, suggests that if the Yabrudian Facies is associated with marked interpluvial conditions most of the artifactual evidence may have been removed and buried by the erosion that characterizes such periods in the Levant.

The abundance of Acheulian artifacts in open-air contexts poses at least one major problem with relation to the sequence at Tabun. In some of these concentrations, such as at Kissufim (Ronen *et al.* 1972) and the Upper level at Evron (Ronen 1979: 301) typical Late Acheulian bifaces occur in association with a Levallois flake industry. While we do have such an association in Unit X at Tabun, it appears to be of relatively short duration. There is little earlier evidence of a well developed Levallois technique at Tabun (or any other Levantine cave or shelter), with the exception of some flakes in the Amudian levels. It seems curious that if this technique were known throughout the Late Acheulian it would not have been practiced to any significant extent in any cave or shelter. There does seem to be good chronological evidence for the presence of this technique in earlier contexts in several other parts of the Old World (e.g. Africa and Western Europe) so that its absence in these sites remains an enigma.

The Mousterian Industries

The Mousterian from our excavations at Tabun is confined to Units I–IX. As mentioned above, the distinctive Levallois elements that characterize the Mousterian increase in frequency gradually from the top of Unit XI to the lower beds of Unit IX (see Table IV).

Table IV
The Transition to Mousterian:
Frequency (%) of Levallois Flakes, Levallois Points, and Prismatic Blades

Major Unit	Bed	Levallois Flakes	Levallois Points	Prismatic Blades	Total %	Total Lithics
IX	67D	5.1	18.2	17.2	40.5	99
IX	68	11.4	10.0	12.8	34.2	290
X	70	13.0	5.4	7.6	26.0	277
X	71S	6.8	5.1	5.1	17.0	176
X	71I	6.5	3.0	3.6	13.1	168
X	72S	6.1	.8	2.5	9.4	476
X	72I	5.6	.5	1.4	7.4	591
XI	73S	4.6	.2	1.4	6.1	637

The Mousterian sequence at Tabun was disrupted by the collapse of sediments into a solution cavity at the base of the interior chamber. The effect of this subsidence was an interruption in the depositional sequence between the top of those beds that Garrod saw as Layer D and the base of her Layer C. We must recognize here that the units defined in her stratigraphic diagrams have idealized boundaries with respect to the natural stratigraphy as a result of her use of arbitrary horizontal levels in excavation and changes in artifact typology to designate each "Layer". Her excavation did not significantly penetrate the edge of the inner chamber collapse and thus the materials she ascribed to Layer D must have been largely derived from deposits that precede this event. While our excavations did expose some Mousterian artifacts in this context in Unit IX, most of our Mousterian below Layer C is derived from the long series of deposits that continued to fill and then subside into the great pit that had opened at the back of the cave. Thus the temporal origin of the artifacts in these deposits must, to some extent, be open to question. While a portion of these materials may have been discarded from the terrace into the back of the cave or utilized in the butchering of game trapped in the pit, others, without question, must have weathered from earlier deposits that formed the steep walls of the pit. We can therefore expect that the artifact inventories in Units VIII to II are to some degree mixed, with a greater chance of mixing in the lower units, and less in the upper units. I believe that the relative lack of continuity in the basic lithic categories (illustrated in Figure 2) through these units in part represents the discontinuity in the depositional sequence, and in part represents differential mixing of older and younger artifacts. The chaos of deposition is so marked in the middle of this sequence that the ordering of a few units (e.g., Units V, VI, and VII) may be said to be based largely on inspired guesswork due to the absence of evidence of clear stratigraphic superposition. Because of these factors, it has been considered reasonable to lump the artifacts between Unit IX and Unit II in most of our comparisons.

There are several aspects of the accepted Mousterian chronology that can be tested, and perhaps clarified by quantification, in our sequence. Of first importance here is the relationship of the Tabun "D" and the Tabun "C" variants of the Mousterian. Here a comparison between the industries of Unit IX and Unit I will be of primary importance. The defining characteristics of these two kinds of Mousterian, as described by Copeland (1975: 329–336) and Ronen (1979: 302–305), are as follows:

(1) Tabun D (or Phase 1, or Abou Sif type) – an abundance of Levallois points, frequently elongate; high proportions of blades, frequently with plain platforms; high proportions of Upper Palaeolithic tool types; and, by inference, relatively low frequencies of broad, radially-prepared, Levallois flakes.

(2) Tabun C (or Phase 2, or Tabun C-B type) – an abundance of broad Levallois flakes from radially-prepared cores; broad, short Levallois points, sometimes very few in number; relatively few blades; and high frequencies of side-scrapers among the retouched tools, with Upper Palaeolithic tools infrequent or absent.

Whether the Tabun C and B Mousterian constitutes a single cultural-temporal phase (re. Ronen) or two distinct temporal phases (re. Copeland), is an issue that is relatively difficult to treat with the evidence from our excavation, since we sampled only a very small portion of the deposits equivalent to the base of Garrod's Layer B. Copeland suggests that an earlier phase (2) is characterized by a near-absence of Levallois points, and a later phase (3) sees a renewal of point manufacture, accompanied by narrow thin flakes and blades.

The Mousterian industries, as a whole (see Figure 2), are distinguished from the preceding Mugharan Tradition by lower percentages of retouched tools and (with the exception of the uppermost beds) cores. Bifaces are confined to the lower Mousterian levels, where they are always very rare. This means that the basic inventory of lithic materials in the Mousterian is dominated by complete flakes, partial flakes, and fragments, which occur in widely varying ratios with respect to each other. There is some continuity with the underlying industries, implied by the relatively higher frequencies of retouched tools in Unit IX at the base of the Mousterian sequence, and in the similar trends leading up to the Early Mousterian and the Amudian that were discussed above. Beyond these ties, the Mousterian industries are quite distinct from the Mugharan Tradition.

The comparison of "D" Mousterian and "C" Mousterian is confined here to a comparison of materials from deposits that can be clearly equated with Garrod's layers on which these distinctions were originally based, i.e., our Units IX (D) and I, below Bed 17 (C). In terms of basic lithic variables (Figure 2), there are some obvious differences between these industries; e.g., Unit IX has consistently higher ratios of retouched tools and complete flakes. All basic lithic variables for these two groups of beds are compared in Table V.

Table V

Basic Lithic Variables in Beds Equivalent to Garrod's "C" and "D"

Variable	"C"; Unit I (18-26)		"D"; Unit IX	
	Number	%	Number	%
Retouched Flake Tools	176	4.47	262	15.87
Complete Flakes	1193	30.32	686	41.55
Partial Flakes	1754	44.58	533	32.28
Fragments	611	15.52	55	3.33
Cores	201	5.11	106	6.42
Bifaces	0	0.00	8	.49
Biface Fragments	0	0.00	1	.06
Total	3935	100.00	1651	100.00

As was noted previously, a few bifaces are found through Unit IX (and above this unit). In general, it might be suspected that those categories present in higher ratios in Unit IX reflect a greater degree of selection (as opposed to all of the debris of lithic manufacture) than do the industries in lower Unit I. This possibility will be pursued further below.

In terms of those defining characters noted by Ronen and Copeland, we can first examine the ratios of Levallois flakes, Levallois points, and prismatic blades as seen in Table VI. I include prismatic blades here as a special class of flakes equivalent in focus on core preparation to the two Levallois categories. I have eliminated from this category those prismatic blades that seem to be "accidents" rather than to reflect deliberate preparation in the form of parallel flake scars. Bordes frequently puts the best defined of the blades that I include here into the typical and atypical Levallois flake categories. I have stated earlier that I would favor recognizing the distinctive nature of these kinds of flakes (Jelinek 1975: 304). The consistent high correlation of prismatic blades with prepared platforms (previously referred to as "Levallois" (*op. cit.*) with those with plain platforms (referred to as "normal") in our Mousterian layers has led me to believe that this distinction may have little definitive value there; I have lumped these blades in Table VI.

The distribution in Table VI conforms to the differences noted by earlier workers based on Garrod's collections; the difference between the two groups of beds is highly significant (χ^2 = 348.73, p is > .001) and the relationships between these variables are clearly distinct in the two stratigraphic units. The relationship just between Levallois flakes and Levallois points is equally clear (χ^2 = 268.14, p = >.001). The question of whether the Levallois points that occur in the later horizon ("C") are less elongate than those in the earlier horizon ("D") is treated in Figure 6. Also included here is the small sample of Levallois points recovered from the upper beds of Unit I, which correspond to the base of Garrod's Layer B. Perhaps the most striking feature of Figure 6 is the suggestion

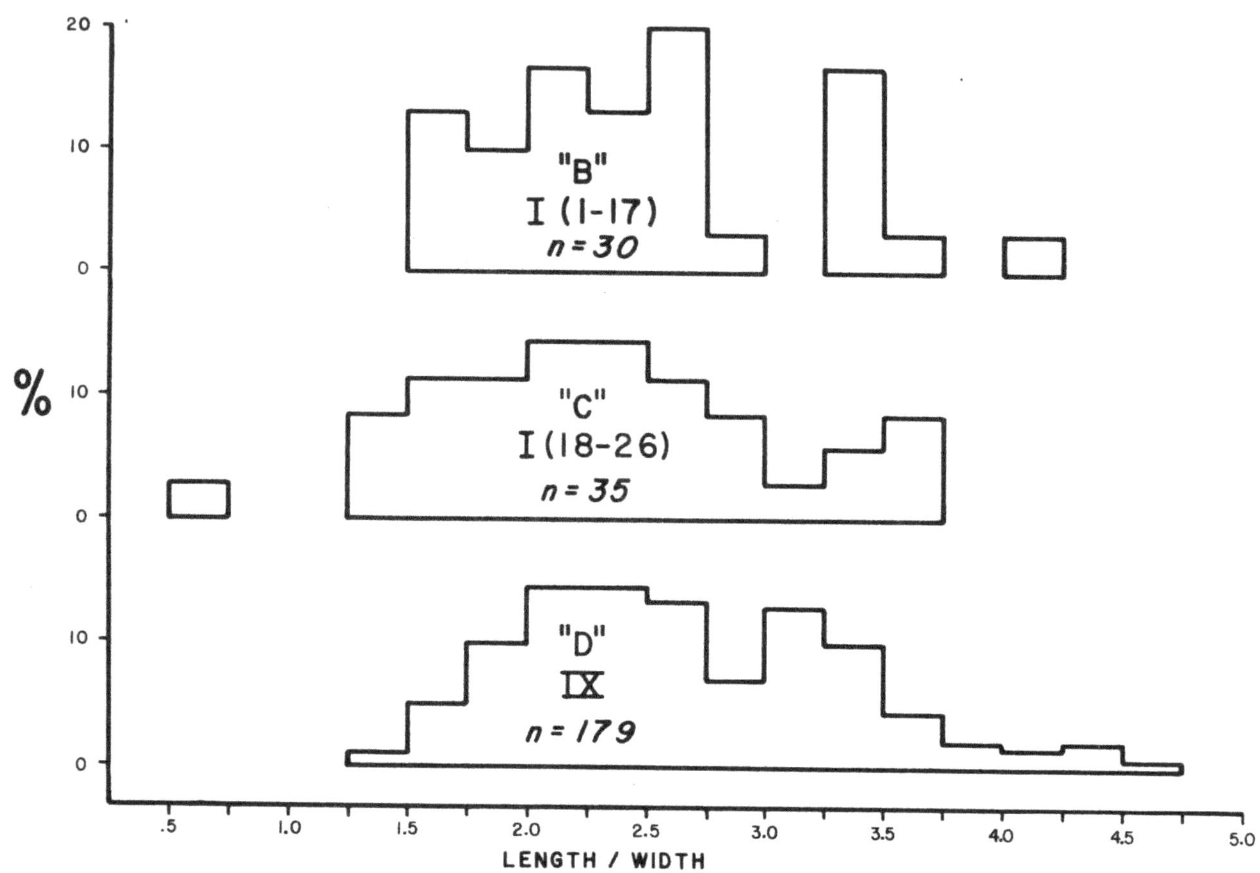

Figure 6 Length/Width ratios for complete Levallois points.

Table VI

Levallois Flakes, Levallois Points, and Prismatic Blades in Beds Equivalent to Garrod's Layers "C" and "D"

Variable	"C"; Unit I (18-26)		"D"; Unit IX	
	Number	%	Number	%
Levallois Flake	507	73.8	154	23.5
Levallois Point	54	7.9	225	34.4
Prismatic Blade	126	18.3	276	42.1
Total	687	100.0	655	100.0

of bimodality in all of the samples, separating points in which length = <2.75 x width from those in which length is >2.75 x width. This indicates that two distinct kinds of points may be present in these collections, one of which is distinctly more elongate than the other. Such a distinction has been recognized in the past (e.g., Hours 1974: 3; Category A4) but generally without objective evidence for determining where to draw the boundary between the two categories. Similar modality is also suggested in the distribution of length/width ratios for points in Units III, VII, VIII and X, however the point of the division between the modes ranges from 2.25 (VIII) to 2.5 (III) to 2.75 (VII). Because of this variation, and the lack of evident bimodality in this distribution in Units II, IV, V and VI, the total distribution of this ratio does not exhibit the clear division seen in Figure 6. The strongest statement that can be made for our evidence is that it suggests that a distinction between normal and elongated Levallois points may be warranted in some industries and that the point of differentiation in the length/width ratio lies between 2.25 and 2.75. Assuming, for the moment, that the division at 2.75 seen in Figure 6 is real, the difference in the two samples can be expressed as follows:

	L/W<2.75		L/W>2.75	
"C" (I: 18-26)	29	82.9%	6	17.1%
"D" (IX)	118	65.9%	61	34.1%

$\chi^2 = 3.156$ $p = .10$

This suggests that while there are relatively more elongated points in the "D" Mousterian represented in our sample from Unit IX than in the "C" Mousterian of Unit I, we cannot on the basis of these samples confirm that the difference between them is statistically significant.

Another means of exploring this relationship is an examination of the means and medians of the distributions, although it should be noted that in our samples these observations can only be treated as descriptive devices. The deviation of our distributions from a normal curve invalidates the use of parametric statistical parameters that imply probability. Table VII presents means and medians for all of the major units in the Mousterian and the Transition.

Table VII

Means and Medians of Length/Width for Samples of Complete Levallois Points

Unit	Mean	Median	*n*
I: 1-17 ("B")	2.27	2.19	30
I: 18-26 ("C")	2.07	2.07	35
II	2.17	2.04	48
III	2.33	2.19	35
IV	2.31	2.13	25
V	2.22	2.22	25
VI	2.13	2.00	75
VII	2.34	2.11	30
VIII	2.25	2.12	69
IX	2.45	2.33	179

Here too, there is a suggestion of more longer points in the earlier levels. The "C" Mousterian has a particularly low mean, however, the higher levels immediately above and below may indicate that this particular low level is not a significant feature in the overall chronology. Table VII does suggest that there may be a general uniformity in Units I through VIII, as opposed to IX, which has higher means for length/thickness than any of the units above. We tested the frequency of points with ratios < 2.75 and > 2.75 in samples from Units I–VIII as opposed to IX with the following results:

	L/T<2.75		L/T>2.75	
Units I-VIII	261	79.6%	67	20.4%
Unit IX	118	65.9%	61	34.1%

$\chi^2 = 10.723$ $p = <.005$

Therefore, we can say with confidence that there is a significant difference in the frequency of elongate Levallois points between the levels with "D" type Mousterian and the levels that follow. This can be demonstrated despite the fact that to some unknown degree artifacts originating in "D" layers were probably introduced into the deposits of Units II through VIII as they filled the subsidence pit in the inner chamber. This conclusion supports the use of higher frequencies of elongated Levallois points (in which the length/width ratio, measuring width at the mid-point of length (Jelinek 1977: 88), exceeds 2.74) in separating "D" type Mousterian from later variants of the Mousterian. About 1/3 of the Levallois points in the "D" Mousterian at Tabun are elongate; only about 1/5 of the points in overlying levels are of this type.

Other typological variables that have been suggested as indicative of the "D" and "C" types of Mousterian are: (1) higher frequencies of Upper Palaeolithic tools in "D", and (2) higher frequencies of side-scrapers (*racloirs*) in "C". Although this difference can be expressed in the customary Group III and IR indices, the relationship can also be seen by directly opposing the number and relative frequency in each category for the two samples:

	Scrapers Types 9-11		U. Palaeolithic Types 30-37, 39	
I: 18-26 ("C")	62	78.5%	17	21.5%
IX ("D")	73	68.9%	33	31.1%

$\chi^2 = 1.662$ $\qquad$ $p = <.20$

The same test applied to all *racloirs* (Types 9-29) yields a lower value for χ^2 (1.100). Thus we would have to say, on the basis of our excavated sample from Tabun, that there is no statistically significant difference in the relative frequencies of side-scrapers and Upper Palaeolithic elements in the Type "C" and Type "D" Mousterian, even though we can see somewhat higher relative frequencies of the elements thought to be diagnostic in each sample. Given the possibility that Garrod's excavations in the terrace deposits may have sampled a wider range of Layer D material, and putting aside the likelihood that the horizontal spits in her excavation mixed some material from the two layers, we can examine the counts made by Skinner (1965) on these two Layers from the original excavation.

	Scrapers Types 9-11		Upper Paleolithic Types 30-37, 39	
Layer C	61	92.4%	5	7.6%
Layer D	29	67.4%	14	32.6%

$\chi^2 = 9.614$ $\qquad$ $p = <.01$

The significant difference between Layers C and D based on Skinner's typological analyses suggest that the use of these typological categories as diagnostic elements for each kind of Mousterian may have validity within the context of Garrod's Tabun collections. The sample here is quite small, however, and is derived from four of the more than 30 collections from the original excavation at this site now distributed around the world; these findings should therefore be treated with caution. While I have examined about 20 of these collections, I have not systematically typed material from layers equivalent to those samples in our excavation, but have concentrated on recording material from Layer B, the Chimney, and Layers F and G in the hope of integrating those materials into our sequence. Because of this I am at present unable to add further data relevant to this problem.

The question of whether the Type C Mousterian is distinct from a succeeding Type B (Phase 3) Mousterian is difficult to resolve with our small sample from those deposits that are assumed to be equivalent to the base of Layer B (our Beds 1–17). The proposed differences (Copeland 1975: 335) include higher frequencies of Levallois points and laminar and narrow flakes. The appropriate frequencies in our samples are shown in Table VIII.

Table VIII

Levallois Flakes, Levallois Points, and Prismatic Blades in Beds Equivalent to Garrod's Layers "B" and "C"

Variable	"B" Unit I: 1-17		"C" Unit: 18-26	
	Number	%	Number	%
Levallois Flake	57	53.3	507	73.8
Levallois Point	30	28.0	54	7.9
Prismatic Blade	20	18.7	126	18.3
Total	107		687	

It is evident in this distribution that the frequency of prismatic blades is virtually identical in the two samples. Therefore, if there is a significant difference it must lie in the relationship between Levallois flakes and points. The difference in the frequencies of these two elements is highly significant ($\chi^2 = 39.07, p = < .001$), which lends support to the classification proposed by Copeland. The question remains, however, of whether these differences reflect only differences in the utilization of the area sampled at Tabun, or whether they have broader temporal implications. We will treat this issue further when we discuss collections from other sites in the Southern Levant. It might be noted here that Skinner's analysis (*op. cit.*) does not show the difference seen above. His counts for the Chimney and Layer B as opposed to Layer C are as follows:

	Levallois Flakes		Levallois Points	
Layer B + Chimney	93	66.9%	46	33.1%
Layer C	116	65.2%	62	34.8%

$\chi^2 = .042$ $p = > .80$

Even if he included all prismatic blades in his Levallois flake count it would not account for this very different result (note the difference in percentages compared to Table VIII for Layer C). All that can be said about this discrepancy is that there is a clear and significant difference in the composition of our "B" and "C" industries as I have classified them, and Garrod's Chimney + B and Layer C industries as they were classified by Skinner.

Several other findings have emerged from our work with the collections from our recent excavation at Tabun that may be relevant to some of the issues treated above and to our following discussion of the Middle Palaeolithic in the whole of the Southern Levant. One interesting finding is a very high correlation between Levallois flakes (with prismatic blades) and all other flakes in the Unit IX ("D") samples ($r = .9553$ for 12 cases, $p = < .001$). In contrast, the Unit I: 18-26 ("C") samples show a markedly lower correlation ($r = .5770$ for 11 cases, $p = < .10$), which is not statistically significant. This apparently indicates that Levallois elements occur in a highly consistent ratio to other flakes in Unit IX, but that this does not hold true for Unit I: 18-26. In fact, in Unit IX in all beds but one, Levallois elements outnumber other flakes ($\bar{x}$ of ratio = 1.32:1), while in Unit I: 18-26 they are outnumbered in all cases ($\bar{x}$ of ratio = .31:1). I would interpret these ratios to mean that most of the flakes in Unit IX were manufactured elsewhere and only a highly selected group of flakes was introduced into the area of the cave that we excavated; in contrast, I would suggest that the flakes in Unit I: 18-26 include more manufacturing debris. This last interpretation is reinforced by the higher percentages of fragments in these upper layers, and lower percentages of complete flakes than are found in Unit IX.

The deposits of Type C Mousterian that we excavated in Unit I: 18-26, by their sloping attitude and heavy mixture of black carbonaceous earth and fragmentary animal bone, suggest a midden of discarded material on the gently inclined edge of the nearly filled pit inside the inner chamber of the cave. If this discarded material reflects the activities that also contributed to Garrod's Layer C collections, we may be looking at extremes in the kinds of tasks that gave rise to the lithic remains in those collections. Thus, while there may be clear differences through time in manufacturing techniques, we might expect that manufacturing sites of "D" type Mousterian will contain a different range of flakes and tools than occurred in the type section, and that facies of the "C" type Mousterian will occur in which there are higher frequencies of Levallois elements and retouched tools, and fewer fragments and broken flakes. This is, in effect, the explanation suggested by Ronen (1979: 303) for the difference between the Layer C Mousterian and the overlying Layer B and Chimney industry (Copeland's Phase 3). The increase in the ratio of complete flakes in the uppermost beds of Unit I (1-17) that we equate with Layer B might indicate a greater selection in what was brought into the cave at that time, as might the higher ratios of Levallois elements to other flakes beginning with Bed 19 (.53:1) and culminating in Beds 1–5 (.94:1) at the top of our excavation. The highest ratio in the 10 samples below Bed 19 is (.44:1).

One final set of relationships discovered in our collections from Tabun must be mentioned before we discuss the Middle Palaeolithic evidence from other sites in the Southern Levant. This is the ratio of width to thickness on complete flakes recovered through the full depth of our section. Relatively early in the study of the flakes from our excavation, a general correlation was noted between this ratio and the stratigraphic position of the samples (Jelinek 1977: 94). Since that preliminary study, all complete flakes with reliable context have been tabulated with the following results:

Table IX

The Ratio of Width/Thickness of Complete Flakes by Major Stratigraphic Unit

Major Unit	$\bar{X}$	Median	Variance	*n*
I	4.633	4.249	5.049	1377
II	4.547	4.169	4.049	334
III–VIII	4.361	4.000	3.956	1275
IX	4.258	3.999	3.128	743
X	3.970	3.643	2.898	424
XI	3.465	3.143	2.709	2373
XII	3.552	3.273	2.570	976
XIII	3.520	3.250	2.391	1905
XIV	3.210	2.917	1.935	766

The clear progression of this ratio through time implies that, on the average, flakes are becoming thinner with respect to their width through the full sequence in our excavation. The gradual increase in the variance of each sample through time seems to reflect the

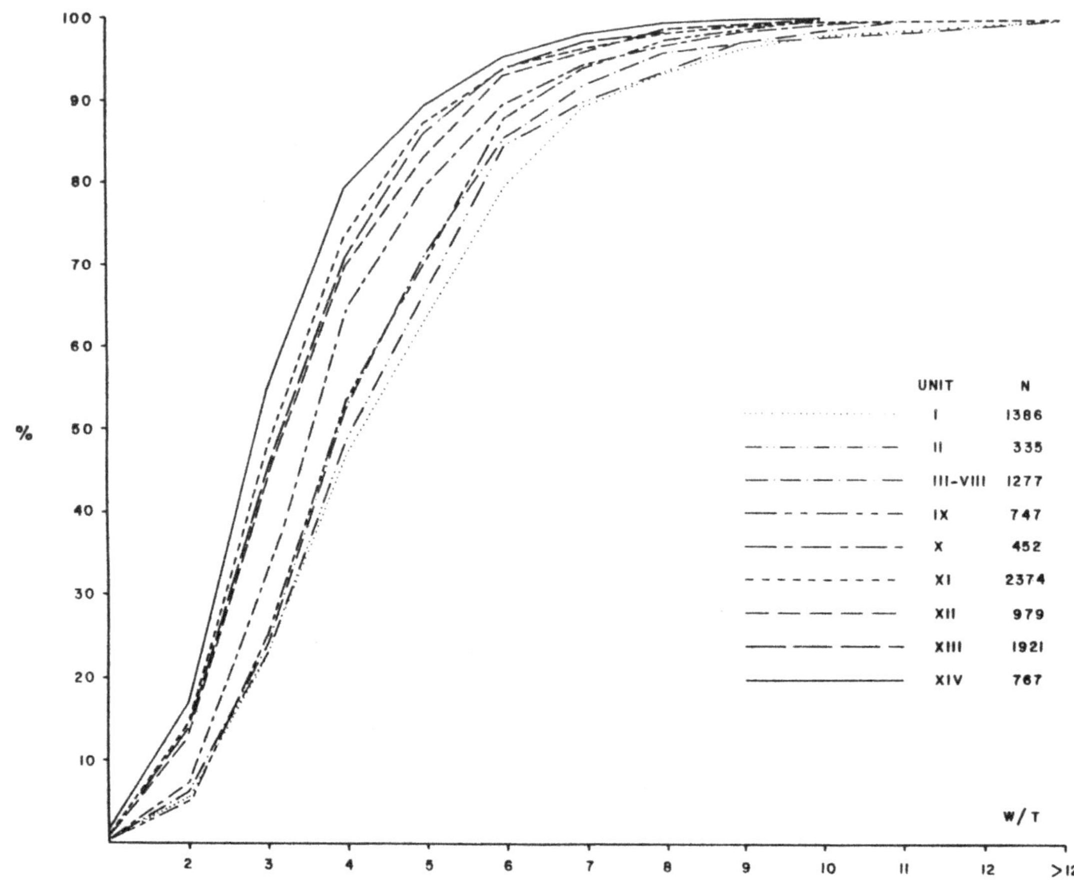

Figure 7 Tabun: Cumulative frequency of Width/Thickness for all complete flakes by stratigraphic unit.

increasing presence of flakes that are more and more extreme in this regard, with the continuing presence of significant quantities of the same kinds of thicker flakes that characterize earlier levels. This relationship is best seen in Figure 7. Here cumulative frequencies at 1.00 intervals for this ratio in each sample show a regular progression of curves through time due to the presence of higher frequencies toward the right side of the diagram (thinner flakes). In other words, there appears to be a continuation in the manufacture of the proportionately thicker flakes that dominate the early levels through the full sequence, with the gradual addition of more thinner flakes. In Figure 8 this trend in increasing variance is calibrated with the chronological sequence proposed in Figure 5. The curve shows a slow increase in the variance through time, which begins to accelerate by the period of Unit II and shows a marked increase over the earlier rate of change by Unit I. The trend appears to be continuous despite the typological differences in the industries, which suggests an underlying continuity in their development at Tabun. A primary question concerning this trend is its applicability to the chronological interpretation of industries outside of Tabun. Another question might relate to the effects on this variance of different kinds of occupations; e.g., might not the stronger emphasis on manufacturing in early Unit I be responsible for greater variance in that sample? Our exploration of these questions must, of necessity, be restricted to collections from other excavations, only a few of which have employed techniques of artifact recovery that can furnish us with samples comparable to those at Tabun. We can also compare samples from the earlier excavations in which artifact retention was selective but the stratigraphic relationship to our sequence is known in order to appraise the possible effects of selection on the kind of variability that concerns us. Probably the best place to begin such a study is in the layers at Tabun that are represented in collections from the original excavation but were not sampled in our work.

The width/thickness ratios of samples from Garrod's excavations at Tabun are shown in Table X, with the uppermost and lowermost samples from our excavation. The observations on the samples from the original excavation were derived from an examination of 20 collections in the United States, Canada, and Western Europe.

Table X

Width/Thickness Ratios of Complete Flakes in the Upper and Lower Levels at Tabun

Layer/Unit	$\overline{X}$	Median	Variance	*n*
Chimney I	6.153	5.750	5.716	74
Chimney II	6.112	5.600	6.993	190
Chimney III & B	6.468	5.667	8.873	74
Chimney I-III	5.667	5.667	7.102	338
Unit I	4.249	4.249	5.049	1377
Unit XIV	3.210	2.917	1.935	766
Layer F	3.610	3.100	2.545	82
Layer G	3.231	3.000	1.581	150

It is apparent in this tabulation that the mean and median ratios, as well as the variance of width/thickness for the Chimney and Layer B samples are markedly higher than our Unit I. Because the samples are small and we assume that deposition was rapid in the

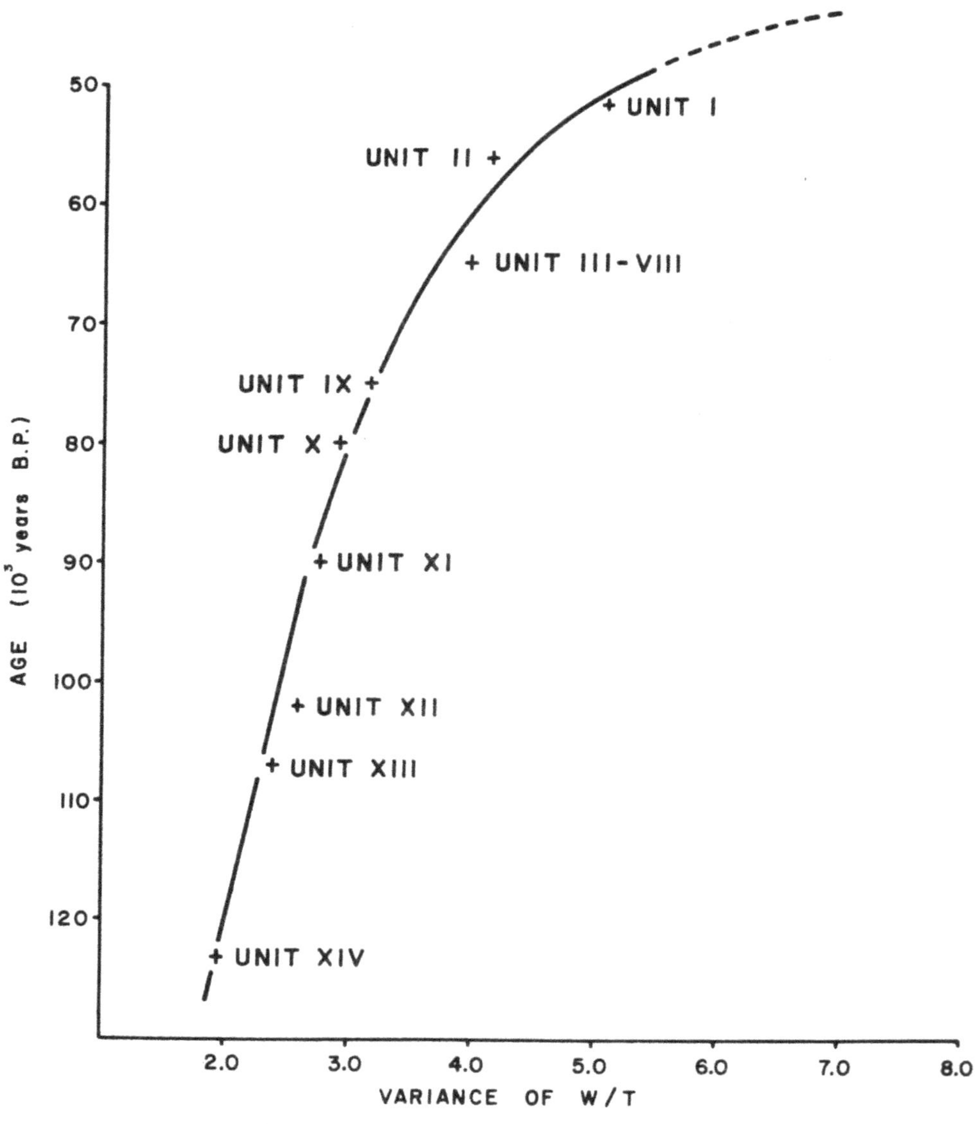

Figure 8 Variance of Width/Thickness for complete flakes at Tabun.

Chimney, the separate levels have been combined for further comparison as "Chimney I–III". Since there is no depositional hiatus between our Unit I and the Chimney Layers, and since the upper portion of Unit I clearly reflects, through abundant roof-fall and *terra rosa* derived from the hillside above the cave (Farrand 1979), the beginning of a rapid filling of the Chimney, we can assume that the temporal position of the Chimney deposits is not very much later than our Unit I. Perhaps an interval of 5,000 to 7,000 years for the filling of the Chimney is not an unreasonable estimate, considering the postulated pluvial conditions at that time. Therefore, the markedly higher variance and ratios seen in Table X is somewhat surprising. Yet, when we project the curve on Figure 7 as indicated by the dashed line, this variance and the time estimate coincide quite closely. At the earlier end of the time scale, the small sample from Layer F is most similar to the predominantly Acheulian Unit XII, while Layer G is quite close to Unit XIV. Considering the uniformity of the Unit XI, XII, and XIII samples and the small size of the Layer F sample this placement is not surprising. If those flakes that were saved from this layer were biased toward somewhat thinner and more regular specimens, only a slight selection would alter the statistical aspects of this stratigraphically earlier layer to resemble the overlying "Layer E". The probable close temporal link between Layer G and Unit XIV (since both precede the major collapse in the outer chamber) seems reinforced by the similar values for these collections.

It may be possible to see some of the kinds of selection of flakes for curation practiced in the early excavations by comparing the distribution of the width/thickness ratio in our collections with collections made by Garrod. In Figure 9 the distributions for Units I, XIII, and XIV are compared with our data on Garrod's collections from the Chimney and Layers F and G. What these cumulative curves show is that in all cases in her samples the very thick flakes (width = thickness, or width = 2 x thickness) are under-represented with respect to thinner categories. This is most marked in the relationship of the Chimney material to Unit I, but is also evident in the cases of Layers F and G with respect to Units XIII and XIV. The small sample of Layer F also seems under-represented with respect to Unit XIII in the middle ranges and over-represented in the higher values (thinner categories). The larger sample from Layer G when compared to Unit XIV shows a clear under-representation in the thicker categories and a corresponding over-representation in the middle and thinner categories. The key to these differences is the relative or complete absence of flakes from the very thick categories in the collections from Garrod's excavation; we can assume that they were discarded as "chunks" or were not recognized as artifacts by the workers. The presence of flakes with very high values (width = 13 to 20 x thickness) in Garrod's Chimney collections is similar to our Unit I; the over-representation of higher values here is seen in the whole higher portion of the curve, following abnormally low values up to width = 4 x thickness. It appears that here a considerable selection was exercised by either the excavator or the aboriginal inhabitants for more than the expected quantity of thinner flakes.

We can conclude this discussion of the Middle Palaeolithic sequence at Tabun with the following summary:

(1) A consideration of the cultural and environmental record at Tabun, in the light of recent knowledge of paleotemperature changes in the Late Pleistocene, suggests a likely time scale for the depositional sequence and particular environmental associations with distinct cultural patterns.

(2) The relatively large and well controlled samples of artifacts from this Middle Palaeolithic type-section have allowed us to examine the kinds of distinctions that have been made in the past for the separation of particular cultural phases and to confirm or question them within limits of probability.

(3) Within the sequence at Tabun we can distinguish time trends in typological relationships and technological factors that suggest an underlying continuity in the sequence. These trends may be of wide, and perhaps fundamental, importance in the interpretation of the Late Pleistocene prehistory of the Levant, and possibly beyond the Levant.

(4) It appears, on the basis of comparison with collections from stratigraphically compatible layers from the Garrod excavation at Tabun, that the technological variability in the relationship of width to thickness of complete flakes may serve as a useful guide to the general temporal position of an industry. These relationships may be further clarified for older collections if we can determine the kind of selection exercised in retaining material from the excavations for curation.

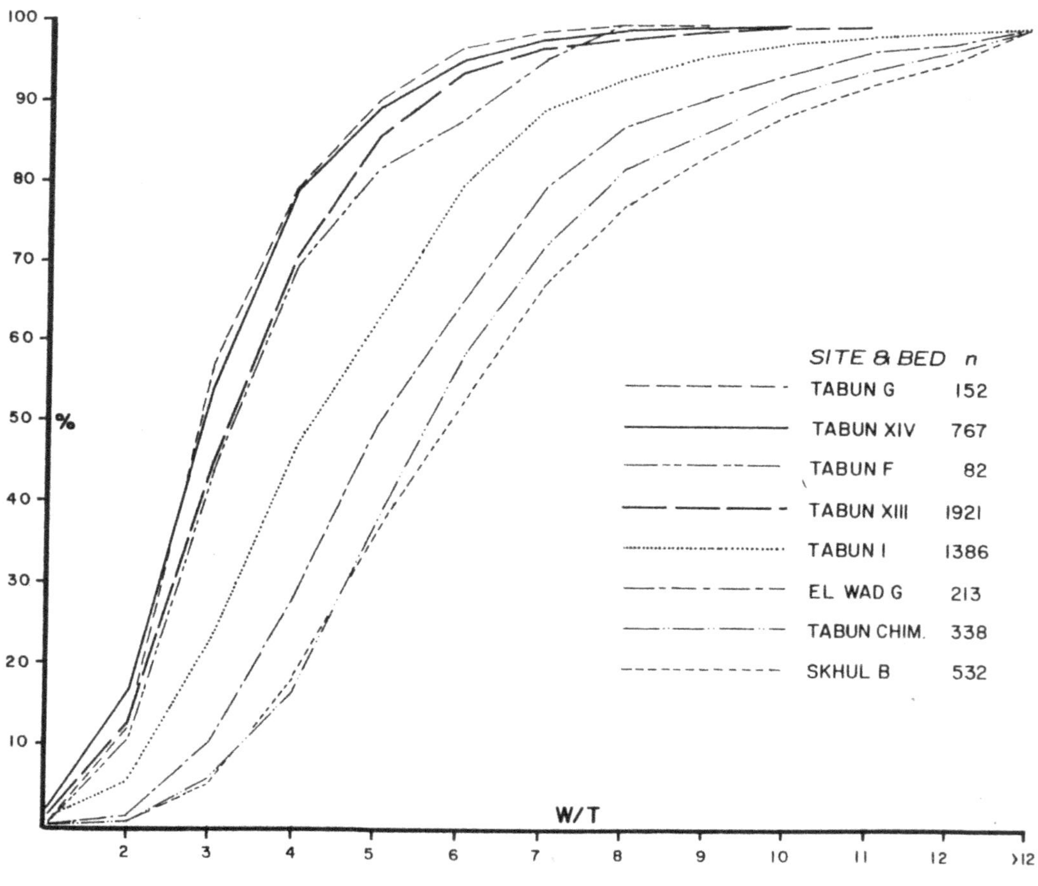

Figure 9 Wadi Mughara: Cumulative percentage of Width/Thickness (complete flakes).

The Middle Palaeolithic Beyond Tabun in the Wadi Mughara

The Wadi Mughara is a reasonable place to begin our examination of other sites in the Southern Levant, using Tabun as a baseline, because the proximity of the sites of El Wad and Skhul insures a uniform access to resources and a uniform environment for the three sites through time.

Garrod remarks on the "unsatisfactory" nature of the materials from the Mousterian deposits (lower Layer F and Layer G) at El Wad (Garrod and Bate 1937: 53). The considerable amount of abrasion on the edges and surfaces of many of the flakes, and the irregular floor of the cave, with underlying solution cavities convinced Garrod that most of the material had been moved to a considerable degree in antiquity; partly by the action of a spring emanating from the back of the cave. The badly crushed nature of many of these pieces encountered in our examination of the collections suggests that additional forces, such as trampling by large or hooved animals probably also played some part in this abrasion. Garrod's interpretation of this Mousterian, based on the high frequencies of Levallois points and radially-prepared cores, was that it was generally similar to the Chimney industry at Tabun. Our examination of material from El Wad was restricted to the Layer G context, since significant amounts of obvious Levantine Aurignacian artifacts were encountered in all of the Layer F collections that we observed. Within the Layer G industry, the frequency of Levallois elements is shown in Table XI, where the frequencies of all late Mousterian collections from the Wadi Mughara are compared on the basis of our (as opposed to Skinner's) count from excavation and museum collections. Since our observations on the museum collections were largely restricted to complete flakes, Table XI and the following tables include counts only on complete flakes from the Tabun collection.

Table XI

Percentage of Levallois Elements in the Late Mousterian in the Wadi Mughara (Complete Flakes)

Site and Bed	Levallois Flake	Levallois Point	Prismatic Blade	*n*
Tabun I: 18-26 (C)	63.6	11.2	25.2	313
Tabun I: 1-17 (B)	42.8	39.0	18.2	77
Tabun Chimney I-III	52.9	32.0	15.1	291
El Wad G	44.0	27.7	28.3	184
Skhul B2	63.3	24.4	12.2	90
Skhul B1	69.6	20.5	9.9	303

The most similar sample in this table to El Wad G (on the basis of comparisons of absolute frequencies) is Tabun I: 1–17 at the bottom of the Chimney-Layer B sequence ($\chi^2 = 3.73$, $p = < .25$). The El Wad G sample differs significantly from all of the other samples, including Tabun Unit I: 18–26 and Tabun Chimney I–III. If just the ratio between Levallois points and Levallois flakes is considered, the El Wad G sample is similar only to the Tabun Chimney I–III sample. They are very close in this respect with 38.6% points in El Wad G and 37.7% points in Tabun Chimney I–III ($\chi^2 = .090$, $p = > .70$). It can be pointed out here that our counts for the Levallois flake-Levallois point ratio for complete flakes from the Chimney and Layer B (our Chimney I–III) museum collections do not differ significantly from Skinner's counts on all flakes from those loci (χ^2 = .616, p = < .50).

The index of length/width of complete Levallois points is another means of placing the El Wad G industry with respect to the Tabun sequence. Table XII lists the basic information on this variable for the late Mousterian industries in the Wadi Mughara, including El Wad Layer G.

Table XII

Aspects of the Length/Width Ratio of Complete Levallois Points from the Late Mousterian in the Wadi Mughara

Site and Bed	$\bar{X}$	Median	% Elongate	*n*
Tabun I	2.163	2.108	20.0	65
Tabun Chimney I-III	1.802	1.742	3.2	94
El Wad G	1.923	1.813	5.8	52
Skhul B2	2.104	2.001	13.6	22
Skhul B1	1.992	1.857	7.9	63

All of these samples are compatible with the later Mousterian at Tabun, and in most respects they are more extreme than Tabun Unit I in those characteristics that separate the later from the earlier industries. El Wad G is more similar to the Chimney I–III industries in this regard than to Unit I.

Because of the difficulty of distinguishing deliberate from post-depositional retouch in Layer G at El Wad (Garrod and Bate 1937: 53), and because those observations that we did make are not yet fully compiled, there is little that can be added at present to previous typological descriptions of this collection. While Garrod's counts (*op. cit.*: 54) show the retouched tool industry to be dominated by *racloirs* (c. 87%), Skinner's counts (1965: 92) show about 46% *racloirs.* If we eliminate the denticulates from Skinner's counts (they seem not to have been recognized as tools by Garrod) the *racloirs* rise to about 56% of the small sample (56 tools) that remain. Skinner appears to have examined only about 27% of the collection reported by Garrod and this may account for the remaining discrepancy.

One additional clue to the position of the El Wad industry in the Tabun sequence may be provided by the index of width/thickness for the collection. This index for the Late Mousterian industries of the Wadi Mughara is presented in Table XIII.

Here El Wad appears to be somewhat less advanced in terms of the temporal trends demonstrated for Tabun than the Tabun Chimney I–III layers, and relatively close to Unit I. The cumulative curve for El Wad G in Figure 9 shows the same very low frequencies for low values of this index as does the Tabun Chimney I–III sample. However, the full curve lies almost midway between Tabun Unit I and Chimney I–III. It seems likely that if the expected amount of thicker flakes were present in the El Wad G sample (based on the assumption that the close similarity of the Tabun Mousterian samples from Unit I to Unit IX for values up to width = 3 x thickness indicates a reliable expected frequency) the curve would lie closer to Tabun Unit I.

Table XIII

The Ratio of Width/Thickness for Complete Flakes in the Late Mousterian Industries for the Wadi Mughara

Site and Bed	$\bar{X}$	Median	Variance	*n*
Tabun I	4.633	4.249	5.049	1377
Tabun Chimney I–III	6.199	5.667	7.102	338
El Wad G	5.616	5.123	5.694	213
Skhul B2	6.532	5.833	11.405	113
Skhul B1	6.705	6.167	8.262	355
Skhul B, B1, B2	6.528	5.889	8.975	496

The sum of this evidence suggests that, as Garrod also suggested, the industry in El Wad Layer G closely resembles her collections from the Chimney at Tabun; however, our technological evidence derived from the complete flakes in the collection suggests an earlier rather than a later date in the Chimney sequence.

Perhaps the most interesting question concerning the Mousterian industries of the Wadi Mughara is that of the chronological relationship between Skhul and Tabun. The relationship between hominid remains from these two sites has been considered to be of crucial importance to the understanding of the transition from the latest of the fossil hominids (*Homo sapiens neanderthalensis*) to modern man (*Homo sapiens sapiens*). Garrod equated the Layer B deposits at Skhul with her Tạbun C industry (Garrod and Bate 1937: 109–110, 115) on the basis of the frequency and character of the Levallois elements and the associated faunal evidence. The faunal evidence has been shown to be irrelevant to this problem (Jelinek *et al.* 1973: 180– 181). The industries that she described shared a dominance of broad Lavellois flakes over points in a ratio of about 2:1, and a strong dominance of *racloirs* in the retouched tools. She felt that Skhul B2 belonged near the base of Tabun C because of the "smaller size of the implements and flakes as compared with B1, and in the larger proportion of unprepared to prepared striking-platforms (B1, 1:10; B2, 1:5)" (*op. cit.*: 110).

When we examine the data that she used to arrive at these conclusions, it is difficult to substantiate them. Table XIV lists the absolute frequencies of tools and of Levallois flakes and points as they are presented in her report (*ibid.*: 54. 72–74, 76, 110–112). The flake tools have been condensed into points, *racloirs*, and other retouched elements. The retouched tools will not be compared directly with the Levallois flakes and points since varying amounts of the latter were saved in the different excavations.

Tests for paired samples based on her data indicate that the only contexts that do not differ significantly in terms of tools are Tabun Chimney B and Skhul B2 ($\chi^2 = 4.239$, $p = < .20$). The relationships based on Levallois points and flakes are all significantly different except for the case of the Skhul B1 and Skhul B2 samples ($\chi^2 = .4618$, $p = > .50$). In no case is there any indication of a strong similarity between the Skhul collections and Tabun C. It is difficult to assess Garrod's evidence on prepared platforms because she gives precise figures only for Tabun and Skhul B1. In El Wad G, a "majority" of the flakes saved have prepared platforms; in Skhul B2, "rather more than half" are in this category. The Tabun Chimney and Layer B each have approximately 80% retouched platforms in her collection, Skhul B1 has 76%, and Tabun C has about 54%. The degree to which symmetrical and more regularly shaped flakes were selected for curation in her excavations could have related to a selection for more prepared platforms. Our observations at Tabun have yielded an IFs of 50.47 on 212 specimens from Unit I: 1–17, and

Table XIV

Retouched Tools and Levallois Elements Reported by Garrod from Late Mousterian Contexts in the Wadi Mughara

	Retouched Tools			Levallois	
Site and Bed	Points	Racloirs	Other	Flakes	Points
Tabun C	83	411	261	267	138
Tabun B & Chimney	149	515	188	202	273
El Wad G	99	184	67	218	184
Skhul B2	88	404	154	865	328
Skhul B1	177	692	156	956	385

an IFs of 35.9 on 1947 specimens from Unit I:18–26, corresponding to her Layers B and C respectively. If dihedral platforms are included, IF =59.9 for I:1–117 and IF = 52.03 for I:18–26. These figures, assuming that Garrod counted mostly multiple faceted platforms, suggest that there was some selection for more regular flakes for curation in her collection at Tabun. Since we know that proportionately more flakes were saved, presumably less selectively, at Skhul (*ibid.*: 109), there is some question of sample comparability for her collections from these two sites.

Much of our data relevant to this comparison has been presented in Tables XI–XIII. In terms of relative frequencies of Levallois elements (Table XI), Skhul B1 differs significantly (on the basis of absolute frequencies) from all other late Mousterian samples except Skhul B2. Skhul B2, on the other hand, is not significantly different from the Tabun Chimney I–III sample ($\overline{\chi}^2 = 3.032$, $p = < .25$), but does differ significantly from the deeper samples at Tabun and from El Wad G. When only the ratio between Levallois flakes and points is considered, the Skhul samples are similar only to each other, and differ significantly from the Tabun and El Wad samples. With regard to the length/width ratio of Levallois points (Table XII), the Skhul B samples show no marked differences with the El Wad G and Tabun Chimney samples other than somewhat higher ratios of elongated points.

In the ratio of width/thickness of complete flakes, the Skhul samples are the most extreme in terms of relative thinness and variance, of all the Mousterian collections from the Wadi Mughara. In Table XIII all of our data for Skhul B (samples from Layer B1, Layer B2, and a small sample of specimens labeled simply "B" in museum collections) have been combined in an attempt to eliminate possible distortion caused by the small sample size of Layer B1. These values suggest that, on the basis of the time trends demonstrable at Tabun, the Skhul occupation was later than that of the other sites in the wadi.

When we consider the evidence of varying frequencies of striking platforms suggested by Garrod as a clue to these relationships, we are restricted to our recent re-examination of a portion of her collection from the Tabun Chimney and Layer B, Skhul, and El Wad G. In this study, all three loci in the Chimney (I, II, III + B) have an IFs of about 90, as does Skhul B1; Skhul B2 and El Wad G have an IFs of about 80. For the more comprehensive index of faceting (IF), Skhul B2 and El Wad G = c. 88, Skhul B1 and Tabun Chimney II = c. 92.5, and Chimney I and Chimney III + B = c. 94.5. The extremely high indices of faceting seem indicative of selection in these samples, either aboriginally or by the excavator. These figures do not indicate any strong differences between the six samples.

On the basis of all the evidence presented above, it appears that the Mousterian of

Skhul B1 and B2 falls very late in the Middle Palaeolithic sequence in the Wadi Mughara. While Skhul B2 may be contemporary with some of the Chimney deposits at Tabun, Skhul B1 probably post-dates those beds. El Wad appears to be contemporary with the early Chimney layers at Tabun and is thus probably somewhat earlier than Skhul. This evidence implies a date of about 40,000 BP for the lower portion of Layer B at Skhul.

The Middle Palaeolithic in the Southern Levant Beyond the Wadi Mughara

Middle Palaeolithic evidence in the Southern Levant outside of the Wadi Mughara is as widely varied in terms of excavation controls and artifactual reporting as the full group of excavations that have been discussed above. Very few sites have been excavated with precise contextual controls, and fewer still have been reported in the framework of a standardized typology. Four sites which are perhaps most comparable in this respect are Rosh Ein Mor and Nahal Aqev in the Negev Desert, Shubbabiq in the Wadi Amud, and Tirat Carmel on the northwest edge of Mount Carmel. There are several additional sites in which carefully controlled excavations were conducted, but where full reports presenting a standardized typological analysis are still lacking (as in the case with the recent excavation at Tabun). These sites include Qafzeh, near Nazareth, the Amud Cave in the Wadi Amud, Rakefet and Sefunim on Mount Carmel, and Hayonim in the northwest Galilee. Beyond the more recent studies, there are considerable collections from sites excavated with somewhat less precise controls and where typological reporting varies from a rather generalized description to the full set of standard types and indices. These sites include Kebara, at the southern end of Mount Carmel; Shukbah, in the hills above the coastal plain northwest of Jerusalem; and the several caves in the Wadi Khareitoun southeast of Jerusalem. In addition, there are such sites as Geula and Abou Uzba whose limited artifact inventory considerably restricts interpretation. We are also aware of a number of unexcavated surface scatters of Middle Palaeolithic artifacts, particularly in the Negev Desert.

Our discussion here will focus first on sites that appear to represent the earlier variants of the Mousterian, as it is seen in Unit IX ("Layer D") at Tabun.

The "Phase 1" Mousterian in the Southern Levant

The industries assigned to this phase have been classified as "Abou Sif type" or "Tabun D type" by recent workers; the defining characteristics are given on page 74.

The known sites that contain examples of this type of Mousterian are concentrated in two areas, the Negev Desert near Avdat, and the Judean Desert in the hills bordering the Wadi Khareitoun. Tabun is the only known site with this industry in the northern portion of the Southern Levant. The earliest collections of these industries in the south were made by Neuville (1951) at Abou Sif, Sahba, and Larikba (Vandermeersch 1969a) in the Judean Desert. More recently the excavations of Marks and his students at Rosh

Ein Mor (Crew 1976) and Nahal Aqev (Munday 1977) and collections from surface scatters (Munday 1976) have been reported from the Avdat area in the Negev. Our comparisons here will treat the expression of some of the defining characters of the early Mousterian as seen in the available samples.

Table XV presents most of the conventional indices for the recently excavated sites in the Negev and Unit IX at Tabun, as well as data on the ratios of flakes, cores, and "*essentiel*" tools.

Table XV

Standard Indices and Basic Lithic Categories for "Phase 1" Mousterian Sites

Index or Category		Tabun Unit IX		Rosh Ein Mor		Nahal Aqev Level 3
IL		56.3		14.6		8.7
ILty		62.2		60.6		47.1
IF		61.4		54.6		57.4
IFs		48.4		34.7		43.1
Ilam		76.2*		19.5		25.3
IRe		36.1		8.6		12.8
IIe		43.4		10.8		13.1
IIIe		19.3		30.4		20.3
IVe		3.2		14.3		9.2
Flakes	1219	73.9%	37540	87.3%	6992	89.7%
Cores	114	6.9%	2886	6.7%	383	4.9%
Tools (ess.)	316	19.6%	2554	5.9%	417	5.4%

* computed with width at midpoint of length.

It is apparent from Table XV that the two Negev sites are quite similar to each other and differ from Tabun Unit IX is several respects. In particular, Tabun IX has a higher Levallois index, a much higher blade index, higher *racloir* and Mousterian indices, and a higher frequency of restricted tools. All of these frequencies suggest a basic difference in activities between Tabun IX and the Negev sites. It is likely that the latter include much more material resulting from the full range of lithic manufacture than does the Tabun sample. In all three sites there is a strong index of Levallois typology and a relatively high ratio of Group III ("Upper Palaeolithic") tool types. This suggests that the three sites may share the same kind of basic tool kit, but that there was much more lithic reduction being carried out at the Negev sites.

We can bring in several more samples for comparison if the categories are confined to specific kinds of items that were likely to be preserved for curation in the earlier as well as the later excavations and collections. One relationship that is pertinent here is the relative frequency of Levallois flakes and points. These ratios are shown in Table XVI for the three sites treated above, as well as for Abou Sif B and C and Larikba. The Abou Sif data are confined to complete flakes from the collections in Paris and Jerusalem; the Larikba data are from the report of Vandermeersch (1969a).

Table XVI

Ratios Between Levallois Flakes and Levallois Points in the "Phase 1" Mousterian Industries

Site and Bed	Levallois Flakes		Levallois Points	
	n	%	*n*	%
Tabun Unit IX	430*	65.5	225	34.4
Rosh Ein Mor	3455	66.0	1780	34.0
Nahal Aqev 3	259	56.2	202	43.8
Abou Sif B	79*	50.0	79	50.0
Abou Sif C	52*	36.9	89	63.1
Larikba	64	72.7	24	27.3

* includes prismatic blades.

The inclusion of prismatic blades with Levallois flakes at Tabun, and at Abou Sif, is necessitated by somewhat inconsistent typological practice with regard to this kind of flake. As mentioned above, Bordes frequently includes these flakes in typical and atypical Levallois categories. In Tabun IX (see Table VI) they account for 42% of the flakes from prepared cores; in the Abou Sif B sample of complete flakes, 38%; in Abou Sif C, 35%. Thus, we know there are very few broad Levallois flakes in Abou Sif C (c. 2%), somewhat more in Abou Sif (c. 12%), and 23.5% in Tabun IX. Only at Larikba does the percentage of Levallois points to other flakes from prepared cores fall below 30% and even there it is close to that figure. In this regard, it is perhaps pertinent to mention a small sample of flakes from the Amud Cave that we had the opportunity to examine in Jerusalem. Forty of the 51 flakes were from prepared cores; they can be categorized as 43% Levallois points, 33% Levallois flakes, and 25% prismatic blades.

Since the frequency of elongate Levallois points (defined above as points on which length = > 3 x width) has been used as a defining character of the early Mousterian, the following data have been compiled for comparison with other industries.

Table XVII

Length/Width Ratios of Complete Levallois Points from "Phase 1" Mousterian Industries

Site and Bed	$\bar{X}$	Median	% Elongate	*n*
Tabun IX	2.45	2.33	34.1	179
Rosh Ein Mor	2.41	2.39	36.4	11
Nahal Aqev 3	2.48	2.48	28.2	39
Abou Sif B	2.70	2.57	43.4	76
Abou Sif C	2.69	2.67	40.0	50

All points were measured in the system described in an earlier publication (as were all other flakes measured in our study) (Jelinek 1977: 87–88); width is measured perpendicular to length at the midpoint of the flake. Only a very small sample of Levallois points was measured in the Negev collections at Southern Methodist University. All of the industries in the table show relatively high means and medians for this ratio, and only one (Nahal Aqev 3) has fewer than 1/3 elongate points. In general, we can say that these

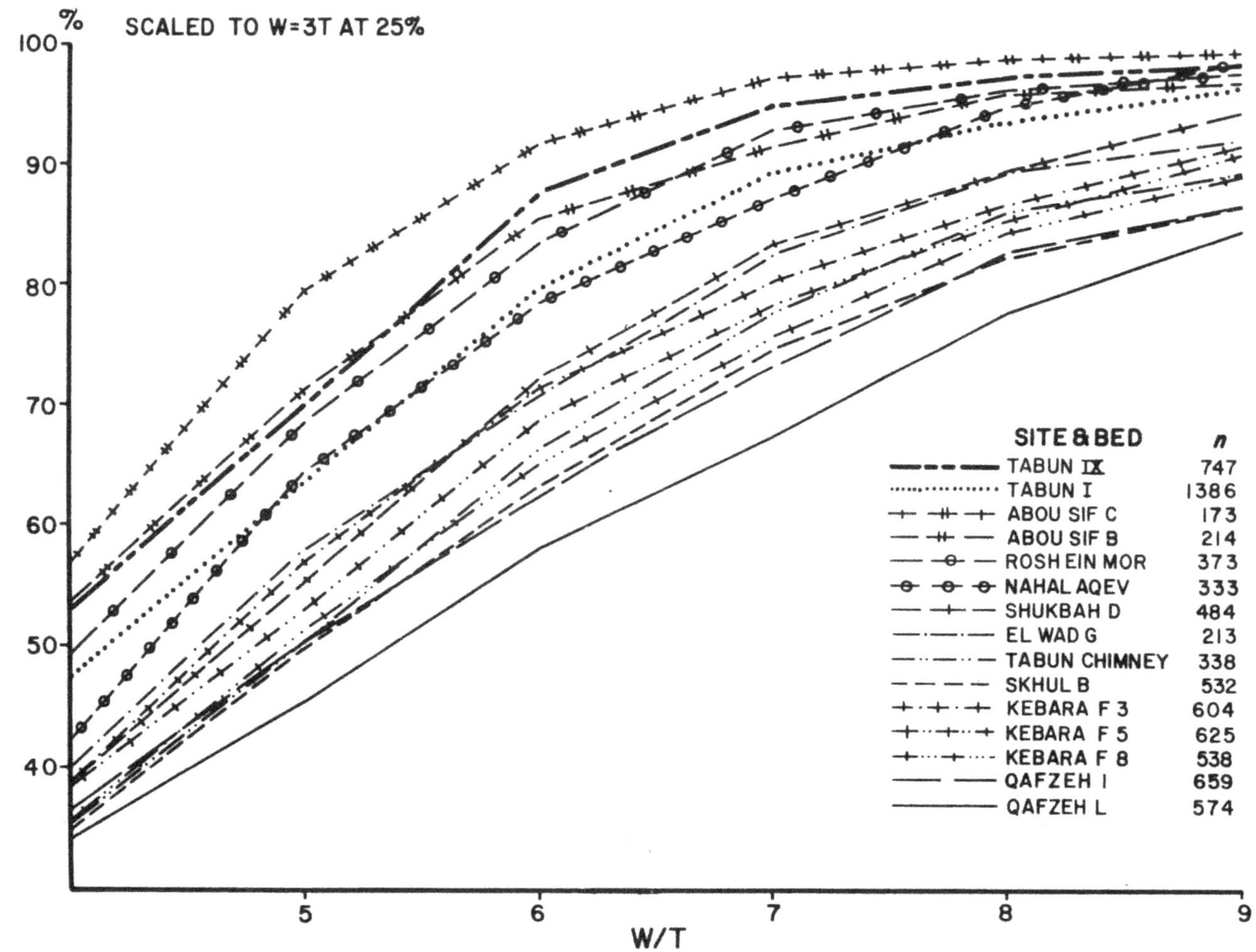

Figure 10 Cumulative percentage of Width/Thickness for complete flakes.

industries are consistent in this respect.

A final tabulation that can be made here concerns the width/thickness ratio for complete flakes. This is presented below for the five sites on which we have data.

Table XVIII

The Ratio of Width/Thickness for Complete Flakes from the "Phase 1" Mousterian

Site and Bed	$\bar{X}$	Median	Variance	*n*
Tabun IX	4.248	3.999	3.128	743
Rosh Ein Mor	4.436	4.111	4.236	373
Nahal Aqev 3	4.925	4.716	3.902	332
Abou Sif B	4.510	4.002	3.342	214
Abou Sif C	4.126	3.875	1.732	173

When we compare these data with the sequence at Tabun (Table IX), most of the values are compatible with a position between Tabun Unit IX and Unit II. One exception is the mean and median for Nahal Aqev 3, which are higher than those from Unit I at Tabun; another is the variance at Rosh Ein Mor, which is higher than that for Tabun II. The variance for Abou Sif C is lower than Tabun XIV. We are not yet in a position to explain the discrepancies in these values. In terms of the most reliable progression in the Tabun material, the variance can be considered the strongest indicator of temporal position. The narrow variance at Abou Sif C might be explained as the result of a poor sample of the full lithic industry; there is very little in the collections that we examined other than prismatic blades and Levallois points. All of these elements are somewhat elongate and rather narrowly patterned in terms of the variation that they exhibit at the midpoint where our measurements were taken. Here the mean and median of the distribution are probably better indicators of temporal position. The sample from Rosh Ein Mor was taken from an area of the site that produced very high ratios of manufacturing detritis with respect to Levallois elements. It is possible that this is related to the higher than expected (on the basis of the mean and median of the distribution) variance there. Figure 10 may clarify some of these relationships. Here a series of cumulative curves are presented summarizing the middle range of the distribution of this ratio for each of the measured museum collections. The lower ratios (1–3) are standardized to a total cumulative percentage of 25 on the assumption that the close correspondence of the samples from Tabun I through IX for these low values signals a kind of standard for the Mousterian industries. The Rosh Ein Mor sample matches this figure almost exactly, but the Nahal Aqev sample falls well below it (16.2%) at the ratio of 3. Even when this sample is corrected to 25% at a value of 3, it appears, with respect to the other curves, that the adjacent lower values are also under-represented. Other data will be needed to explain this anomaly. The other sites all cluster quite closely along the top of the series of curves.

Our conclusion, on the basis of all of the above observations, is that in most respects the collections classified here as "Phase 1" Mousterian (or Abou Sif type, or Tabun D type) share particular features of typology and technology that are similar to Unit IX at Tabun ("Layer D") and are distinct from Unit I ("Layer C"). Probably the best indication of the absolute age of these industries that is now available comes from the palaeotemperature correlation with the Tabun sequence. If these industries are contempoary, this correlation indicates an age of about 60,000 years for Unit IX, which coincides with a maximum cold phase. If these cooler temperatures also brought more moisture to the marginal areas of the region, the Negev and Judean deserts would have been most attractive to hunter-gatherers at that time. The absence of later kinds of Mousterian in the Negev suggest that the area was abandoned by the time these later industries appeared, or that the Negev Mousterian represents a more conservative tradition which persisted in this marginal area while a wider range of developments took place elsewhere. This is one possible explanation for some of the anomalous features in the industries at Rosh Ein Mor and Nahal Aqev with respect to Tabun IX and Abou Sif.

The "Phase 2-3" Mousterian in the Southern Levant

The reported sites outside of the Wadi Mughara which appear similar to the Tabun "C" and later Mousterian there include Shukbah, Shubbabiq, Kebara, Tirat-Carmel, Qafzeh, and Geula. Among these, it is possible to present standard indices with some reliability for Shubbabiq (Binford 1966), Kebara (Shick and Stekelis 1977), and Tirat-Carmel (Ronen 1974). Garrod's Shukbah collections and Neuville and Stekelis's Qafzeh collections appear too highly selected for this kind of treatment and the Geula collection is not sufficiently large.

Table XIX

Standard Indices and Basic Lithic Categories for "Phase 2-3" Mousterian Sites

Index or Category	Kebara Layer F		Tabun I:1-17		Tabun I:18-26		Tirat-Carmel Excavation		Shubbabiq Groups I-IV	
IL	66.5*		36.0		22.0		6.4		41.0	
ILty	58.4		64.7		66.1		8.0		43.3	
IF	71.7*		59.9		52.0		35.1		65.3	
IFs	63.7*		50.5		35.9		28.5		56.8	
IRe	37.7		31.5		23.6		26.5		25.6	
IIe	40.1		31.5		25.0		26.5		29.3	
IIIe	9.0		13.0		3.6		22.4		15.8	
IVe	7.3		7.4		8.2		22.4		15.2	
Ilam	28.6*		64.2**		35.7**		14.6		13.0	
Flakes	9709*	73.7%*	272	72.0%	2947	82.7%	485	68.3%	2081	62.9%
Cores	992	7.5%*	52	13.8%	253	7.1%	127	17.9%	416	12.6%
Tools (ess.)	2478	18.8%*	54	14.3%	364	10.2%	98	13.8%	812	24.5%

* not reliable because many flakes discarded.
** width measured at midpoint of length.

A comparison of Table XIX with Table XV shows that there is no standard index or basic lithic category that clearly separates the "Phase 1" from the "Phase 2-3" Mousterian. The site which seems most different from all others in Table XIX is the excavation collection from Tirat-Carmel. Its very low IL, ILty, IF, and IFs and relatively high frequencies in Group III and IV set it apart from the other samples. Perhaps these unique features are in some way related to the fact that it is the only open-air site in the group. The very high ratio of *essentiel* tools at Shubbabiq, along with somewhat higher ratios of IF and IFs than the other samples suggest a somewhat selected assemblage on the part of the aboriginal inhabitants of the site. Perhaps the same may be said of Tabun I: 1–17 in relation to I: 18–26.

The ratio of Levallois flakes to Levallois points in these and several other samples does seem to show differences with the "Phase 1" Mousterian. The Tirat-Carmel site has been omitted from this and subsequent comparisons because of small sample size. All of the comparisons presented in Table XX are the result of our observations except those from Shubbabiq. As in previous comparisons, we have lumped prismatic blades with Levallois flakes in the samples we recorded.

Table XX

Ratios Between Levallois Flakes and Levallois Points in the "Phase 2-3" Mousterian Industries

Site and Bed	Levallois Flakes		Levallois Points	
	n	%	*n*	%
Tabun I: 1-17	77	71.8	30	28.2
Tabun I: 18-26	633	92.1	54	7.9
Kebara F (3, 5, 8)	876	84.0	167	16.0
Shukbah D	270	60.5	176	39.5
Shubbabiq I-IV	814	93.0	61	7.0
Qafzeh I	470	79.5	121	20.5
Qafzeh L	399	82.4	85	17.6

Here the majority of cases are quite distinct from those postulated earlier Mousterian sites in Table XVI. The major exception is Shukbah D which shows a frequency of Levallois points comparable to the earlier sites. Tabun I: 1–17 and Larikba are intermediate between the "early" and "late" groups in this ratio.

The length/width ratios of samples of complete Levallois points in all of these collections except Shubbabiq (which we have not yet studied) are presented in Table XXI.

Table XXI

Length/Width Ratios of Complete Levallois Points from "Phase 2-3" Mousterian Industries

Site and Bed	$\bar{X}$	Median	% Elongate	*n*
Tabun I: 1-17	2.27	2.19	24.5	30
Tabun I: 18-26	2.07	2.07	14.1	35
Kebara F (3)	2.12	2.03	12.5	56
Kebara F (5)	1.94	1.80	10.8	37
Kebara F (8)	2.07	1.91	14.3	63
Shukbah D	2.29	2.18	22.0	173
Qafzeh I	2.17	2.03	16.1	112
Qafzeh L	2.18	2.19	17.3	81

A comparison of these data with Table XVII shows that there is no overlap in any of these values between the two groups. This suggests that, with adequate sample size, the length/width ratios of complete Levallois points (with width taken at midpoint) may provide a reliable way to distinguish between collections of Phase 1 and Phase 2-3 Mousterian in the Southern Levant.

The ratio of width/thickness of complete flakes can also be examined here as a possible chronological indicator. This index is presented in Table XXII for those collections that we have examined.

Table XXII

The Ratio of Width/Thickness for Complete Flakes from the "Phase 2-3" Mousterian

Site and Bed	$\bar{X}$	Median	Variance	*n*
Tabun Unit I	4.633	4.249	5.049	1377
Kebara F (3)	5.891	5.332	7.129	604
Kebara F (5)	5.962	5.496	7.604	628
Kebara F (8)	6.293	5.668	8.166	539
Shukbah D	5.630	5.204	6.302	484
Qafzeh I	6.476	5.802	7.974	661
Qafzeh L	7.034	6.502	9.508	573

A comparison of Table XXII with the Phase 1 Mousterian in Table XVIII shows that all of the values for the collections from Kebara, Shukbah, and Qafzeh are markedly higher than those for Phase 1 collections. In appraising these values it must be remembered that the flakes in the collections from Kebara, Shukbah, Qafzeh, and Abou Sif are all a selected portion from the full range of material that was present in the site. Those from the Negev and Tabun are presumably unbiased samples of flakes > 2.5 cm in maximum diameter from those contexts penetrated by the excavations. The position of these sites in the cumulative curves in Figure 10 (scaled to compensate for discarded thick flakes) shows them clustering on the late side of the Tabun I sample, in the same general range as the Tabun Chimney, Skhul, and El Wad G samples. One curious feature in Table XXII is the reversal of the trend in values for the stratified samples at Qafzeh and Kebara as opposed to Tabun. Whether this is indicative of a trend toward more narrowly patterned behavior at the end of the Middle Palaeolithic or is an artifact of the kinds of selection practiced in retaining flakes for curation is a question that will have to be left to future analysis.

What can be said at this point, is that the industries that are generally classified as "Phase 2-3" Mousterian in the Southern Levant do seem to share particular features that can be quantified by objective observation and that differentiate them from industries classified as "Phase 1". The fact that the Tabun Unit I and Chimney samples fit into the Phase 2-3 group strongly suggests that these are chronologically distinct phases of the Mousterian. This view seems supported by the sequence at the Yabrud site in the Central Levant, which shows a similar succession. Internally in the Southern Levant there is additional support for such a sequence at Abou Sif, where Layer B contains more broad Levallois flakes than the underlying Layer C. The dominance of the Phase 1 Mousterian in the south of the Southern Levant, and the abundance of Phase 2-3 Mousterian in the Central Levant and the north of the Southern Levant may be an indication that these distinctions will not hold true on the absolute time scale throughout the whole of the Levant. These kinds of questions point up the need for more refined or completely new techniques for chronological placement and environmental reconstruction so that we can better understand the relationships of the Middle Palaeolithic industries to each other and their relevance to man's cultural and biological development.

Conclusions

The evidence presented here for the interpretation of man's development in the Southern Levant strongly supports continuity in both the cultural and biological succession. The best evidence for cultural continuity is seen in technological aspects of the lithic sequence. Here, the gradual changes in flake dimension (W/T ratio) and the gradual appearance of Levallois technique at Tabun provide strong evidence for a continuous sequence without evidence of intrusive elements. The typological changes, particularly in the Mugharan Tradition, suggest cyclical patterns of change in activities that may relate to the shifting emphasis on different kinds of resources that would be expected to accompany the climatic oscillations of the early portion of the Last Glacial/ Interglacial cycle (Oxygen isotope Stage 5).

The time scale suggested by the evidence from Tabun in relation to the Oxygen isotope climatic curves can be used to calibrate the changes in flake dimension. This technological chronology can be extended to other sites for which samples of flakes are available. When this evidence is applied to sites that have produced hominid fossils, it suggests that the hominid that shows the most extreme neanderthaloid features (Tabun I) is markedly earlier than the more "modern" hominids (Skhul and Qafzeh). Here our evidence points to an *in situ* development of modern man in the Levant between about 50,000 and 40,000 BP. These conclusions are supported by the dimensions of a small sample of flakes (54) that were available from the Amud Cave in the Israel Museum. Both in dimensions and in typology these flakes are similar to Unit IX at Tabun ("Layer D"), which places the well defined Neanderthal from Amud at an early point in the technological chronology of the Mousterian.

I believe that the implications of this record of continuity in human biological development in this region are very important. A strong case can be presented for the concomitant development of manipulative and conceptual abilities of the hominids, as shown in the progression toward broader and thinner flakes and in the increasing specialization in assemblages, and for a chronological succession of fossil men that places the Tabun I, II, and Amud Neanderthals antecedent in time to the more modern Skhul and Qafzeh hominids.

References

Bar Yosef, O., P. Goldberg, and T. Leveson 1974 Kebaran and Natufian sites in the Wadi Fazael, Jordan Valley, Israel *Paléorient* 2: 415-428.

Binford, S.R. 1966 Me'arat Shovakh (Mugharet es Shubbabiq) *Israel Exploration Journal* 16: 18-32, 96-103.

Bordes, F. 1955 Le Paléolithique inférieur et moyen de Jabroud (Syrie) et la question du préaurignacien *L'Anthropologie* 59: 486-507.

Copeland, L. 1978 The Middle Palaeolithic of Adlun and Ras el Kelb (Lebanon): First results from a study of the flint industry *Paléorient* 4: 33-57.

Copeland, L. 1975 The Middle and Upper Palaeolithic of Lebanon and Syria in the light of recent research. *In:* F. Wendorf and A. Marks (eds.) *Problems in Prehistory: North Africa and the Levant:* 317-350.

Crew, H.L. 1976 The Mousterian Site of Rosh Ein Mor. *In:* A. Marks (ed.) *Prehistory and Palaeoenvironments in the Central Negev, Israel, I: The Avdat Aqev Area, Part 1:* 75-112.

Emiliani, C. and N.J. Shackleton 1974 The Brunhes Epoch: isotopic palaeotemperatures and geochronology *Science* 183: 511-514.

Farrand, W.R. 1979 Chronology and Palaeoenvironment of Levantine Prehistoric Sites as seen from Sediment Studies *Journal of Archaeological Science* 6: 369-392.

Garrod, D. 1942 Excavations at the Cave of Shukbah, Palestine, 1928 *Proceedings of the Prehistoric Society* 8: 1-20.

Garrod, D. and D. Bate 1937 *The Stone Age of Mount Carmel, I.*

Gisis, I., and O. Bar Yosef 1974 New excavations in Zuttiyeh cave, wadi Amud, Israel *Paléorient* 2: 175-180.

Horowitz, A. 1979 *The Quaternary of Israel* Acedemic Press.

Hours, F. 1974 Remarques sur l'utilization de listestypes pour l'etude du Paléolithic supérieur et de l'Epipaléolithique du Levant *Paléorient* 2: 3-18.

Hours, F., L. Copeland and O. Aurenche 1973 Les Industries Paléolithiques du Proche-Orient, Essai de Corrélation *L'Anthropologie* 77: 229-280, 437-496.

Howell, F.C. 1959 Upper Pleistocene Stratigraphy and Early Man in the Levant *Proceedings of the American Philosophical Society* 103: 1-65.

Jelinek, A.J. 1977 A Preliminary Study of Flakes from the Tabun Cave, Mount Carmel *Eretz-Israel* 13: 87-96.

Jelinek, A.J. 1975 A Preliminary Report on some Lower and Middle Palaeolithic Industries from the Tabun Cave, Mount Carmel (Israel). *In:* F. Wendorf and A. Marks (eds.) *Problems in Prehistory: North Africa and the Levant:* 297-316.

Jelinek, A.J., W.R. Farrand, G. Haas, A. Horowitz, and P. Goldberg 1973 New Excavations at the Tabun Cave, Mount Carmel, Israel: a preliminary report *Paléorient* 1: 151-183.

Marks, A. and H. Crew 1971 Preliminary report on Rosh Ein Mor, an open air mousterian site in the Central Negev, Israel *Current Anthropology* 13: 591-593.

Munday, F.C. 1977 Nahal Aqev (D35): A Stratified, Open-Air Mousterian Occupation in the Avdat/Aqev Area. *In:* A. Marks (ed.) *Prehistory and Palaeoenvironments in the Central Negev, Israel, II: The Avdat/Aqev Area, Part 2 and the Har Harif:* 35-60.

Munday, F.C. 1976 Intersite Variability in the Mousterian of the Central Negev. *In: Prehistory and Palaeoenvironments in the Central Negev, Israel I: The Avdat/*

Aqev Area, Part 1: 113-140.

Neuville, R. 1951 Le Paléolithique et le Mésolithique du désert de Judée *Mémoire des Archives de l'Institut de Paléontologie Humaine* 24.

Perrot, J. 1968 La Préhistoire Palestinienne *Supplément au Dictionnaire de la Bible* VIII, fasc. 43: 286-446.

Porter, S.C., M. Stuiver, and I.C. Yang 1977 Chronology of Hawaiian Glaciations *Science* 195 (4273): 61-63.

Ronen, A. 1979 Palaeolithic Industries: Upper Acheulian, Early Palaeolithic Flake Industries, Middle Palaeolithic. *In:* A. Horowitz *The Quaternary of Israel:* 300-305.

Ronen, A. 1974 Tirat-Carmel: A Mousterian Open-Air Site in Israel *Institute of Archaeology* Tel Aviv University, Publication 3.

Ronen, A. 1968 Excavations at the Cave of Sefunim (Iraq el Baroud, Mount Carmel) *Quartär* 19: 275-288.

Ronen, A., D. Gilead, E. Shachnai, and A. Saull 1972 Upper Acheulian in the Kissufim Region *Proceedings of the American Philosophical Society* 116 (1): 68-96.

Rust, A. 1950 *Die Hohlenfunde von Jabrud (Syrien).*

Shick, T. and M. Stekelis 1977 Mousterian Assemblages in Kebara Cave, Mount Carmel *Eretz-Israel* 13: 97-149.

Skinner, J.H. 1965 *The Flake Industries of Southwest Asia: A Typological Study* (Doctoral Dissertation, Columbia University).

Solecki, R.S. 1970 A Sketch of the Columbia University Archaeological Investigations at Yabroud (Syria) *Fundamenta, R.A.,* 2: 199-211.

Stekelis, M., and G. Hass 1952 The Abu Usba Cave (Mount Carmel) *Israel Exploration Journal* 2: 15-47.

Suzuki, H. and F. Takai 1970 *The Amud Man and His Cave Site.*

Turville-Petre, F. 1927 *Researches in Prehistoric Galillee, 1925-26.*

Vandermeersch, B. 1969 Les nouveaux squelettes moustériens découverts à Qafzeh, Israël, et leur signification *Comptes Rendus de L'Académie des Sciences* 268.

Vandermeersch, B. 1967 C.R. de fouilles sur Har ha Qefisan (Jebel Qafzeh) *Israel Exploration Journal* 17: 264-266.

Vandermeersch, B. 1966 C.R. de fouilles sur Har ha Qefisan (Jebel Qafzeh) *Israel Exploration Journal* 16: 267-269.

Vandermeersch, B. 1966a L'industrie Mousterienne de Larikba *L'Anthropologie* 70: 123-130.

Watanabe, H. 1970 A Palaeolithic Industry from the Amud Cave. *In:* H. Suzuki and F. Takai (eds.) *The Amud Man and His Cave Site:* 77-114.

Watanabe, H. 1968 A Palaeolithic Industry from the Amud Cave, Israel *Proceedings of the VIIth International Congress of Anthropological and Ethnological Sciences, Moscow* 3: 320-327.

Wreschner, E. 1967 The Geula Caves, Mount Carmel: excavations, finds and summary *Quaternaria* 9: 69-90.

Zeuner, F., D. Kirkbride, and B.C. Park 1957 Stone Age Exploration in Jordan I *Palestine Exploration Quarterly:* 17-54.

Discussion

M. Lamdan: It does not appear likely to me that the only distinction between the Yabrudian and the Acheulian is in the proportions of scrapers and handaxes. I think that unit XIV (layer 90, or former layer 80) – from where you have taken the samples – may be correlated to lower layer E, or more probably, to upper layer F. So there is yet another possibility that layer F, from which you have no sample, is Acheulian, and layer E, including your unit XIV, is Yabrudian.

A.J. Jelinek: We will look at this section tomorrow and I will show you the evidence which indicates that our unit XIV is quite early. It is interesting to note that I have looked at most of the available samples from Dorothy Garrod's layer G, which I would correlate with our unit XIV, both in terms of its stratigraphic situation and in terms of the character of the industry, and I find that the two industries are typologically similar. There are, in fact, two bifaces from layer G in Miss Garrod's material. One has been restruck and a large flake has been removed from one face; the other is a sort of Abbevillian biface. Both of these are in the collection at Cambridge. The flake dimensions from Miss Garrod's layer G, that is the proportions of width/thickness, are virtually identical to those from our unit XIV. The proportions of flakes from her layer F, where she saved relatively few flakes, so the sample is small, fall between layer G and the base of our layer E samples.

I think that this is strong evidence that, in fact, our unit XIV does correlate with the deeper deposits rather than with layer F or layer E. Perhaps when we look at the section I can explain this further.

P. Goldberg: Just to add more mud to the water; sedimentologically, as far as I could tell, there is no real difference between layer F and bed 80. I collected a few samples from bed 80, or unit XIV, and they look much more like layer F than they do like layer G.

A.J. Jelinek: The layer G samples that you have drawn on were taken from just above the bedrock and they have a lot of weathered material from the bedrock, as might be expected, as well as other compounds that have accumulated down next to the limestone. Our unit XIV and layer F might well be expected to look similar sedimentologically, since they are both mostly windblown sand. They lie higher up above the bedrock and there is not very much alteration that could be expected there. Perhaps if a sample of unit XIV were taken next to the limestone wall of the cave it might look similar to what you saw in layer G. That is something that could be tested. It would take some excavation because I do not think that unit XIV is now exposed near the cave wall. But I do not think that very much excavation would be required.

It would also be interesting to excavate some of unit XIV near the cave wall for another reason. The depositional environment there might be basic enough for bone preservation. It would be extremely interesting for problems of faunal correlation to have some bone from this deposit (as well as for amino acid analysis).

G. Bosinski: A few questions from the Central European point of view. I do not know the Tabun material, but the Yabrud collection is housed in my university. This concerns first of all the proposed dating – you showed in your paper a very typical Pradnik knife, including the sharpening blow.

A.J. Jelinek: That is the best one, of course.

G. Bosinski: In Central Europe this type is surely placed into Early Wurm, and I learned it comes from unit XIV, the Yabrudian. Everybody knows the Yabrudian is almost the same as the Charentian of Quina type. Most of these sites, there are exceptions I know, date from Early Wurm. You and others always refer to the bifaces as belonging to the Acheulian. But there is a similarity with the Mousterian of Acheulian Tradition. You yourself showed very regularly trimmed bifaces, and in the Yabrud collection they are together with backed blades. There are indications that in Europe such a material would be placed into Early Wurm. If I understood the paper of Patricia Masters, there is nothing against such a date. Is it excluded that this material is younger than you propose?

A.J. Jelinek: If we are going to correlate things typologically, I think that we are going to end up on shaky ground. If we are going to look for an industry that was virtually identical to the Amudian we would be left with the Chatelperronian in Europe; I do not think that anyone would correlate these two industries in absolute time anymore. I do not see any serious problem with the possibility of the same tool types arising at different times in different places, and especially in places as far removed environmentally and physically as the Levant and Central Europe.

There is another factor to consider here and that is the traditional view of interglacial/glacial relationships in the period since the penultimate glaciation. In particular, I am talking about the view that is emerging from the study of such variables as the deep-sea oxygen isotope series, other kinds of Palaeoclimatic evidence, and the correlation of Central and Western European deposits.

At a conference of the American Quaternary Association about a month ago, the question of what should be included as "Glacial" or "Interglacial" was raised and Shackleton, the man who was responsible for the isotope curve that you just saw, said that the only thing that he would call "Last Interglacial" was the first marked warm peak on the curve (isotope stage 5e). Everything after that would fall within the last glaciation in his terminology.

This could, in fact, be one of the explanations for the apparent lack of Last Interglacial deposits in caves and shelters in Western Europe. There one commonly finds only a relatively thin weathered layer assigned to that period and the interpretation is that there was not much deposition in sites during that period. But if the Last Interglacial is confined to isotope stage 5e, it had a time span of peak warm conditions of only about 12,000 years. The colder conditions that followed were more analagous to the succeeding glacial phases; the whole episode is a succession of increasingly cooler oscillations, climaxed by the cold peak at the end of the Last Glaciation.

If we tend to think of things that way, some possibility exists for correlating "early Wurm" materials in Europe with our later isotope stage 5 Yabrudian-Acheulian remains. What we really need now are systems of absolute dating within that time range to give us dates for that part of the Last Glacial/Interglacial cycle that will resolve these questions. In the absence of such evidence it is possible to try to correlate deposits on the basis of typological similarities, but my personal view is that such correlations are questionable over long distances.

A. Tuffreau: I am not a specialist of the Near East, and I wish to have precision on your Late Acheulian. Do you have Levallois flakes in the Late Acheulian, and what is the percentage?

A.J. Jelinek: We have very few Levallois flakes in that whole part of the sequence. It is a curious thing that there is no cave or shelter in the Levant that has a significant level of Levallois technique associated with an Acheulian industry. There is no question about these people being capable of making good Levallois flakes. We have a few cores and a few flakes that suggest that this was done occasionally in the caves. We do not know whether they made and used more Levallois artifacts outside of the caves or whether those surface sites with Levallois and Acheulian artifacts span a long period of time that includes separate occupations that could be assigned to separate Acheulian and Levallois industries. In Tabun there is very little evidence of Levallois technique (except in the Amudian facies) earlier than the "Transitional" unit X. In unit X there is a gradual, but rapid, rise in Levallois manufacture leading into the heavy Levallois industry of unit XI ("Layer D" Mousterian).

A. **Tuffreau**: Another question: Are the flake tools in the late Acheulian in your site and other sites in this country different from the Mousterian tools which are not made on Levallois flakes?

A.J. **Jelinek**: Typologically they are not very different, but their frequency and diversity is markedly different. There are very few different kinds of tools in the later Mousterian: most of the retouched pieces are side scrapers.

In the Acheulian of unit XI there are frequent burins, including some rather peculiar special types, and scrapers, side scrapers, and some curious small tools that occur with some frequency but do not correspond to any widely recognized typological categories. The industry as a whole is somewhat peculiar in the general context of the Acheulian. When I first examined it, as it was coming out of the excavation, my reaction was, 'This does not look like Acheulian; it looks like something else". Then I asked myself, "Where is a good, carefully excavated sample of Acheulian from the Levant in which all of the artifact materials were saved?" In the industry at Oumm Qatafa, for instance, we have many large artifacts, but very few small specimens and almost no flakes; presumably the smaller materials and flakes were discarded at the time of the excavation. My conclusion was, "Well, that is what the Acheulian looks like at this level in the Levant".

This is not really such a simple issue. The Acheulian from Isimila in East Africa, for instance, has numbers of small well shaped tools and I think that it may be that the more carefully we excavate the later Acheulian sites, the more of this kind of thing we are going to see. Our notion of the Acheulian as an industry composed primarily of large tools is probably due for a change.

ENVIRONMENTAL CONDITIONS DURING THE LOWER/MIDDLE PALEOLITHIC TRANSITION IN THE NEAR EAST AND THE BALKANS

William R. Farrand

If, by way of definition, the transition from Lower to Middle Paleolithic in this area is considered to be the transition from Late Acheulian industries (and variants such as Yabrudian, Amudian, Samoukian, etc.) to (Levalloiso-) Mousterian industries, then the period of time we are concerned with is more or less centered on the time of the last interglaciation. Traditional terms for that time period, such as Riss/Würm or Eemian, cannot be applied directly here because of the lack of means of correlation, on one hand, and the major uncertainties in the significance of those terms even in their type areas, on the other. The most direct type of evidence, in the Near East, for an interglacial period in northern latitudes is that of sea levels very close to or above the present sea level. In our area this period of high sea levels is called the Enfean-Naamean (Sanlaville 1977) and its evolution and chronology fit quite reasonably with reconstructions of sea-level history from the western Mediterranean, Barbados and New Guinea, where better radiometric dating is available. Fortunately, numerous prehistoric sites are associated directly or indirectly with these former shorelines (Farrand 1979), providing us with good stratigraphic control in many cases, e.g., at Tabun, Adlun, Ras Beirut, Ras el-Kelb and Nahr el-Kebir (Figure 1).

Unfortunately, however, sea-level history does not provide us abundant or clear environmental information. We can establish, of course, that certain peoples were living in and exploiting a coastal environment, but the former concept of interglacial high sea levels being equated with interpluvial dry climates can no longer be considered valid. Worldwide evidence shows clearly that there were at least three major sea-level peaks during the last interglaciation (*sensu lato*) and that those peaks and the intervening low sea levels correlate in time with oxygen 18/16 fluctuations in the world oceans. The latter reflect primarily the volume of glacier ice stored on the continents, but they also are influenced by other (direct?) climatic conditions, as can be seen from Figure 1 where the amplitudes of contemporaneous peaks and troughs of the two curves are different. The isotope curve shows a progressive enrichment in oxygen 18 from peak 5e into the beginning of stage 3, but the sea level peaks (therefore, ice volumes) rise to nearly the same level repeatedly throughout this same interval. In brief, interglaciations were times of fluctuating climate as well as fluctuating sea level, and these were not parallel.

Moreover, the climate of the Near East during the time of the last major glaciation was by no means uniformly "pluvial," as shown by recent pollen studies throughout the area (Figure 2). The interval between c. 80,000 and 10,000 yrs ago was interrupted by dry episodes in which arboreal vegetation dropped to levels well below that of the

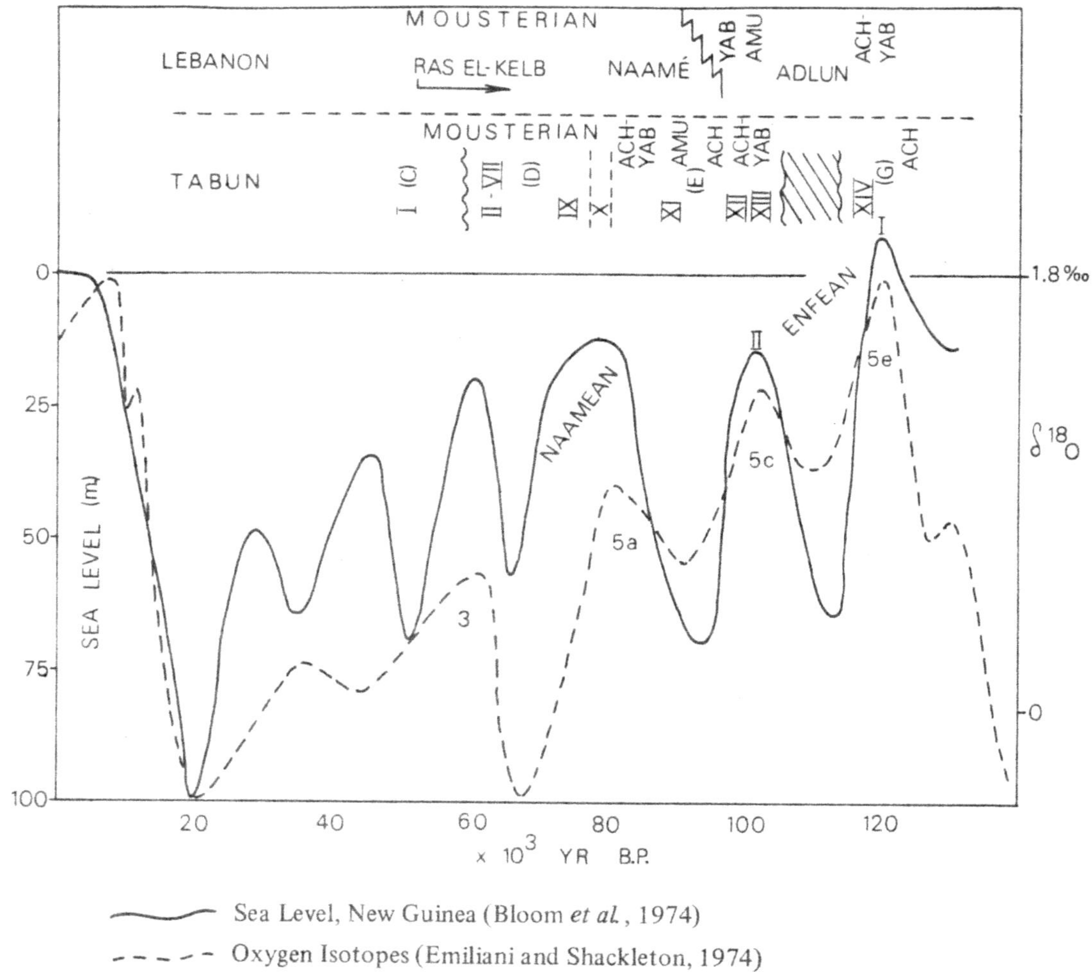

Figure 1 The Lower/Middle Palaeolithic transition in the Near East relative to sea levels and oxygen isotopes.

present (Holocene) interglaciation. Unfortunately, few pollen studies are available for the last interglaciation in this area, and even fewer for earlier times. Exceptions are the long core from Tenaghi Philippon in northern Greece (van der Hammen *et al.* 1971) and pollen from marine sediments along the Israel coast (Rossignol 1969). However, the latter are difficult to interpret and lack pollen from critical stratigraphic intervals, and the former (Philippon) is a long distance from the Levant and probably does not apply directly here. It is becoming clear from various studies of the last glacial interval that the northern perimeter of the Mediterranean was experiencing relatively dry conditions while the southern perimeter was wetter than usual (Farrand 1971). Israel and Lebanon are situated in the tension zone between those two areas of opposing climatic tendencies, and therefore extrapolation of paleoclimates even across relatively short distances is not straightforward.

Direct evidence of environmental conditions from Near Eastern prehistoric sites is unfortunately quite limited. Fossils have not been particularly informative, especially because they are rare to absent in last interglacial sites. Pollen is not well preserved in general, although at Tabun and at Nahr Ibrahim abundant tree pollen characterizes Middle Paleolithic strata (Figure 2). Frost action appears to be absent in the coastal zones; no cryoclastic debris has been identified at Tabun, Ksar 'Aqil, Franchthi or other sea-level sites where detailed sediment analysis has been done, and is of questionable importance at Qafzeh. Cryoclastic eboulis is characteristic of inland sites, such as Yabrud, but these sites experience freezing climates during present-day winters.

Preserved paleosols tell us that the landscape was stabilized at certain times in the past, e.g., the hamra paleosols on the Israel coastal plain that indicate an absence (or marked diminution) of eolian input, which can be interpreted only indirectly in terms of environmental change. However, from another point of view, the fact that all the paleosols are of a similar kind and similar to presently forming soils indicates that climatic change throughout the later Quaternary time was minimal in this area (Dan and Yaalon 1971). This interpretation is reinforced by Tchernov's (1975) study of fossil rodents which do not reflect a strong cycling of climate that could be correlated with glacial-interglacial cycles. These fossils, instead, show a gradual, progressive drying trend from some time within the last interglaciation straight through the last glaciation into the Holocene.

Information from the Balkans is even more limited. Excavated Middle Paleolithic sites are few (e.g., Asprochaliko, Kastritsa) and Lower Paleolithic ones almost unknown (Petralona). Franchthi Cave at sea level in the southeastern Peloponnesus provides a model, however, of full-glacial conditions, although during Upper Paleolithic occupations.

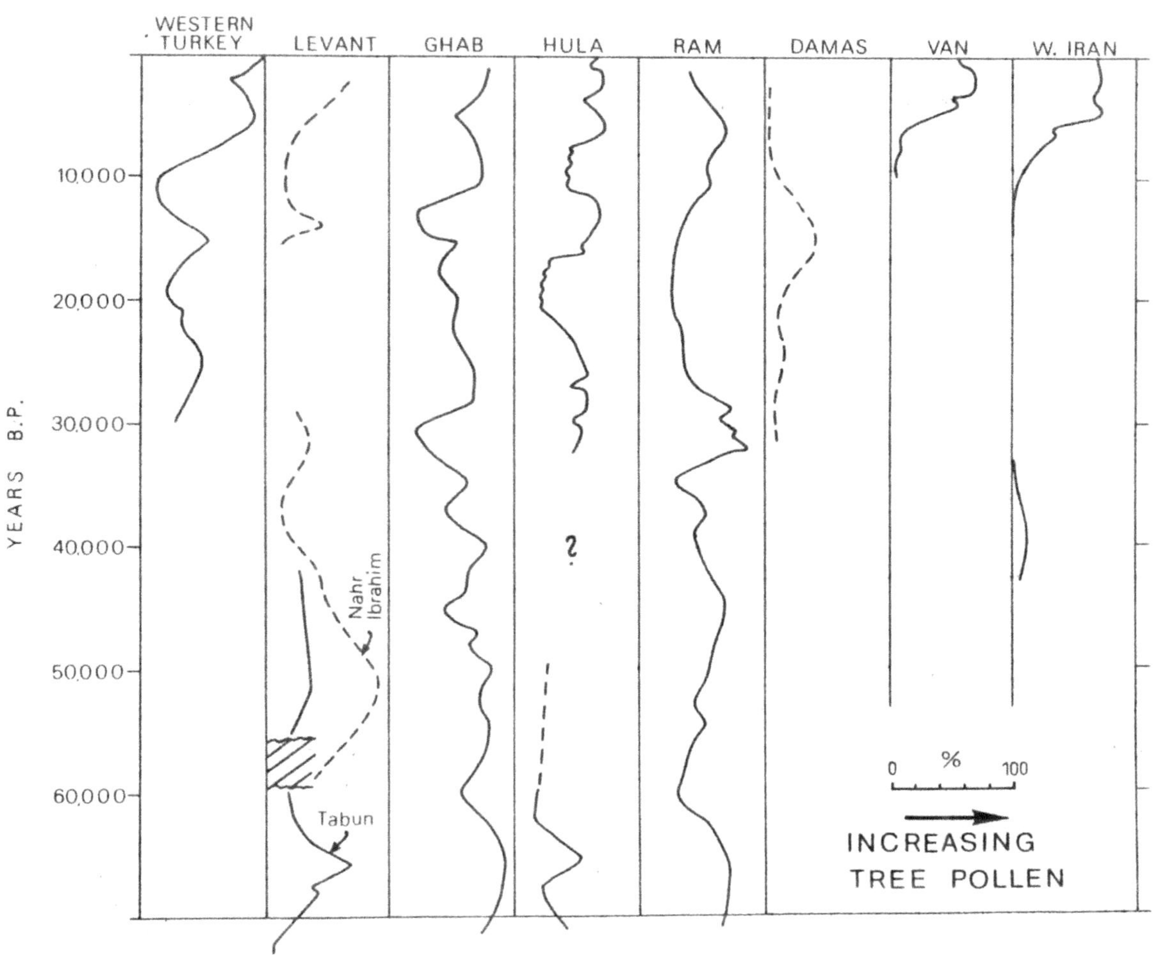

Figure 2 Pollen diagrams of the Near East.

Here, as in the Levant, "glacial" conditions do not seem to have differed greatly from those of the Holocene, other than those changing environments conditioned by the proximity to the fluctuating marine shoreline.

References

Bloom, A.L., W.S. Broecker, J.M.A. Chappell, R.K. Matthews and K.J. Mesolella 1974 Quaternary sea level fluctuations on a tectonic coast: new 230Th/234U dates from the Huon Peninsula, New Guinea *Quaternary Research* 4: 185-205.

Dan, J. and D.H. Yaalon 1971 On the origin and nature of the paleopedological formations in the coastal desert fringe areas of Israel. *In:* D.H. Yaalon (ed.) *Paleopedology* Jerusalem. pp. 245-259.

Emiliani, C. and N.J. Shackleton 1974 The Brunhes Epoch: isotopic paleotemperatures and geochronology *Science* 183: 511-514.

Farrand, W.R. 1971 Late Quaternary paleoclimates of the eastern Mediterranean area. *In:* K.K. Turekian (ed.) *The Late Cenozoic Ages* New Haven. pp. 529-564.

Farrand, W.R. 1979 Chronology and palaeoenvironment of Levantine prehistoric sites as seen from sediment studies *Jour. Archaeological Science* 6: 369-392.

Hammen, T. van der, T.A. Wijmstra and W.H. Zagwijn 1971 The floral record of the late Cenozoic of Europe. *In:* K.K. Turekian (ed.) *The Late Cenozoic Glacial Ages* New Haven. pp. 391-424.

Rossignol, M. 1969 *Sedimentation palynologique dans le domaine marin quaternaire de Palestine.* Notes et Memoires sur le Moyen-Orient, Museum National d'Histoire Naturelle, Paris, 10: 1-261.

Sanlaville, P. 1977 *Etude geomorphologique de la region littorale du Liban* Publication de l'Universite Libanaise, Beyrouth, 3 vol.

Tchernov, E. 1975 Rodent faunas and environmental changes in the Pleistocene of Israel. *In: Rodents in Desert Environments.* pp. 331-362.

Discussion

P.M. Masters: I have a comment about the age of the beachrock. As I pointed out in my lecture, there is a relationship between temperature and the rate constant for aspartic acid racemization. Using the D/L ratio that I experimentally determined on the shell (alloisoleucine/isoleucine = 0.388) and the age of 80,000 years from Professor Brunnacker's Uranium/Thorium work, it is possible to calculate the temperature over the amount of time since the deposits. The average mean temperature estimated by means of the Arrhenius equation (eq. 2 in my paper) comes out to be 23°C, which is rather high for a Pleistocene environment here. Therefore it is somewhat more reasonable in my opinion to take the older date of 130,000 years which I have determined using a well-controlled rate constant from a southern California marine terrace. The error for that rate constant is ± 10,000 years, so I would not distinguish between 130,000 and 120,000 years. Such an age would make the Hahotrim beachrock contemporaneous with oxygen isotope stage 5e. Several problems with these dates should be pointed out here. The beachrock shell dated by either method is reworked material and may present a problem of interpretation. An additional problem for racemization dating of shell is that there are generic differences in racemization rates, and the shell is so fragmentary as to preclude analysis of a single molluscan species. A difficulty with uranium-series dating of molluscs is that the shells represent an open system, which cannot be as accurately dated as a closed system such as coral.

A. Ronen: Prof. Farrand said that we do not know the stratigraphy of the extreme western ridge. That is generally correct, but recently Dr. A. Raban told me that on the Islands off Achziv and Rosh Hanikra a thin layer of red soil was exposed by ancient quarries between two sandstone layers, the same as we find on the coastal ridge. We do not have as yet any tools from that red soil, but I would guess it is a Mousterian soil, like everywhere else on the northern coastal plain of Israel.

W.R. Farrand: That is the point. Is it Mousterian or is it Epi-Palaeolithic?

A. Ronen: There is no Epi-Palaeolithic soil in the north of Israel, except for scarce signs of it. This includes the Carmel coastal plain and the Western Galilee coastal plain. So the odds are that this particular soil, some 10 centimeters thick, is Mousterian too.

P. Goldberg: A few related comments: I think one of the problems with rock shelters – if you like, with archaeological sites in general – is that we do not really understand the effect of human or biological influence on the substratum. With the little study I did in Hayonim Cave – which I believe is quite applicable to caves with similar lithological types (Kebara, Zuttiyeh, Rakefet) – you can see in thin section that the last observable sedimentological "effect" is that of biological or anthropological activity. I think that the climatic signal, if it is there at all, has been more or less modified by the biological effects, by leaching, or whatever.

Consequently, I think that using classical sedimentological methods such as those which work quite well in France, require great care in their application to Israeli caves, if they will work at all. I suspect that more often than not you will have to find a totally different methodology to evaluate past environments and past human activities using cave sediments. Furthermore, I am quite skeptical at this point of interpretations attached to former cave sediment studies – in spite of the fact that some of these were done by myself (Jerf 'Ajla, Yabrud) – now that I have seen these biological effects.

W.R. Farrand: Your comments are very well stated, but I think they do not apply very greatly to most of the sites involved because the human influence has been really pretty slight in all of these. You do not see habitation layers, you do not see living floors, you do not see organics, you do not see much in the way of hearths even.

P. Goldberg: Take for example the Qafzeh terrace deposits, beds 7 to 10 in the new excavations by Vandermeersch. In the field they look like 'eboulis' or calcareous debris. But on an area basis, 50 per cent of what you see in the thin section is organic debris: Wood, pieces of bone cemented by manganese. There is no way you can evaluate this type of mixture by looking at it in the field or doing any other sedimentological analysis. Take again the terrace deposits of Qafzeh, for example: You can see very distinct sedimentological units which appear in thin section but do not appear in the field. And these repeat themselves: The top is very organic. Layers 10–15 are very pedological, probably with some kind of soil development going on; below that again there is a large component of organic material and finally the eboulis sec. Classical sedimentology won't get at these sedimentological types.

W.R. Farrand: Let's not throw the baby out with the bath water.

P. Goldberg: Well, no, but I think that one has to be very careful. The interior Qafzeh sediments seen in thin section have a lot of cultural disturbance. Yet a size analysis may have a tendency of confounding the geogenic effect with the anthropogenic ones, and you could never see it in the sieve shaker, or in the hydrometer.

A.J. Jelinek: I would like to introduce another note of caution in accepting Farrand's second interpretation. I think that the correlation between the deposits in the Tabun Cave and the beach formations is at present a very tenuous one. There simply is no direct way of correlating the two. The beach rock may well be dated to 120,000, in which case it would equate with Enfean I, but that does not necessarily mean that the deposits that you say are equivalent in the cave are really of the same age.

We return again to the fact that we are not going to understand the time relationships until we have a good system of absolute dating that can be applied to both situations. We can speculate about these things and sort of juggle them back and forth to look for the most probable correlations, but all of this kind of correlation may be highly questionable.

With regard to the Amudian correlation in your second step, I would point out that the Amudian does in fact lie directly on a beach and is in part incorporated in a beach at Adlun. We do not, however, know how long after that beach was abandoned by the sea the Amudian industry was deposited there. The Amudian artifacts are not rolled at Adlun, which suggests that the beach was not active at the time the industry was made. We do not know how long the sand was soft enough for the artifacts to become incorporated in it. All we know is that the beach had been abandoned by the sea at the time the Amudian was laid down. This sea retreat, we would assume, was associated with cooler conditions, in a global sense.

Let me approach this in another way. We assume that these marine retreats correlate with the apparent cool peaks on the oxygen isotope curve. To be more specific, we assume that they may correlate with isotope Stages 5b and 5d, rather than with 5a, 5c or 5e.

What I am trying to point out is that the Amudian industry on the beach at Adlun does not necessarily correlate with a period of high sea level that we would equate with isotope Stages 5a, 5c or 5e. Until we have some method of absolute dating we are simply juggling in the dark; we can only take the most highly probable solutions to this problem of correlation.

D.L. Inman: I simply want to add an observation regarding the dating and its implication for sea level and chronology in general. Actually 15 years ago we were very careful with radio carbon analysis and what it meant, and how clean the samples were, and so forth. Now there seems to be a general assumption that the radio carbon or Uranium-Thorium date is good. This just does not hold. Anything in Carbon 14 much beyond 30,000 years, even in the best conditions, has limited value. 50,000 years of Carbon 14 – I would not put much belief in it in any way. So that it is a little bit like the racemization situation now. It is a new technique and people have been very careful with their calibrations and how they do them. It seems to me that we are now forgetting that the same things apply to radio carbon and other isotope techniques. And in the case of Uranium-Thorium, the worst possible thing is to get some recrystallization and diagenesis which always gives a younger date than the one the sample really has. So that we have to be very careful about these so-called absolute dates – they are not all that absolute unless the samples dated were carefully tested for contaminants and this factor was considered in the analysis.

O. Bar-Yosef: Within the discussion of the relative chronology of certain lithic assemblages I would like to mention an alternative viewpoint. Two days ago, Jelinek presented an evolutive curve demonstrating the thinning of flakes through time. In the old terminology this means through the "Late Acheulian", the "Acheulo-Yabrudian" and the Mousterian. If the evolution of stone technology is a good chronological scale, then there is no way to accept the correlation, just presented by Farrand, between the lower levels at Qafzeh and layer D at Tabun. Neither can we accept the late continuation of the Acheulo-Yabrudian in Yabrud Rockshelter I relative to its correlates in Tabun. Tabun is, for the time being, the only complete stratigraphic sequence where assemblages were carefully measured. Perhaps we should, in correlating these sites in the Levant, go by archaeology alone and abandon the contradictory evidence that comes from geological observations.

A.J. Jelinek: I would like to respnd to that. I would be very cautious about doing this. The only sample from Yabrud that I was able to study, because of the limits of the range of the collection imposed on the enterprising Alfred Rust by what he could carry on his bicycle, was from the Pre-Aurignacian. He seems to have made a deliberate effort to collect everything possible from these levels (Yabrud I, layers 15 and 13), including a large sample of flakes. I was able to measure these flakes and, according to all the kinds of information that our observations can yield, they seem to comprise a good representative sample of lithic debitage. The interesting thing is that the width/thickness ratio for this sample is quite close to Unit XI at Tabun, in which we find the Amudian. This little bit of information suggests that the two industries are quite similar and may be temporally equivalent. This increases the probability that the typological sequences at Tabun and Yabrud I are temporally equivalent, both in terms of the evolution of the Mugharan Tradition and the later Mousterian developments.

However, I would like to also say that this system needs a lot more testing before we should give it priority over other kinds of evidence in establishing chronological relationships. I personally have a fair degree of confidence that it will hold up, but we do not yet have enough comparable samples from deep sites in the Levant that can be correlated through the use of independent dating techniques to be sure about this. I hope to make an effort to apply this technique to sites in Western Europe, where we have more deep sections, more material that is comparable, and more of a chance to tie it to an absolute chronology.

Right now I would say that this is simply another line of evidence that we have to consider in relation to all of the other kinds of evidence that we can apply to problems of correlation. We have to see how each works within the system of overall weight of evidence; I would not give the flake dimension data priority in a correlation when it was in conflict with several other strong lines of evidence. I think that the general similarity of the industrial sequences at Adlun, Yabrud I, and at Tabun suggests very strongly that this sequence holds true for much of the Levant. I think that it is a strong counter argument for the postulated early position of the Tabun B/C type of Mousterian in the north of Lebanon and Syria.

However, without dependable absolute dating techniques that can be applied to these materials we will never be able to be really sure about these time relationships. It is possible that the new linear accelerator techniques of C 14 analysis may be of some help here, as well as the use of thermoluminescence analysis on heated lithic materials.

We will have to rely on all of these things taken together to demonstrate whether or not this system of correlation is going to work, but right now we do not yet have enough evidence. We can now see what we have to do and some of the ways that it can be done, but, much as we wish we had the answers, we do not.

THE UPPER ACHEULIAN ON THE ISRAELI COASTAL PLAIN

Modi Lamdan

Before beginning my brief lecture, I would like to note that most of the data on which I base my remarks was compiled by the late Dr. Gilead from the collections of many amateurs (Gilead 1970). The upper Acheulian sites in the coastal plain are open-air sites and the lithic assemblages were laying on the surface of red sandy loam soils called "Hamra" – or in the upper few cm. of that soil.

The lithic industry of upper Acheulian coastal sites contains handaxes, flake-tools, Levallois cores and Levallois flakes. The industry is a typical Mousterian if one excludes the handaxes. This can be seen, for example, in some indices from two sites, Evron-Zinat in the northern coastal plain and Kissufim in the south:

	Kissufim[1]	Evron[2]
IL (real)	55	33
IF (real)	73	47
ILty	49	27
IB	30	70
Cordiform & Amygdaloid handaxes	45%	50%
Discoidal & Ovaloid handaxes	28%	23%

[1] After Ronen *et al.* 1972.
[2] After Gilead and Ronen 1977.

Most of the handaxes in the upper Acheulian open-air sites are Cordiforms and Amygdaloids, contrary to what we have in the Cave of Tabun – mainly Micoquian in layer E or Discoidal and Ovaloids in layer "90" (Unit XIV).

The coastal plain may be divided by the Yarqon river (which enters the sea near Tel-Aviv) into two major parts: the northern, which gets its raw flint from the Samarian hills, Mt. Carmel or western Galilee – it is a good quality homogeneous flint; and the southern part, which gets its raw flint from the Judean hills – a largely brecciated flint.

The sites of the eastern zone of the coastal plain seem to be larger and more important than the sites of the western zone.

The distribution of the upper Acheulian sites on the littoral (Figure 1) is limited to the east of the third "Kurkar" ridge (fossil sand dunes), counting from the sea. The west flanks of this ridge fit the +20 m. contour line. The suggestion can be made that all the coastal upper Acheulian sites date from a +20 m transgression.

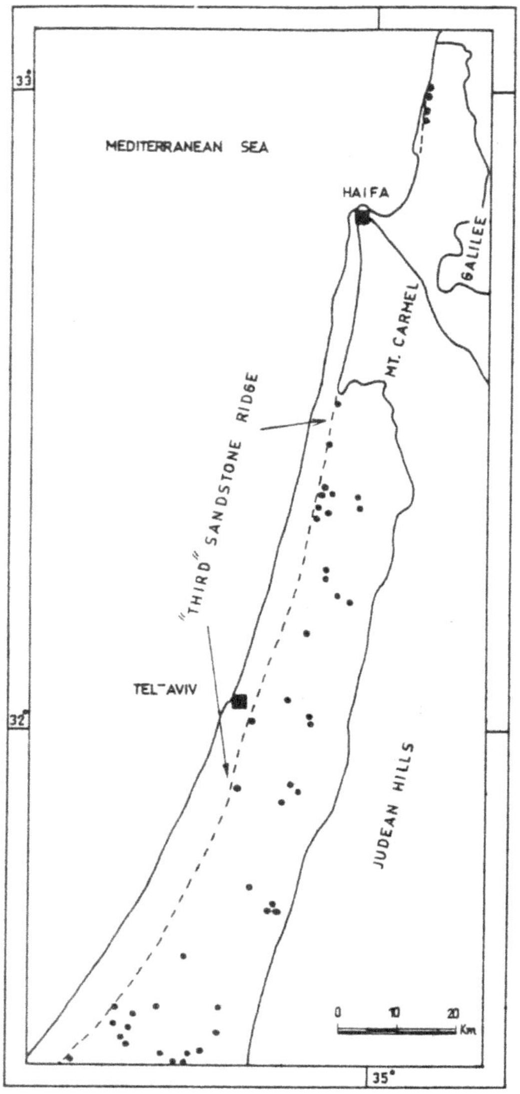

Figure 1 Distribution of Acheulian sites on the Coastal Plain of Israel (after Gilead 1970, with additions).

References

Gilead, D. 1970 *Early Palaeolithic Cultures in Israel and the Near East* Unpubl. Ph.D. Thesis, Hebrew University, Jerusalem.

Gilead, D. and A. Ronen 1977 Acheulian Industries from Evron on the Western Galilee Coastal Plain *Eretz-Israel* 13: 56-86.

Ronen, A., D. Gilead, E. Shachnai and A. Shaul 1972 Upper Acheulian in the Kissufim Region *Proc. of the Am. Phil. Soc.* 116: 86-96.

Discussion

A.J. Jelinek: These collections that you are talking about now and that you mentioned earlier, were they all collected by amateurs or were any sites collected methodically? I know that the Kissufim material was collected more or less methodically, but mostly by amateurs. The question that arises here is how reliable, for instance, is the Levallois Index in such a collection? The fact that we have 500 handaxes, 11 cores, and 69 flakes suggests that these people were really picking up handaxes and not paying much attention to flakes. They picked up the flakes that were nice and looked like Levallois. I think that in our comparisons we have to treat these indices with caution. The main problem here is one that you pointed out very well: in the cave and shelter sites, such as Oumm Qatafa, Bezez, Tabun and Yabrud I, through the whole early part of the sequence where we have a lot of handaxes we have very little Levallois. It suggests that if the quantity of Levallois that is reported for the open sites and surface sites is really a valid ratio, there are two most likely explanations for the differences between these sites and the cave sites:

One explanation is that the people behaved very differently once they got out of the caves and made a lot more Levallois flakes. Perhaps they did this because of better access to raw material or some other related reason. The other explanation is that since this is all surface material or derived from a few centimeters below the surface, it is quite possible that we are looking at material that represents widely separated time periods. We may be looking at Middle Palaeolithic industries mixed with Lower Palaeolithic industries. This is always going to be a problem in the interpretation of surface collections.

What we really need are some good *in situ* deposits of this material that will rule out mixing. Until we have this evidence we have what may be a "phantom" of an Acheulian industry with heavy Levallois technique up and down the Levantine coast. This would not look out of place in Western Europe where these things are demonstrably associated, but here all of the *in situ* deposits that we have with handaxes are without much Levallois technique. So, I would have to say that up to now we have a "phantom" Acheulian of Levallois Facies whose actual existence still has to be demonstrated in an *in situ* context of demonstrably restricted temporal duration.

M. Lamdan: The criticism about the possible mixture of Lower and Middle Palaeolithic material is well known. But the picture we receive from this entire zone of the country is of an industry in which a flake element, including Levallois technique, always goes together with the handaxes, as I have shown. The number of handaxes is very large. This is contrary to what we heard from Bosinski and Tuffreau, who reported having a low number of handaxes in this period in Europe.

A. Tuffreau: It is possible that your handaxes and Levallois flakes are mixed. You are in the situation in which we in northern France were in the 50's. It will be very interesting to make a real excavation in a site on the coastal plain and compare it with the caves.

M. Lamdan: There are very few sites that can be excavated in the coastal plain. The lithic material is normally spread on the surface.

G. Bosinski: Are the Tabun handaxes of the same type as the Late Acheulian handaxes from the surface?

M. Lamdan: In Tabun we have more Micoquian handaxes, specially in layer E, and more ovaloid and discoidals, as in unit XIV. On the coastal plain we have more cordiforms and amygdaloids.

A. Ronen: There is no doubt that we need to excavate some good sites on the coastal plain. But it should be stressed that there is no one concentration of handaxes there with no Levallois elements mingled with it, and it is at least strange to assume that every time Middle Palaeolithic people saw a handaxe, they settled near it and made their tools.

The possibility raised by Jelinek that perhaps people behaved differently outside of caves cannot be questioned, in my opinion, at least as far as the Yabrudian is concerned. We do not have anywhere an open-air site with a Yabrudian industry like that of Tabun E. While the matter is open to question, for the time being we have to assume that the Levallois technique goes with the open-air Acheulian.

THE ACHEULIAN SITE OF BREKHAT RAM

Naama Goren

The Northern Golan Heights are characterized by basalt flows, volcanic cones made of scoria, and sporadic appearance of limestones inliers (Mor 1973, 1981). In the spring of 1980, artifacts were found at three different localities south of Mt. Hermon, in a clayey horizon embedded between two basalt flows. The artifact-bearing horizon was exposed as a result of agricultural activities along the western edge of the Brekhat Ram volcano crater (ca. 1,000 m a.s.l.). A visit to the find spots indicated that excavation might elucidate the archaeological horizon (Goren 1981a). The quarrying of several terraces has exposed many sections. Observations in the western part of the volcano have revealed two distinct basalt flows, named the "Kramim Basalt" (Mor 1981). The following stratigraphic units are exposed (bottom to top):

1) Basalt flow ("Lower Kramim Basalt") – 3 m.
2) Clayey unit (palaeosols) with artifacts – ca. 1.5 m.
3) Basalt flow ("Upper Kramim Basalt") – 4 m.
4) Volcanic tuff, palaeosols, with artifacts.
5) Basalt flow ("Saar Basalt").

Units 1–3 were exposed at one of the find spots. An area adjacent to a modern track, leading from the rim of the crater to the lake, was chosen for excavation. Heavy machinery quarried away an overburden, ca. 4 m thick, of the "Upper Kramim Basalt", thus exposing unit 2, which later led to its erosion and the appearance of flint artifacts on the surface.

In September 1980 and 1981, a total of about 30 sq. meters was excavated (Grid system of 1 sq. meter), resulting in the exposure of a distinct archaeological horizon 0.3 m thick at the bottom of unit 2. The artifacts, mostly of flint, are found above, among and below basalt boulders, cobbles and pebbles, and are overlaid by ca. 1.2 m of red clay. This upper horizon deviates from the general characteristics of unit 2. A micromorphological study (Goldberg, in press) indicates a volcanic tuff origin for this unit, and distinguishes two distinct phases of pedogenesis, one on top of the archaeological horizon, and another below the contact between units 2 and 3. The sedimentological character-

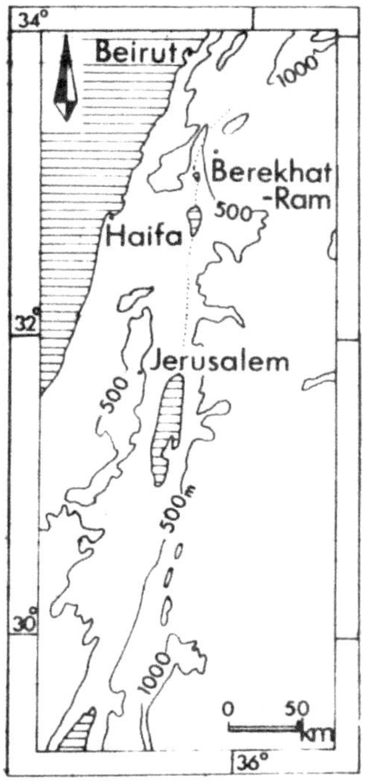

Figure 1 Location of the Brekhat Ram Acheulian site.

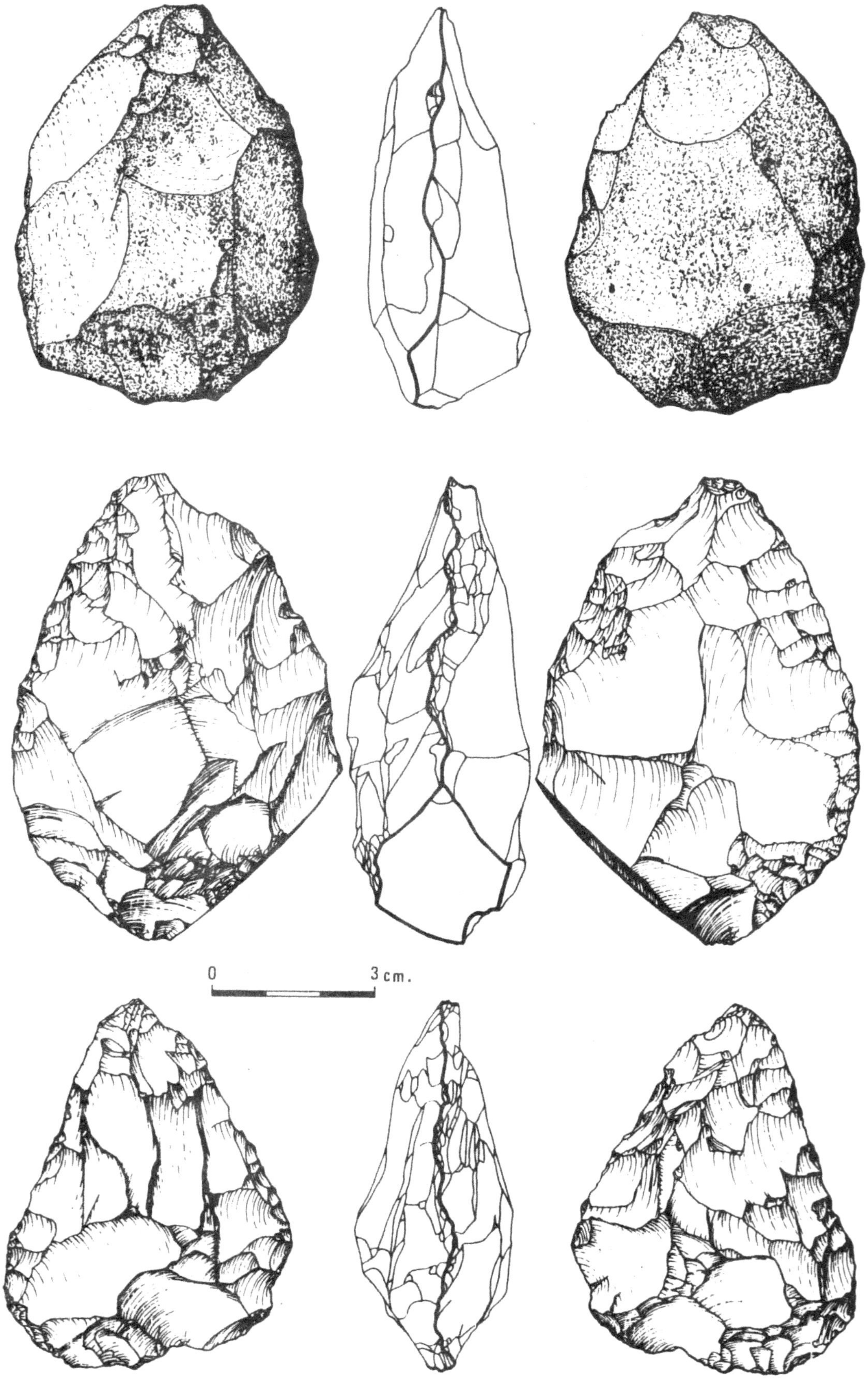

Figure 2 Basalt (top) and flint bifaces from the excavation.

istics of unit 2 are presently undergoing further analysis.

The lithic assemblage comprises, typologically, small bifaces (Figure 2), flake tools (scrapers, burins, awls, denticulates, etc.), and cores. Technologically, it is characterized by the Levallois technique, and the assemblage can be assigned to the Acheulian Industrial Complex.

The evidence clearly demonstrates that volcanic activity in the site vicinity did not prevent early prehistoric communities from occupying this region (Goren 1979, 1981b). The Brekhat Ram Acheulian site has no stratigraphic parallels in the Levant and thus is considered as a key site for resolving the issue of absolute chronology of the many Acheulian facies known from this region.

References

Goldberg, P. In press Application of Micromorphology in Archaeology *Proceedings Sixth International Working Meeting on Soil Micromorphology* London, 1981.

Goren, N. 1979 An Upper Acheulian Industry from the Golan Heights *Quartär* 29/30: 105-131.

Goren, N. 1981a An Acheulian Site in the Northern Golan Heights. *In: Geology of The Golan Heights Proceedings* Annual Meeting, Israel Geological Society, pp. 22-23.

Goren, N. 1981b The Lower Palaeolithic in Israel and Adjacent Countries. *In:* Sanlaville, P. and J. Cauvin (eds.) *Prehistoire du Levant* C.N.R.S., Paris.

Mor, D. 1973 *The Volcanism of the Central Golan Heights* T.H.L. Ltd., Tel Aviv (in Hebrew).

Mor, D. 1981 Volcanism of the Golan Plateau. *In: Geology of the Golan Heights Proceedings* Annual Meeting, Israel Geological Society, pp. 30-32.

Discussion

A. Ronen: Were any of the basalts dated?

N. Goren: Samples were sent for dating.

P.M. Masters: Are you certain that the handaxes are coming from the same horizon as the Levallois flakes?

N. Goren: I am certain.

3

EUROPE

ENVIRONMENTAL CONDITIONS IN MIDDLE EUROPE DURING THE LOWER/MIDDLE PALAEOLITHIC TRANSITION

Karl Brunnacker

In the northern Alpine region the classification of the younger Quaternary is based on the glacial series as defined by Penck. In the periglacial area it is classified by terraces, loess covers and palaeosols. At the moment, it is not possible to clearly correlate both systems, as the problem of Upper Riss = Lower Würm has been taken up again.

The glacial before the last, Riss ("Saale" or "Warthe"), is documented by loess in the periglacial area. It is, in contrast to the recent arid regions, in relative humid areas, i.e. South Bavaria, replaced by dust loam. This dust loam indicates, in most cases, a more humid area. The Last Interglacial (R/W–I) is documented by more or less pseudo-gleyed parabrownsoil, which can also be observed on the recent land surface. Typical loess has a greater expansion during the last glacial (Würm). Only in the recent and glacial more humid Alpine area is dust loam again present.

The detail structure of both loess layers is in parts very complicated (Figure 1). Facies areas as well as the development of the respective palaeorelief modify the special sequence. In general, one can say that in the beginning of both glacials at first washout and solifluction have dominated. Sooner or later the loess-facies follows, depending on the different areas and localities. In both cover layers weaker palaeosols also appear in different number and development. They are soils of a tundra or cold steppe.

This principle is also relevant for the profile of Rheindahlen (Lower Rhine Basin), where loess of the Würm-glacial, in the special form as spot loam, overlies the dust loam of the older glacials.

In cave sediments near the entrance, the profile is also modified by the effects of resedimentation during the beginning of the glacials and by congelifraction in the upper part of the profile. Debris is accumulated chiefly at the beginning and at the end of the glacial (Figures 2, 3).

High-terrace can be determined, which belongs to the Riss-glacial, in areas with fluviatile activity. Until now older lower-terrace of the high glacial and younger lower-terrace of the young Tundra period could be classified in the valleys belonging to the Würm-glacial. Recently the number of indications for a further oldest lower-terrace, which belongs to the early glacial, increase.

For Middle Europe we therefore have the following palaeoclimatic results:

1. There is a polarity between warm humid and cold arid conditions during the transition of warm and cold periods.
2. Climax of the Riss-glacial: cold (congelifraction in the cave of Hunas), relative humid (dust loam), the fauna in the (Riss-glacial) rubble of Hunas (Figure 2) is

characterized through late glacial biocoenosis rather than a high glacial (Prof. Groiss, Erlangen, pers. comm.).

3. R/W–I: as today.
4. The transition between this interglacial and the Würm-glacial took place in a very short time.
5. Würm-glacial: in the earliest Würm, humus zones indicate warmer phases. Three cold climaxes during the early high and late glacial can be determined with the help of terraces. In the upper Würm weak palaeosols indicate somewhat humid rather than warmer conditions (mollusc fauna).
6. Regional differentiation during the Würm-glacial: a) SE–NW: here the facies in the loess layers change from a continental loess type to an oceanic wash-out type. b) N–S: in the Mediterranean area the relation between pedogenesis and extra-periglacial loess sedimentation is modified. Here the genesis of intensively developed soils reaches up to the early Würm-glacial. Also the genesis towards the end of the glacial starts earlier than in Middle Europe.

Until now the chronostratigraphy is based on C^{14}–data and its extrapolations. Absolute datings of cave-sinter have, in the meantime, shown that for the last glacial older dates must be assumed. But still the correlation between the cave profiles and other profiles has to be undertaken. A comparison of the environmental situation between the two discussed glacials show:

1. The Riss-glacial was in general slightly more humid than the Würm-glacial; this results from the loess-facies.
2. The regional differentiation from the SE to the NW of Middle Europe during the Riss-glacial was in the same way present as in the Würm-glacial. The number of tundra wet soils during the Riss increases towards the NW but the degree of development decreases. Ice wedges and especially cryoturbations increase towards the NW.
3. One can assume that the Riss-glacial was not so cold as the Würm-glacial during their cold maxima, but this still has to be further investigated.
4. The question as to when the transition between the lower and middle palaeolithicum took place remains open.

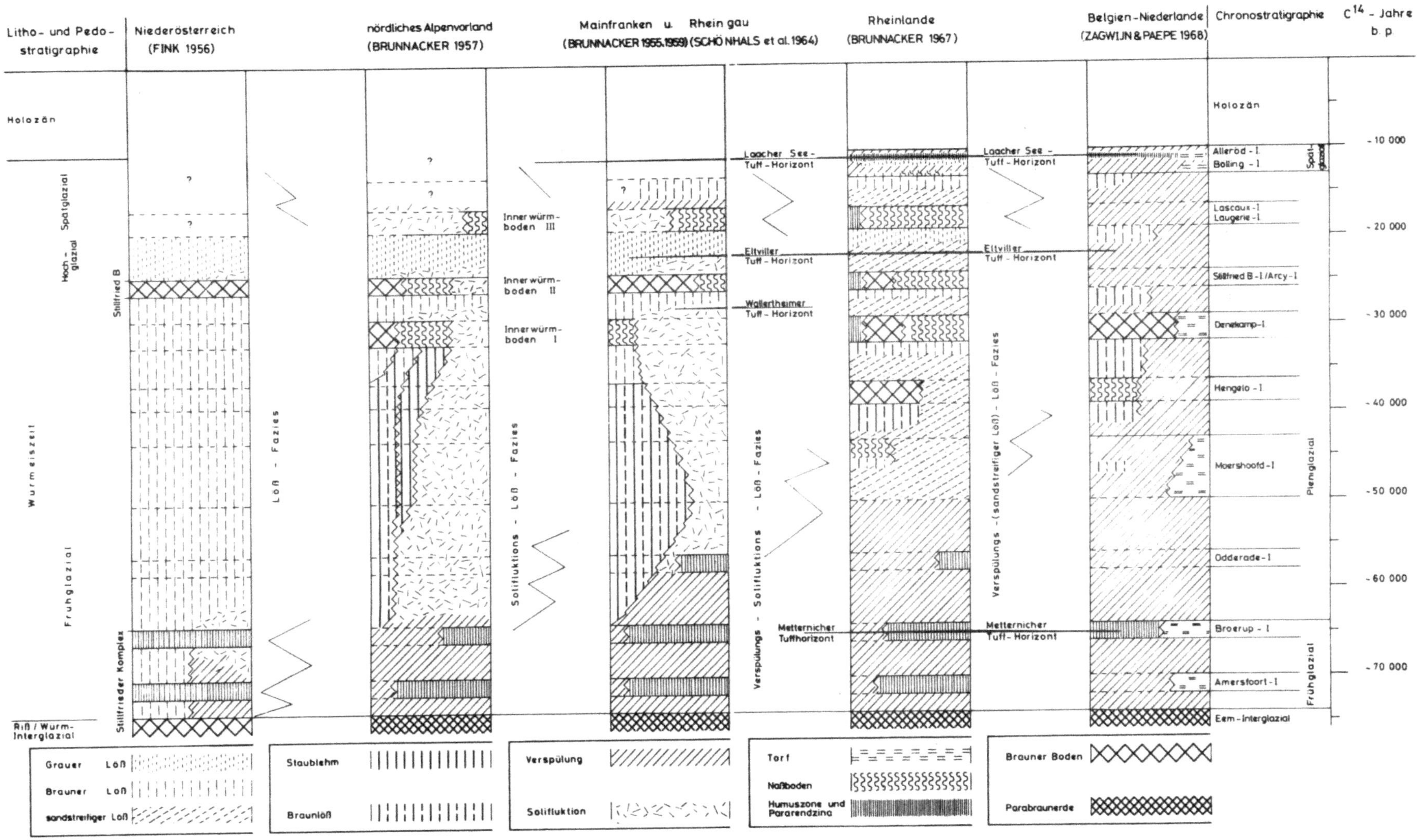

Figure 1 Regional and temporal changes in facies in Middle Europe in the central area of the loess provinces.

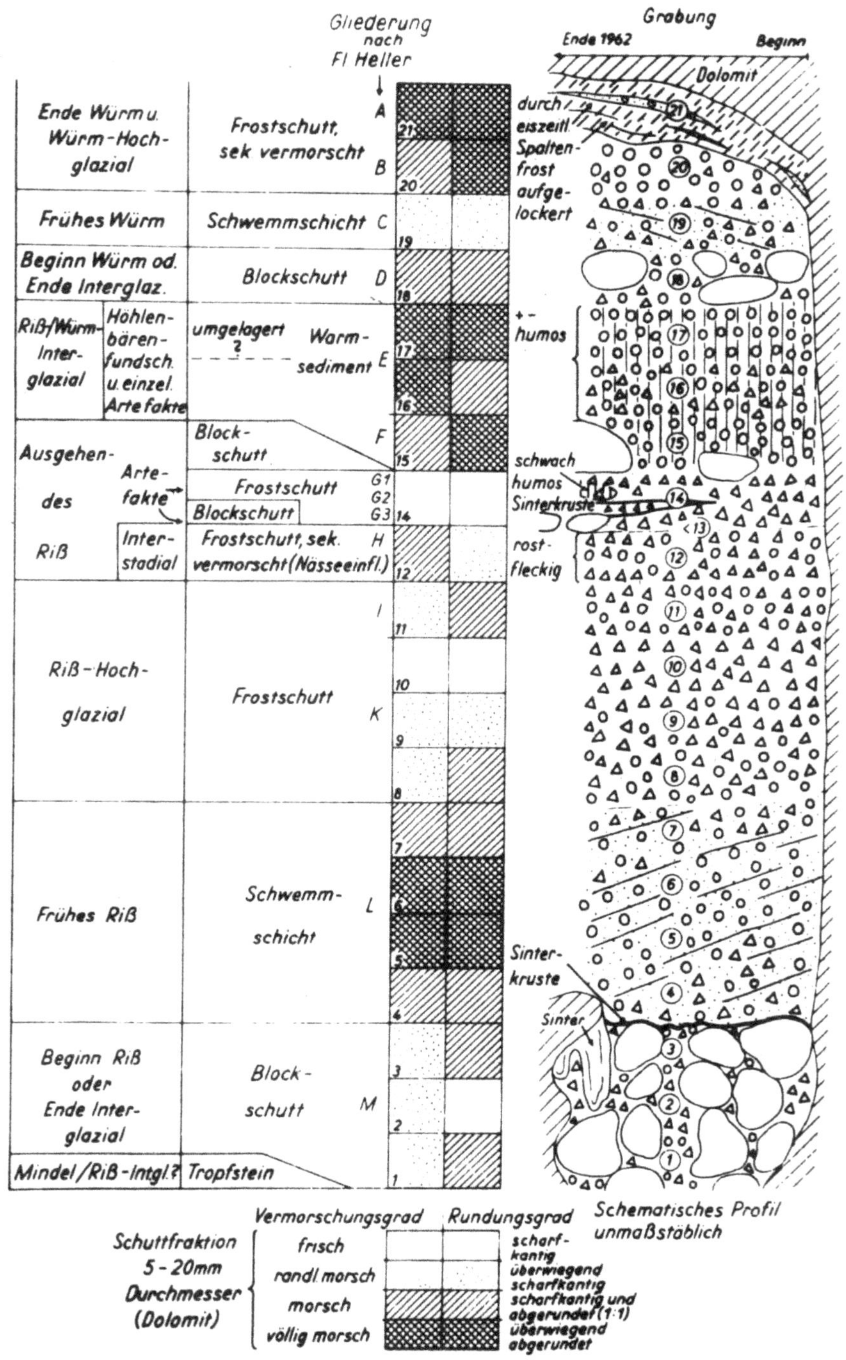

Figure 2 The build-up and classification of the sediments in the cave of Hunas, E of Nürnberg. Layers M–G are Riss-glacial.

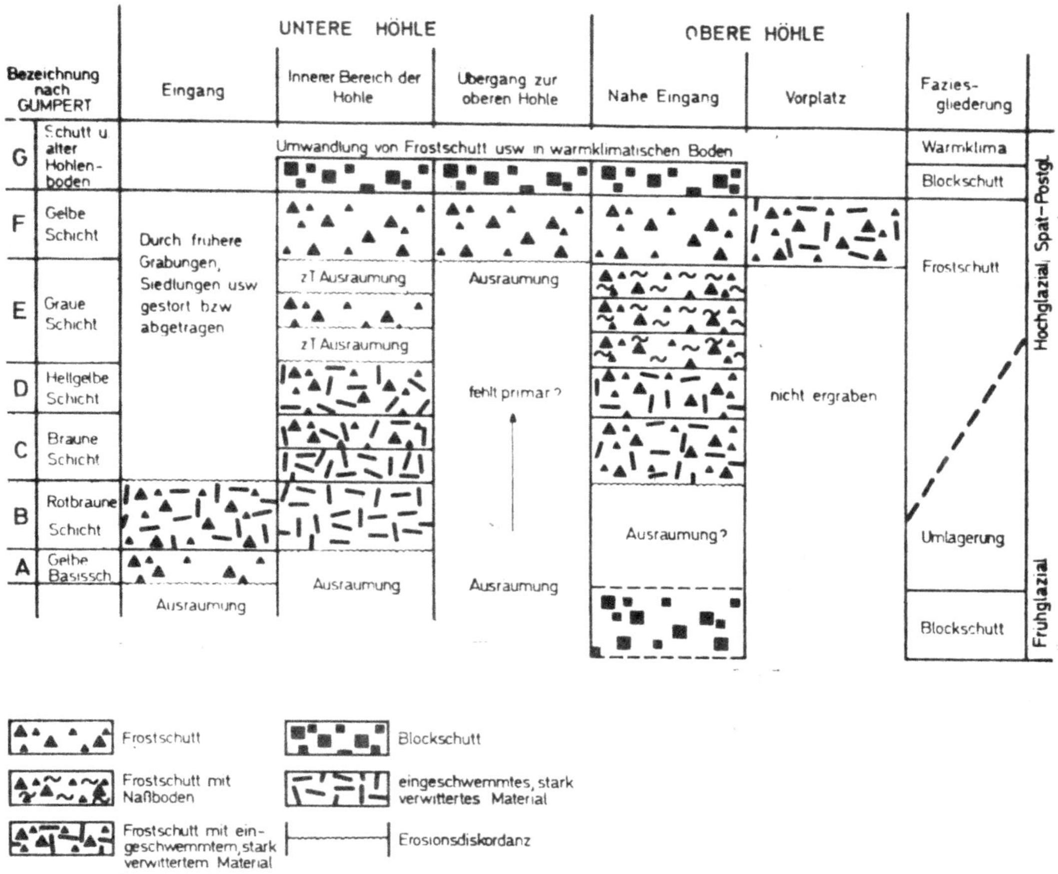

Figure 3 Facies build-up of the Würm-glacial sediments in Hohle Stein, S of Nürnberg.

References

Brunnacker, K. 1963 Die Sedimente in der Höhlenruine von Hunas (Nördliche Frankenalb) *Eiszeitalter u. Gegenwart* 14: 117-120, Öhringen.

Brunnacker, K. 1979 Die Sedimente im Hohlen Stein bei Schambach (Südliche Frankenalb) *Geol. Bl. NO Bayern* 29: 89-112, Erlangen.

Brunnacker, K and J. Hahn 1978 Der jungpleistozäne Loess samt paläolithischen Kulturen in den Rheinlanden als Glied einer zeitlichen und räumlichen Faziesänderung *Beiträge zur Quartär-u. Landschafts-forschung.* pp. 37-51, Wien (F. Hirt).

Hennig, G.J. 1979 *Beiträge zur Th-230/U-234-Altersbestimmung von Höhlensintern sowie ein Vergleich der erzielten Ergebnisse mit denen anderer Absolut-datierungs-methoden* Diss. Univ. Köln.

Discussion

H. Laville: I am very astonished by the length of time you give to the last Interglacial.

K. Brunnacker: The Interglacial Palaeosol of the type of the para-brown soil is in every case thicker than the present soil. On the other hand, I believe that Interglacial conditions were a little wetter and somewhat warmer than today. On the basis of diatomites, pollen analysis and the different stages of the succession of the flora, we can say that this soil was formed no longer than 12,000 or 15,000 years, perhaps 20,000 years at most.

H. Laville: What you say seems to me to contradict the known evidence of the manifestations of the Last Interglacial in various regions. For example, in the south-west of France stratigraphical, sedimentological and palynological studies all point towards an extremely complex period, with multiple major climatic events. We do not have the evidence for the onset of the Last Interglacial in our region, because it was destroyed by subsequent erosion. However, a series of subsequent events can be described. The climatic optimum led to a formation of very thick and well-developed soils, both in open air sites and in rock shelters, and of thick stalagmitic layers inside caves. Then occurred a period of considerable moisture, which promoted the erosion processes that I mentioned, led to the renewal of Karstic activity in the caves and the partial or complete removal of some sediments. Finally, the palynology indicates that at the end of the Interglacial, after the period of erosion, five temperate climatic phases have succeeded before the major Wurmian cold. We attribute to the Interglacial the global down-cutting of the hydrographic network and its lowering by 8 to 10 meters. All these events could hardly have occurred in a period of only 15,000 years. The real problem, I think, is of the boundaries which we accord to the Last Interglacial.

K. Brunnacker: We have in the Interglacial periods weathering by frost. The problem is the velocity of the sedimentation, that is, the difference between sedimentation and weathering, which constitutes an indicator for Interglacial periods. Sedimentation is quicker than weathering in rock shelters.

H. Laville: Personally, I do not know of any cryoclastic deposits from the Interglacial period in rock shelters. Another argument which may show the magnitude of the Riss-Wurm Interglacial in south-west France is the complete disappearance of cold fauna at this time. In particular the reindeer, which comprised more than 95 per cent of the fauna at the end of Riss, disappeared completely during the entire Interglacial and reappeared only long after the onset of Wurm.

K. Brunnacker: At the cave of Hunas the results of the faunal determination and sedimentological analysis concurred. Prof. Heller, who studied the fauna arrived independently at the same conclusions as I did.

W.R. Farrand: I am not quite sure I understand exactly the point Laville wanted to make. I thought you were asking about the duration of the Interglacial on one point and also its characters. But as to the question of duration and that the soils may appear to represent a longer period of time than Brunnacker has suggested, I have another question which I think links with that. You have indicated, Prof. Brunnacker, that the Eemian is represented by one marine ingression. I understand from other people who have interpretations that perhaps there are three different things being called Eemian. That is one problem. I also think, with respect to the isotope curve and sea levels determined around the world, that there are three sea levels which should be included in the Eemian. Now each of those warm and high sea stands was only some 10,000 years long, but the Last Interglacial soil that we see may really encompass all three peaks. This is particularly true if you are away from the sea.

K. Brunnacker: The problem is that we have only one ingression in the Eemian. But on top of it we have two interstadials, the Amersfoot and the Brorup, and perhaps another one. We have a high sea level in the Mediterranean area, but it is not possible for us to incorporate it in the Eemian, it is a question of definition. The Loess profiles too have a dividing line there and we cannot change our definition.

P. Goldberg: I did a little work on loess sections in the Paris Basin. One sees very clearly within the profile the Last Interglacial soil which is very similar, in terms of microscopic properties, to what you find today. However, superimposed upon this soil, one witnesses a degradational phase (textural degradation) which represents a rapid climatic deterioration, and it is very hard to tell whether it is the Eemian as you have it here, or the Amersfoot or Brorup; they are all related to stage 5. As for the Middle East, your picture does not seem to be the one I have come up with, based on the work I have done so far. I think that the climate tends to change much more, and there is localized soil development during the Last Interglaciation. I think it is a little more detailed than you have it indicated.

K. Brunnacker: The Mediterranean soil, for instance in Kytros, Greece, where we have a well developed profile, is the same as in Germany – two layers of Loess with a weathering in the upper layer alone. By micromorphology, it is more a brown loam type of soil than a para-brown loam. Judging by the topography, it must be a Wurm soil. In any case it is very difficult to explain this early Wurm soil.

ON THE TRANSITION FROM "LOWER" TO "MIDDLE" PALAEOLITHIC IN SOUTH-WEST FRANCE

Henri Laville

According to the accepted frame of reference, the "Lower" Palaeolithic includes all the industries preceding the Last Interglacial, whereas the "Middle" Palaeolithic corresponds to the complex of industries in the first half of the Last Glaciation.

This division, based exclusively on chronostratigraphic criteria leads to the grouping of true Acheulian industries in the Lower Palaeolithic with others already characteristic of the Mousterian. In the same way, it includes in the "Middle" Palaeolithic industries which are truly Mousterian with others still retaining Acheulian characters such as the Micoquian or even the Mousterian of Acheulian Tradition.

These considerations have led several authors to propose new criteria of definition and classification: criteria that are not only chronological, but also involving the technological and typological characteristics of lithic industries.

F. Bordes (1977) considering the composition and characteristics of the flake tool industries, has defined a "Mousterian stage of evolution" which would be reached at different periods by both true Acheulian industries or by pre-Würm industries without handaxes.

A. Tuffreau (1979) considers "Lower" Palaeolithic as a stage where flake tool industries are generally rough and not much evolved from a typological point of view, whereas Middle Palaeolithic is characterized by the systematic production of flakes and an abundance of flake tools that are typologically quite evolved and well differentiated, of Levallois and non-Levallois debitage. Recent results in chronostratigraphy and in the knowledge of lithic industries (new digs, new analyses of old collections) allow a reconsideration of the problems of the "transition from the Lower Palaeolithic to the Middle Palaeolithic" in Southwest France in view of these new classification criteria.

A detailed palaeoclimatic and chronological frame, based on corcondant results of sedimentological, palynological and palaeontological research in caves and rockshelters in the Périgord area serves as a reference to define the relative position of archaeological levels with regard to the two climatic phases of Riss I, the four phases of Riss II and the seven oscillations of Riss III.

The archaeological levels of the following caves and rockshelters, Pech-de-l'Azé II, La Micoque and Combe-Grenal (Dordogne) and La Chaise (Charente) are thus replaced with precision. We can also replace levels of Nantet and Bouheben (Landes) and of Les Tares (in the Isle basin), those of the alluvial terraces of the Libournais and finally, more difficult to situate, those of the sites of Cantalouette, Barbas and Les Pendus (Bergeracois).

A consideration of the principal characteristics of these industries and details concerning the chronological position of the corresponding archaeological levels, confirm the contemporaneity of industries with handaxes and of industries without handaxes during different times of the Rissian glacial. The former are called, by definition, "Acheulian" and the later "Pre-Mousterian", "Proto-Mousterian" or "Primitif Mousterian". Industries without handaxes have been found with all certainty at the top of Riss II deposits of La Micoque (layer 4) and in the silty-clay deposits from Riss III in Les Tares in the Dordogne region. The attribution of levels 2 and 3 of La Micoque to the group of industries "without handaxes" is not unanimously accepted. Both of them are contemporaneous with Riss I and probably, more precisely, with the second climatic phase of this stadial. No handaxe has been found so far in level 2, but the whole industry is completely altered by cryoturbation. Level 3 has, in fact, some handaxes but, according to Bordes, they are not typical, they are not of "Acheulian style" and are associated with a context resembling a "primitif Mousterian" probably related to the Quina type. But " . . . s'il est vrai que l'Acheuléen se définit par des bifaces (. . .) il faut inclure la couche 3 de La Micoque dans l'Acheuléen . . . " (Guichard, 1976a)[1].

In any event, excepting level 2 of La Micoque which is completely spoiled, the "Mousterian level of evolution" or the "Middle Palaeolithic" stage of all these industries (levels 3 and 4 of La Micoque, level 1 of Les Tares) is especially certified by the high cumulated percentage values of side scrapers, notches and denticulates, which, according to J.-M. Le Tensorer (1978), could be significant of a certain affinity with Quina type Mousterian. Keeping these values in mind, the Mousterian or Middle Palaeolithic stage of the group of industries without handaxes would have already been attained as early as Riss I if level 3 of La Micoque really belongs to this group, or at least at the end of Riss II, as demonstrated by the chronological position of level 4 of La Micoque. Industries with handaxes, which are by definition true Acheulian, begin during Riss I and stake out the three rissian stadials.

Levels 9 and 8 of Pech-de-l'Azé, whose industry is classified by Bordes as Acheulian of southern facies, belong to Riss I, and consequently are approximately contemporaneous with levels 2 and 3 of La Micoque.

To Riss II are attributed, apart from level 4 of La Micoque, levels 7 (Acheulian of southern facies) and 6 of Pech-de-l'Azé II, the Acheulian of Chalossian facies of Nantet in the Chalosse region and, probably, the industries of Barbas and Les Pendus in Bergeracois.

Riss III includes levels 64 to 56 (Acheulian of southern facies) of Combe-Grenal and level 5 of La Micoque, 54 to 50 and VI to III of the Abri Suard in La Chaise, the industry of Bouheben, those of the middle terrace of the Isle in Libournais and, probably, that of Cantalouette in Bergeracois. The characteristics of those different industries cannot be presented here. In spite of their great technological and typological variations, they show a certain number of common features, as the unusual proportion of rough handaxes, the presence of certain particular types such as backed handaxes, beaked handaxes, etc. . . , the existence of cleavers in many levels and in many sites: Pech-de-l'Azé, Cantalouette, Les Pendus, Vallée de l'Isle.

Nevertheless, the characteristics and especially the composition of the flake tool group also give a basis to situate the precise time when the Mousterian or Middle Palaeolithic level of Evolution was reached. In view of criteria such as the relative importance of the sidescraper, notch and denticulate groups, this stage seems to have been reached

[1] "If it is the rule that Acheulian is defined by the presence of handaxes, level 3 of La Micoque should be included in the Acheulian".

at the very beginning of Riss III. The composition of flake tools from different Riss III levels of Combe-Grenal and from Bouheben suggests just this, these flake industries being highly comparable to that characteristic of the Mousterian of Acheulian Tradition. It is important to remember, however, the existence of the industry of level 3 at La Micoque which, if identified as Acheulian, would mean that the Mousterian stage was reached in the first rissian stadial by the group of corresponding industries. The new division of the "Lower" and the "Middle" Palaeolithic which, as we have seen, can be reached with the use of technical and typological criteria, tends to mask the enormous void represented by the Riss-Würm interglacial in the evolution of the Palaeolithic industries, the Interglacial which, as we know, constitutes a chronological limit between the "Lower" Palaeolithic and the "Middle" Palaeolithic in the accepted frame.

The data drawn from sedimentological and paleopedological studies attribute to the interglacial stalagmitic floors, gully erosion and highly evolved but abruptly truncated weathered soils to this Interglacial, the later corresponding not only to an extensive halt in the deposition of sediment but also to a highly intense period of erosion. For these reasons, no industry dating back to the Riss-Würm Interglacial has ever been found in the South-West France.

Taking these data into account, if the link between certain industries at the end of the Riss and at the beginning of the Würm is certain (the Rissian Acheulian/the Mousterian of Acheulian Tradition and Micoquian relationship; the industries without handaxes of Les Tares/Mousterian of the Quina type relationship), the progressive "evolution" which some authors have perceived in view of the technical and typological characters of industries before and after this Interglacial, remains very difficult to explain.

References

Bordes, F. 1961 Mousterian cultures in France *Science* 134: 803-810.

Bordes, F. 1966 Acheulian cultures in Southwest France *In:* Sen, E. and A.K. Ghosh (eds.) *Studies in Prehistory. Robert Bruce Foote Memorial Volume*, Calcutta. pp. 50-54.

Bordes, F. 1968 *The Old Stone Age* World University Library, McGraw Hill, New York, Toronto.

Bordes, F. 1969 Livret-guide de l'excursion A5 (Landes-Périgord) (en collaboration avec Cl. Thibault, H. Laville et M.–M. Paquereau) *Union internat. pour l'Etude du Quaternaire*, VIIIème Congrés de l'INQUA.

Bordes, F. 1971 Observations sur l'Acheuléen des grottes de la Dordogne *Munibe* XXII(1): 5-23.

Bordes, F. 1972 *A tale of two Caves* Harper & Row, New York.

Bordes, F. 1973 On the chronology and the contemporaneity of different palaeolithic cultures in France *Research Seminar Univ. of Sheffield.* pp. 217-226.

Bordes, F. 1977 Time and space limits of the Mousterian. *In: Stone tools as cultural markers* Australian Institute of Aboriginal Studies, Canberra. pp. 37-39.

Bordes, F. 1979 *Le Paléolithique en Europe* Cours de Préhistoire, Institut du Quaternaire, Université Bordeaux I (non publié).

Bordes, F. 1980 Hommes et cultures du Paléolithique moyen *Science et vie,* no. hors série: *"Les ancêtres de l'Homme".* pp. 90-102.

Bordes, F., A. Debenath, B. Kervazo, H. Laville, J.M. Le Tensorer, J.P. Texier et Cl. Thibault 1981 Les dépôts quaternaires en Aquitaine. *In:* Problèmes de stratigraphique quaternaire en France et dans les pays limitrophes. *Suppl. Bull. Assoc. Franç. Et. Quat.* N.S.(1): 250-267.

Bourgon, M. 1957 *Les industries moustériennes et pré-moustériennes du Périgord* Archives de l'Institut de Paléontologie humaine 27.

Debenath, A. 1974 *Recherches sur les terrains quaternaires des Charentes et les industries qui leur sont associées* Thèse de Doctorat d'Etat ès-Sciences naturelles.

Debenath, A. 1976 Les civilisations du Paléolithique inférieur en Charente. *In: La Préhistoire française* I(2): 929-935.

Delpech, F. et F. Prat 1980 Les grands mammifères du Sud-Ouest de la France. *In: Problèmes stratigraphiques du Quaternaire en France et dans les pays limitrophes* Dijon, Suppl. Bull. Assoc. Franç. Et. Quat. N.S.(1): 268-281.

Guichard, J. 1965 Un faciès régional de l'Acheuléen: Cantalouette (commune de Creysse, Dordogne) *L'Anthropologie* 69(5-6): 413-464.

Guichard, J. 1976 Les civilisations du Paléolithique inférieur en Périgord. *In: La Préhistoire française* 1(2): 1053-1069.

Guichard, J. et G. Guichard 1966 A propos d'un site acheuléen du Bergeracois (Les Pendus, commune de Creysse). Bifaces, hachereaux et hachereaux sur éclat. Aperçu typologique *Actes de la Soc. linéenne de Bordeaux* 10,B,(5).

Guichard, G. 1976 Les civilisations du Paléolithique moyen en Périgord. *In: La Préhistoire française* I(2): 908-928.

Laville, H. 1972 The relative position of Mousterian industries in the climatic chronology of the early Würm in the Périgord *World Archaeology* 4(3): 321-329.

Laville, H. 1975 *Climatologie et chronologie du Paléolithique en Périgord. Etude sédimentologique des dépôts en grottes et sous abris* Etudes quaternaires 4, Université de Provence.

Laville, H. 1976 Le remplissage des grottes et abris sous roche du Sud-Ouest. *In: La Préhistoire française* I(1): 250-270.

Laville, H., J.Ph. Rigaud and J. Sackett 1980 Rock Shelters of the Périgord. *Geological Stratigraphy and Archaeological Succession* Academic Press, New York.

Le Tensorer, J.-M. 1978 Le Moustérien type Quina et son évolution dans le sud de la France *Bull. Soc. préhist. franç.* C.R.S.M. 75(5):143-149.

Moisan, L. 1978 *Recherches sur les terrasses alluviales du Libournais et leurs industries préhistoriques* Doctorat d'Université, Université Bordeaux I.

Paquereau, M.-M. 1970 Flores et climats paléolithiques dans le Sud-Ouest de la France *Revue de Geographie physique, géologique et dynamique* XII(2): 109-116.

Paquereau, M.-M. 1976 La végétation au Pléistocène moyen dans le Sud-Ouest. *In: La Préhistoire française* I(1): 491-493.

Paquereau, M.-M. 1976 La végétation au Pléistocène supérieur et au début de l'Holocène dans le Sud-Ouest. *In: La Préhistoire française* I(1): 525-530.

Peyrony, D. 1938 La Micoque, les fouilles récentes, leur signification *Bull. Soc. préhist. franç.* 6: 257-283.

Rigaud, J.-Ph. and J.-P. Texier 1981 A propos des particularités techniques et typologiques du gisement des Tares, commune de Sourzac (Dordogne) *Bull. Soc. préhist. franç.* 78(4): 109-117.

Texier, J.-P. 1979 *Recherches sur les formations superficielles du bassin de L'Isle* Thèse de Doctorat d'Etat ès-Sciences naturelles, Université Bordeaux I.

Thibault, Cl. 1968 Un gisement paléolithique inférieur et moyen de plein air en Châlosse: Nantet (commune d'Eyres-Moncube, Landes). *In: La Préhistoire. Problèmes et tendances* ed. C.N.R.S. pp. 427-438.

Thibault, Cl. 1970 *Recherches sur les terrains quaternaires du bassin de L'Adour* Thèse de Doctorat d'Etat ès-Sciences naturelles, Université Bordeaux I.

Thibault, Cl. 1976 Les civilisations du Paléolithique inférieur dans le Sud-Ouest (Pays Basque, Landes, Gironde). *In: La Préhistoire française* I(2): 905-908.

Tuffreau, A. 1979a Les débuts du Paléolithique moyen dans la France septentrionale *Bull. Soc. préhist. franç.* 76(5): 140-142.

Tuffreau, A. 1979b Recherches récentes sur le Paléolithique moyen dans la France septentrionale *Bull. Soc. royale belge d'Anthrop. et de Préhist.* 90: 161-177.

THE TRANSITION LOWER/MIDDLE PALAEOLITHIC IN NORTHERN FRANCE

Alain Tuffreau

Problems of Definition

The problem of transition from Lower Palaeolithic to Middle Palaeolithic depends first of all on the definition of Middle Palaeolithic and on its meaning. If by Middle Palaeolithic it is meant (as usually in France and as I myself did before) the complex of industries (Mousterian, Micoquian) from the first part of the Last Glacial (Weichselian, Wurm), then the problem does not even exist. In this case, the identification of Middle Palaeolithic is implicitly based on essentially chronostratigraphical tests. But recent research conducted in Northern France (Tuffreau 1979a) with the excavations of Biache-Saint-Vaast site (Pas-de-Calais) of saalian age constituting the most spectacular example, shows that such a concept would place in Lower Palaeolithic industries which do not differ technically and typologically from more recent industries of Middle Palaeolithic, understood in a chronological meaning. That is why it is better to define Middle Palaeolithic in terms of technological and typological characteristics of the lithic assemblages. Other data (evidence concerning living floors, bone industries) are too poor to be used.

Consequently, Middle Palaeolithic is defined (Tuffreau 1979b) as a stage anterior to Upper Palaeolithic. It is characterized by a vast production of flakes and flake-tools, typologically evolved and well standardized, the debitage of which is often, but not always, done by the Levallois technique. It is possible to consider Lower Palaeolithic as being the oldest stage of Stone Age where the unifacial and bifacial tools (handaxes of Acheulian type, choppers, chopping-tools, chopper-cores) are numerous but sometimes missing. The flake-tools are most of the time rough and typologically little evolved.

Recent excavations and research in Northern France (North of France and Picardy) show that industries responding to the typological and technical definition of Middle Palaeolithic exist from the beginning of the Saalian glaciation. Before examining these industries, the oldest of Middle Palaeolithic, it is advisable to study the latest industries of Lower Palaeolithic.

The End of Lower Palaeolithic

In the area under consideration, the industries of the end of Lower Palaeolithic are preserved in the fluvial formations of the Middle terrace of the Somme basin (Atelier de Cagny-la-Garenne, sites of the rue de Boves and of the rue de Cagny in Saint-Acheul) or at the base of the loessic cover (Atelier Commont, Upper level of the rue de Cagny in Saint-Acheul; base of the old loesses in Cagny l'Epinette pit).

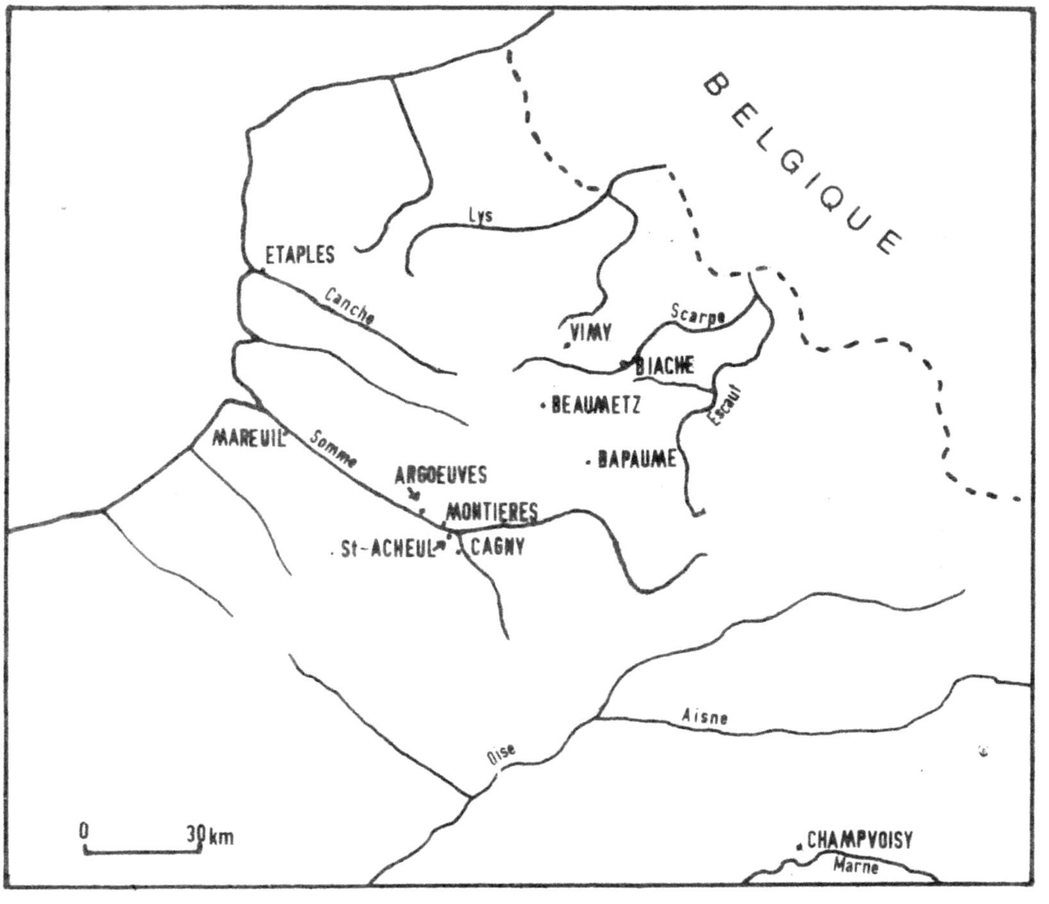

Figure 1 Sites mentioned in the text.

Chronostratigraphical problems

The age of the fluvial formations of the Middle terrace of the Somme basin depends on the interpretation of its loessic cover. This is difficult to establish on the one hand because the chronostratigraphy of anteweichselian loesses is not as well-known as the one of Last Glacial; on the other hand, because the loessic cover of the Middle terrace of the Somme basin seems clearly poorer than, and not so well preserved as that of the Middle Seine terrace, in Saint-Pierre-les-Elbeuf (Lautridou et Verron 1970). The most complete section, visible in Cagny-la-Garenne (Bordes 1954; Bourdier *et al.* 1974a) shows recent loesses overlying the Eemian interglacial soil, surmounting a sequence of old loesses attributed to the three phases of the Rissian glaciation. According to Bourdier, an interglacial soil (chocolate colored clay) of Mindel-Riss age lays on the Middle terrace attributed to a late phase of Mindel (Mindel II). But Bordes maintained that the chocolate colored clay is only an "argile à silex" soliflued, and the fluvial deposits of the

Middle terrace date of the beginning of the Rissian glaciation. Thus, the fragility of the alpine and nordic chronostratigraphical systems now appears more and more as the pre-Eemian phases are concerned.

Climatic data

Except for the industry of Atelier Commont, preserved in its primary context, flint artifacts are replaced in some fluvial deposits or in gravels. Therefore, the pollinic or faunistic (molluscs, microfauna) studies may be misleading. The macrofauna of Atelier Commont is rather temperate (big horse, bovidae and *Elephas antiquus*).

The lithic industries

Formations of the Middle terrace in the Somme Valley

Lithic assemblages, more or less important, have been discovered in the gravels of the Middle terrace in the Somme basin. The most important is Cagny-la-Garenne, included in the gravels near the chalky slope. Known by old collections (Tuffreau 1978a), the industry contains many handaxes (more than 50% of the tools). Between these, elongated handaxes (lanceolates and amygdaloids) dominate. Cortical bases are abundant. The originality of Cagny-la-Garenne industry resides in the presence of varied flake-tools (side-scrapers, notches, denticulates and also flake-tools of Upper Palaeolithic type). On the whole, their classification into well-defined types is difficult. The Cagny-la-Garenne industry also comprises some Levallois flakes, the percentage of which has been multiplied by the conditions of collecting. Some can be the result of the transformation of handaxes into cores. According to Callow (1976), after deducting Bordes' collection the typical Levallois flakes would be about 1% (fig. 2).

Our excavations in Cagny-Cimetière, also in the gravels of the Middle terrace, have shown that the Levallois flakes there are very rare and not typical. They may have been made accidentally because no Levallois core is present.

In the rue de Cagny site (Saint-Acheul), the upper part of the gravels and the sandy levels contain an assemblage of 220 non-patinated handaxes. From the publication of Commont (1908), the handaxes have often thick butts, sometimes with cortex. There are many "ficrons" and lanceolates but the limandes are rare. The flake-tools (side-scrapers, denticulates, notches) seem less numerous than the handaxes, but this material does not come from an excavation. This assemblage also includes some Levallois flakes. Important assemblages have been discovered not far from there, in the rue de Boves site. We can see some Levallois flakes (fig. 3). The possibility of mixing of flint artifacts in this sector of Saint-Acheul, where the fluvial formations are not protected by an important loessic layer and may have been eroded by periglacial phenomena, is not to be excluded.

The lower part of the old loesses covering the Middle terrace formations of the Somme basin

The most interesting assemblage is that of the Atelier Commont in Saint-Acheul (Tellier Pit), which comes from a layer of "sables roux" and includes two archaeological levels. It is known by the publication of Bordes and Fitte (1953) and by the flint artifacts kept in the Musée de l'Homme. The only numerical indications about this assemblage have been left by Commont. At the end of 1905, he had collected 968 flakes, 92 cores, 20 hammerstones and 70 tools including 20 handaxes. After that, thousands of artifacts were gathered in the layer of "sables roux". Among the handaxes, there are amygdaloids, lanceolates, cordiforms and elongated cordiforms. There are no Levallois flakes and the faceted striking platforms are rare. The majority of flake-tools show thick

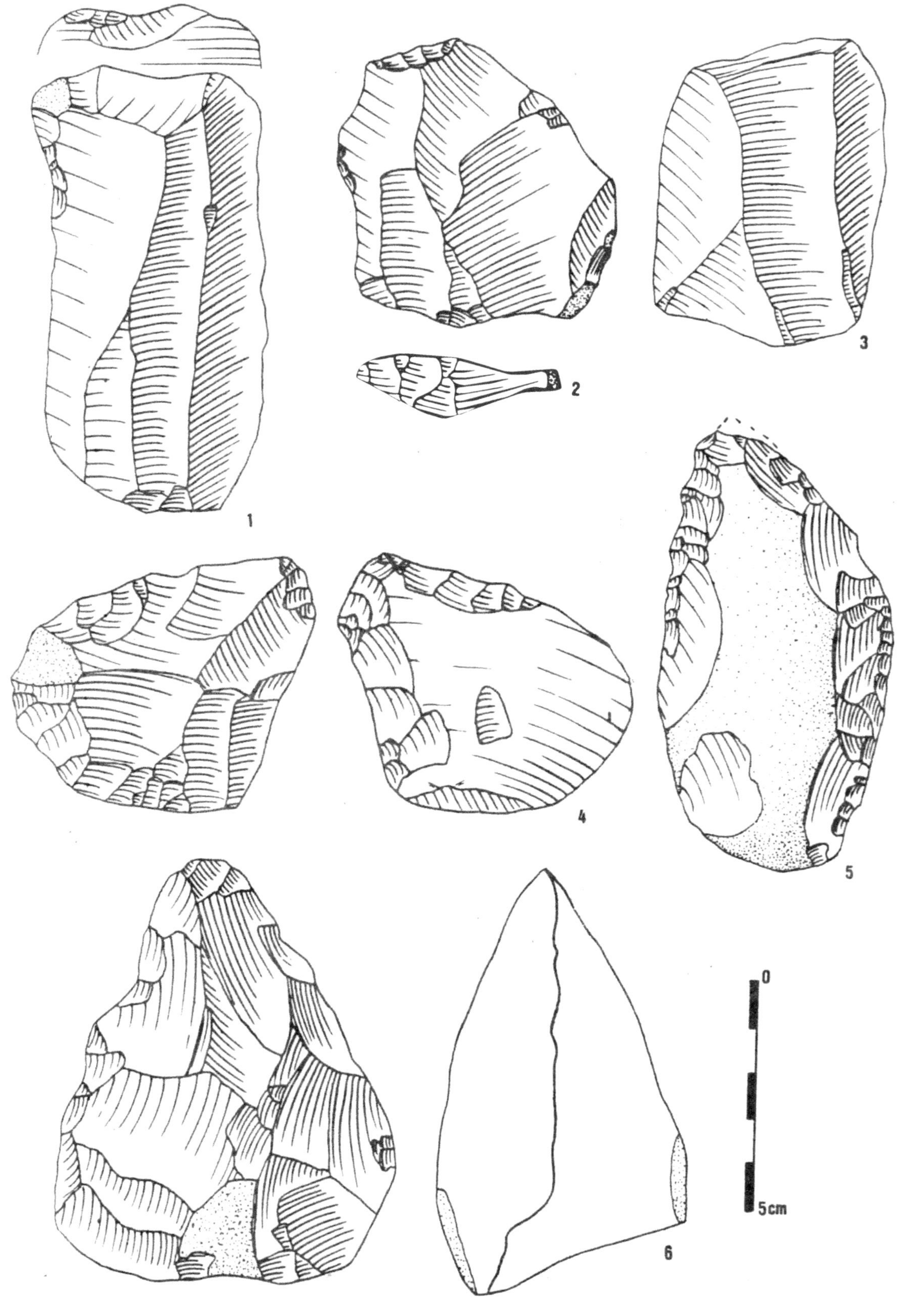

Figure 2 Cagny-la-Garenne. Industry from the gravels of the terrace. 1, notch; 2, truncated flake; 3, atypical Levailois flake; 4, canted side-scraper; 5, convergent side-scraper; 6, handaxe.

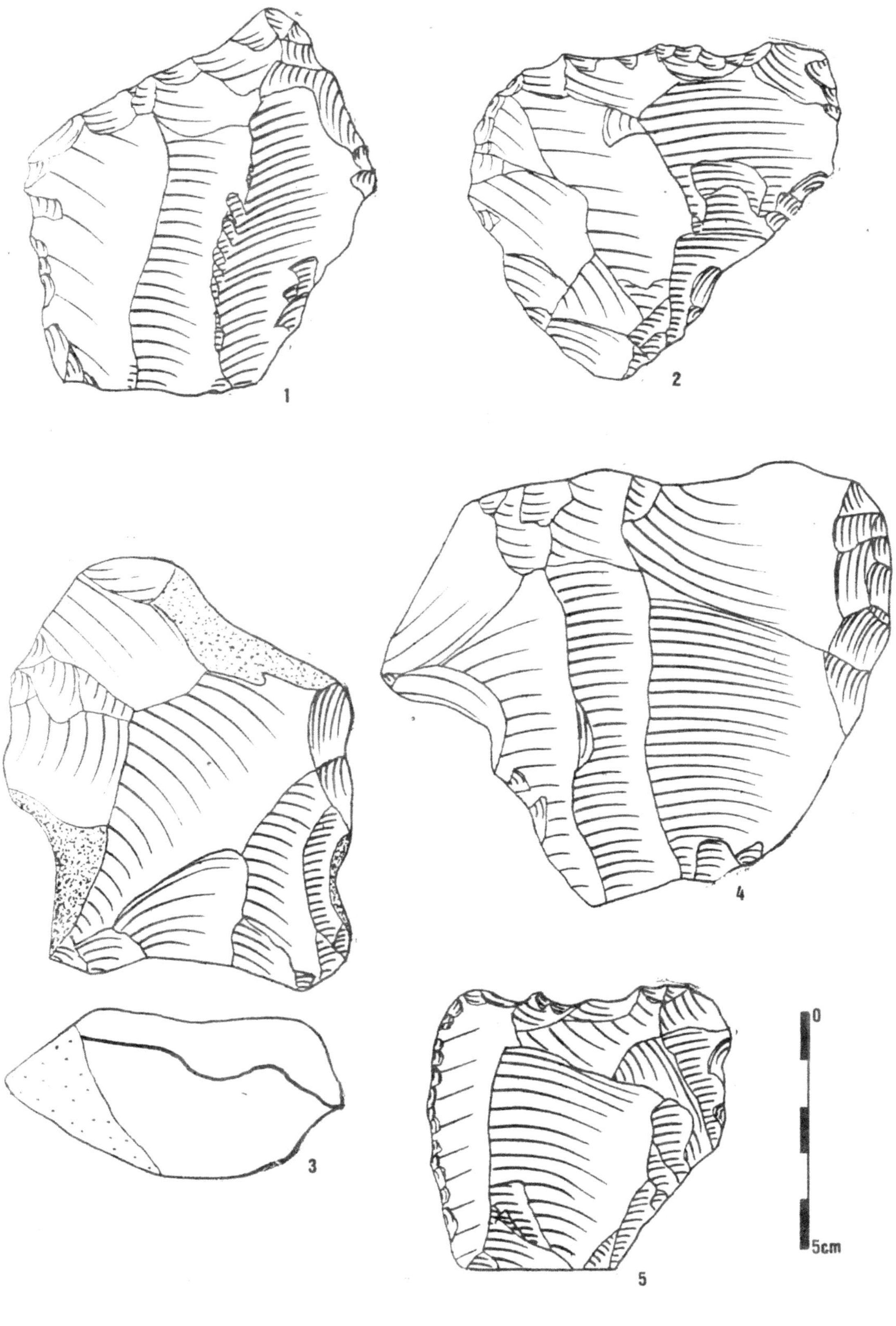

Figure 3 Saint-Acheul, rue de Boves. Industry from the gravels of the Middle terrace. 1, 4, side-scrapers; 2, retouched Levallois flake; 5, denticulated Levallois flake; 3, core.

bulbs. The flake-tools are composed mainly of side-scrapers and denticulates; there are some end-scrapers, backed knives, notches and also some borers and burins.

An assemblage resembling that of the Atelier Commont has been discovered at the base of the loamy cover overlying the high terrace of Mareuil (described as Middle terrace by Breuil).

In the rue de Cagny site, Commont (1908) has also gathered in a layer of "sables roux" numerous handaxes, especially limandes (271 for 300 handaxes). Nearly one half of them have a twisted edge; the others are elongated.

In Cagny l'Epinette, where we have begun an excavation, the gravels of the low part of the loessic cover contain small handaxes, especially short amygdaloids. The flakes are numerous but the Levallois technique seems unknown. The same holds true for the gravels in the Cagny-Cimetière Pit (Tuffreau 1980).

General characteristics of the industries of the Late Lower Palaeolithic

Although lithic assemblages which could be gathered in good stratigraphical conditions are rare, it is possible to make the following observations. All the assemblages belong to Acheulian industries with numerous handaxes, their morphology varies in the different assemblages. Flake-tools are well represented and occasionally well refined from a typological point of view. Rare Levallois flakes are present in the gravels of the Middle terrace in Cagny-la-Garenne; but Ohel's opinion (1979) that the industries at Cagny-la-Garenne and Cagny-l'Epinette belong to an Acheulian with Levallois technique does not agree with reality. It is possible that the idea of Levallois technique was born from the transformation, perhaps accidental, of handaxes into cores. The abundance of raw material has made easier the discovery of Levallois technique. On the other hand, in later assemblages (Cagny l'Epinette, Atelier Commont) Levallois technique is unknown (Tuffreau 1981).

The Beginning of Middle Palaeolithic

All the low terraces of the Somme basin and Northern France date prior to the Last Interglacial. They are covered by a sequence of old loesses surmounted by the Eemian interglacial soil: Montières, low terrace of "10 m"; Argoeuves; Etaples; Biache-Saint-Vaast (Bourdier 1973; Sommé 1978; Tuffreau 1979a). In the loess sequences the lithic assemblages come from the gravels at the base of the last old loess, or are still older like that of Champvoisy, which is found in gravels covered by loess deposits including two interglacial soils.

Palaecological data

Most of the lithic assemblages are reworked or have been collected long ago (Montières, terrace of "10 m": "Moustérien chaud" of Commont 1912); the palaecological data can be reinterpreted with the results of the Biache excavations, where living floors have been discovered (Tuffreau *et al.* 1977; Tuffreau 1978b). The archaeological sequence of Biache is located in the upper part of fine calcareous fluvial formations and at the base of the loess cover. Palynological analysis (by Munaut) and determinations of malacological fauna (Puisségur) have shown that the human occupation took place during an interstadial, the climate of which was not very different from that of today. The landscape was in part wooded with a preponderance of *pinus* and *betula*. The macrofauna also suggests a fairly mild climate. This data seem to match Commont's "Moustérien chaud" in Montières.

No lithic assemblage has been discovered in primary context in loessic series, except for gravels. It is not certain that Northern France was frequented during the coldest phases of the end of the Middle Pleistocene.

The lithic industries

All lithic assemblages prior to Eemian and belonging to the Middle Palaeolithic are typologically evolved. It is possible to distinguish the following industries:

Upper Acheulian

Upper Acheulian is only known from old collections, where the percentage of handaxes is too high due to the conditions of gathering. The most important assemblage, gathered in good conditions, is the one of Vimy (Tuffreau 1979a), from a thin layer of gravels surmounted by a saalian loess. The lithic assemblage presents a clear Levalloisian facies. The flake-tools are varied with numerous side-scrapers (IRess. = 40.82) (percentages of the Bordian method), especially simple side-scrapers, abundant backed knives with cortex (16.10% ess.) and denticulates (IVess. = 13.85). The flake-tools of Upper Palaeolithic type are not rare. The percentage of handaxes, mainly amygdaloids and lanceolates, is very important (IBess. = 24.57).

Epi-Acheulian

We call Epi-Acheulian lithic assemblages where the handaxes, of Acheulian type, are very rare. The most important assemblages of this type are those of the Osiers site in Bapaume (Tuffreau 1976). The percentage of handaxes is less than 1% in the B2 assemblage. The Levallois flakes are well represented. The flake-tools are varied: side-scrapers (35.80% ess.), numerous denticulates (IVess. = 15), notches (28.32% ess.) and tools of Upper Palaeolithic type (IIIess. = 10.83).

In some assemblages, as the so-called "Moustérien chaud" (Commont 1912) of Montières, blades are numerous. This characteristic is also present in assemblages of the Oise Valley.

Biache complex

The Biache excavations, which are still continuing, have already permitted to gather several lithic assemblages, the study of which is not finished. However, it is possible to give their most important technical and typological characteristics (figs. 4 and 5). Levallois technique is present in all the assemblages and the Levallois flakes are numerous. Side-scrapers are abundant, especially the convergent ones (convergent side-scrapers and Mousterian points) of a very high quality. The denticulates are well represented, like the cortical backed knives, but not the notches. The flake-tools of Upper Palaeolithic type are very rare. No handaxe has been discovered in Biache. It is still too early for a detailed study of typological variations among the different assemblages.

Ferrassie Mousterian

The Champvoisy lithic assemblage (Hinout and Tuffreau, to be published) of Levalloisian facies contains many side-scrapers (73.66% ess.), including abundant convergent pieces: 1/4 of the side-scrapers. These characteristics are those of the classical Ferrassie Mousterian of the Last Glacial; consequently, there is no reason not to put the Champvoisy assemblage in the Ferrassie Mousterian.

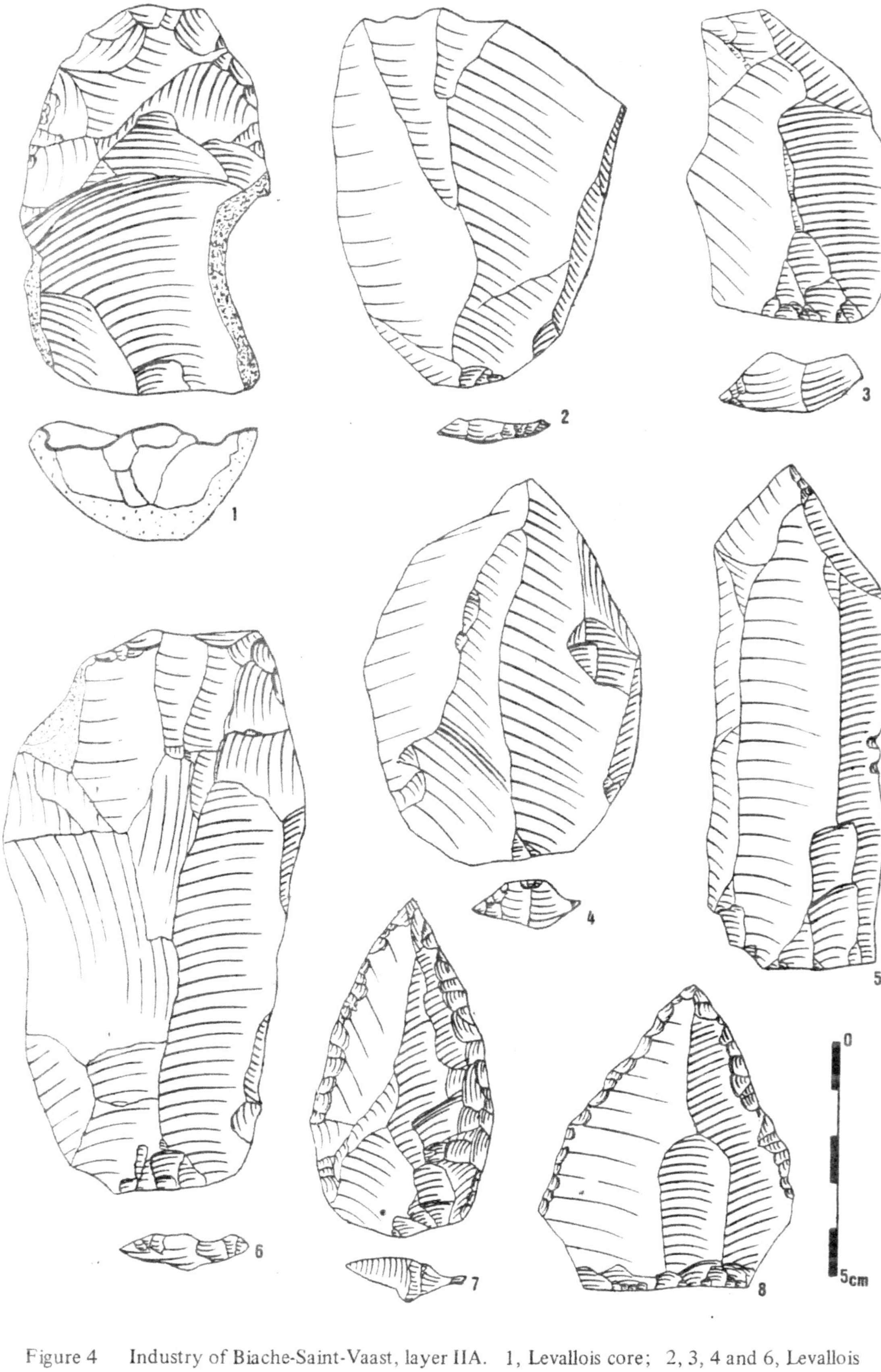

Figure 4 Industry of Biache-Saint-Vaast, layer IIA. 1, Levallois core; 2, 3, 4 and 6, Levallois flakes; 5, Levallois blade; 7, Mousterian point; 8, retouched Levallois point.

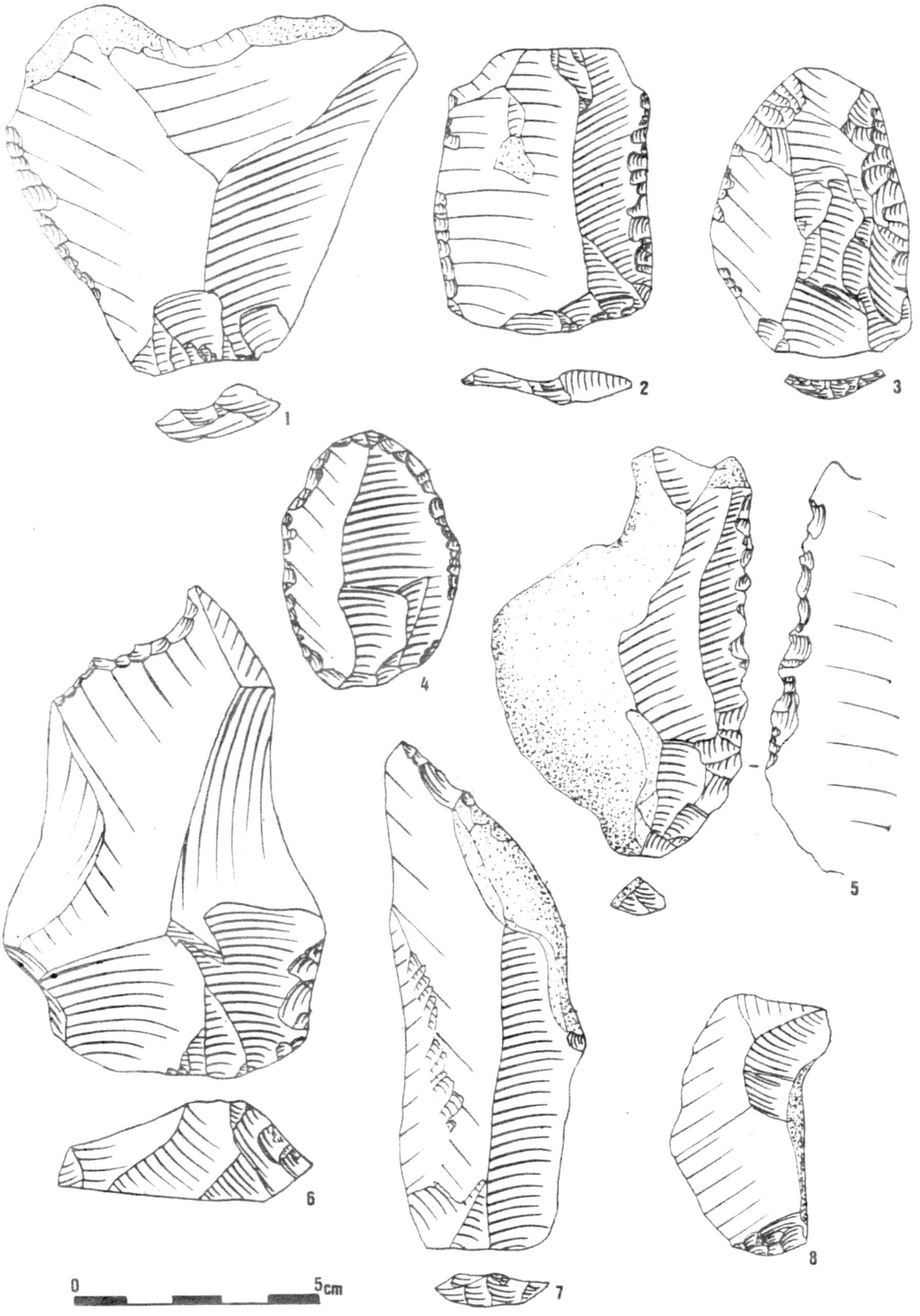

Figure 5 Industry of Biache-Saint-Vaast, layer IIA. 1, simple side-scraper; 2, 3, double side-scraper; 4, end-scraper; 5, 6, denticulates; 7, backed knife; 8, backed knife with cortex.

Other assemblages without handaxes

Beaumetz-les-Loges (Tuffreau 1979a), an assemblage from a thin layer of gravels subdividing the last saalian loess, contains numerous Levallois flakes. The flake-tools are highly evolved: side-scrapers (42.96% ess.) and numerous Upper Palaeolithic tools (17.03% ess.).

An assemblage without handaxes is also present in the fluvial gravels of Etaples low terrace. The material is essentially composed of Levallois flakes, Levallois blades and Levallois cores.

The material from the low terraces of the Somme has been described as "Levalloisian" by Breuil (Breuil et Kelley 1954; Sommé et Tuffreau 1978). In fact, the assemblages of the gravels of the Montières low terraces include some handaxes and a great quantity of large Levallois flakes and Levallois cores. According to Bourdier (1973), the Levalloisian exists also in Argoeuves. In reality, the lithic material of Argoeuves collected by Agache (1976) contains also Acheulian handaxes which are as rolled as the Levallois flakes in the fluvial sequence (fig. 6). Breuil's Old Levalloisian appears to be a mixture of various assemblages in the classical sites of the Somme Valley.

General characteristics of the industries of the Early Middle Palaeolithic

During the penultimate Glacial different industries coexisted. They were all of Levalloisian facies. Some belong to the end of the Acheulian complex: Upper Acheulian and Epi-Acheulian. Others are very similar to certain classical Mousterian industries of the Last Glacial. It appears that the complexity of the Mousterian industries exists since the Saalian in Northern France. In reality, the only difference between Acheulian and Mousterian assemblages in the Saalian is the presence, in low percentage, of Acheulian handaxes, or their absence. We propose to call this early phase of Middle Palaeolithic the Artesian stage of Middle Palaeolithic, where Late Acheulian and Early Mousterian industries coexist. In this stage, the assemblages are quite evolved and very different of those of Late Lower Palaeolithic. At present it is difficult to know if transitional industries are absent in Northern France, or if this absence is the result of a gap in the research.

Human remains

The only human remains discovered in Northern France are the fragments of the Biache skull (Vandermeersch 1978). This fossil belongs to the Preneanderthalians and resembles the Swanscombe skull.

Conclusions

It appears that the Levallois technique is occasionally present in some Acheulian assemblages of the Somme Middle terrace. Its use was very limited and became forgotten in more recent assemblages (Atelier Commont, for instance). During the Saalian, Middle Palaeolithic already exists in Northern France. In the early stage of Middle Palaeolithic (the Artesian stage), Late Acheulian industries coexisted with Mousterian industries. The flake-tools were very evolved and did not differ from those found in the first part of the Last Glacial.

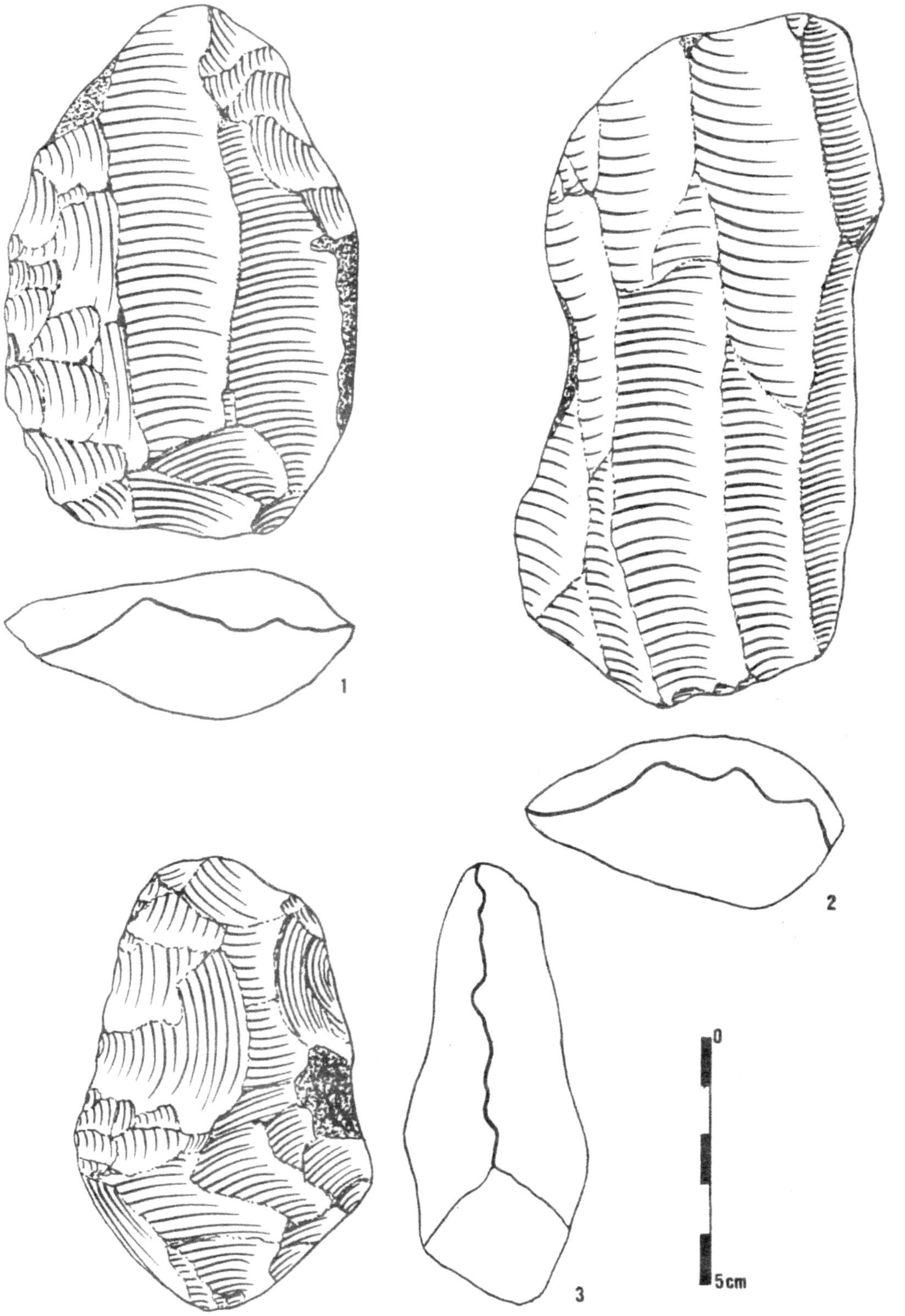

Figure 6 Industry of Argoeuves from the gravels of the terrace. 1, 2, Levallois cores; 3, handaxe.

At present transitional industries more typical than the assemblage of Cagny-la-Garenne are unknown. It seems certain that Northern France was not frequented during the coldest phases of the Late Middle Pleistocene. We must not forget also that the nature of the transition from Lower to Middle Palaeolithic depends only on lithic artifacts because, unlike the transition Middle/Upper Palaeolithic, data concerning features, living floors, bone industries and human remains are scarce.

References

Agache, R. 1976 Argoeuves (Somme). Livret-guide de l'excursion A 10. *IX Congr. U.I.S.P.P.* Nice: 140-145.

Bordes, F. 1954 *Les limons quaternaires du bassin de la Seine* Archives Inst. Paléont. Hum. mémoire 26, 472 p., 175 figs.

Bordes, F. and P. Fitte 1953 L'Atelier Commont *L'Anthropologie* LVII: 1-45, 6 figs., 29 pl.

Bourdier, F. 1973 Dépôts rissiens de la vallée de la Somme. Le problème du Pré-Riss. La datation de l'Acheuléen et du Levalloisien *Le Quaternaire: Géodynamique, Stratigraphie et Environnement, IX Congrès INQUA* Christchurch, pp. 51-53.

Bourdier, F., J. Chaline, A.V. Munaut and J.J. Puisségur 1974a Le complexe Mindelien: II–La moyenne terrasse de l'Avre *Bull. Ass. Fr. Et. Quatern.* 11: 168-180.

Bourdier, F., A.V. Munaut, F. Prat and J.J. Puisségur 1974b Les dépôts du complexe rissien de la Somme *Bull. Ass. Fr. Et. Quatern.* 11: 219-224.

Breuil, H. and H. Kelley 1954 Le Paléolithique ancien: Abbevillien, Clactonien, Acheuléen, Levalloisien. *In:* "Les grandes civilisations préhistoriques de la France". *Bull. Soc. Préhist. Fr.* LI: 9-26.

Callow, P. 1976 *The Lower and Middle Palaeolithic of Britain and adjacent areas of Europe* Ph.D., Cambridge.

Commont, V. 1908 Les industries de l'ancien Saint-Acheul *L'Anthropologie* XIX: 527-572, 75 figs.

Commont, V. 1909 Saint-Acheul et Montières. Notes de Géologie, de Paléontologie et de Préhistoire *Mém. Soc. Géol. Nord.* VI, mémoire 3, 68 p., 52 figs.

Commont, V. 1912 Moustérien à faune chaude dans la vallée de la Somme à Montiéres-les-Amiens *Congr. Intern. Anthr. Arch. Préhist.*, Genève: 291-300.

Lautridou, J.P. and G. Verron 1970 Paléosols et loess de Saint-Pierre-les-Elbeuf (Seine-Maritime) *Bull. Ass. Fr. Et. Quatern.* 7: 145-165.

Ohel, M. 1979 Technological efficiency in Prehistoric ALT assemblages from North-western and Southeastern England *Lithic Technol.* 8(2): 17-20.

Sommé, J. 1978 Chronologie et lithostratigraphie du gisement paléolithique de Biache-Saint-Vaast (Pas-de-Calais) *Bull. Ass. Fr. Et. Quatern.* 15: 27-34.

Sommé, J. and A. Tuffreau 1978 Historique des recherches sur le Quaternaire de la région du Nord de la France *Bull. Ass. Fr. Et. Quatern.* 15: 5-13.

Tuffreau, A. 1976 Les fouilles du gisement acheuléen supérieur des Osiers à Bapaume (Pas-de-Calais) *Bull. Soc. Préhist. Fr.* LXXIII: 231-243.

Tuffreau, A., J. Sommé, J. Chaline, A.V. Munaut, J.F. Piningre, J.J. Puisségur and B. Vandermeersch 1977 Der altpaläolitische Fundplatz Biache-Saint-Vaast (Nordfrankreich) *Archäologisches Korrespondenzblatt* 7: 1-7.

Tuffreau, A. 1978a Les industries acheuléenes de Cagny-la-Garenne (Somme) *L'Anthropologie* 82: 37-60.

Tuffreau, A. 1978b Les fouilles du gisement paléolithique de Biache-Saint-Vaast (Pas-de-Calais): année 1976 et 1977; premiers résultats *Bull. Ass. Fr. Et. Quatern.* 15: 46-55.

Tuffreau, A 1979a Recherches récentes sur le Paléolithique inférieur et moyen de la France septentrionale *Bull. Soc. roy. belge Anthr. Préhist.* 90: 161-177.

Tuffreau, A. 1979b Les débuts du Paléolithique moyen dans la France septentrionale *Bull. Soc. Préhist. Fr.* LXXVI: 140-142.

Tuffreau, A. 1980 Les fouilles de Cagny-Cimetière (Somme): rapport préliminaire *Cahiers archéologiques de Picardie* 7: 5-16.

Tuffreau, A. 1981 L'Acheuléen dans la France septentrionale *Anthropologie* Brno XIX/2: 171-183.

Vandermeersch, B. 1978 Etude préliminaire du crâne humain du gisement paléolithique de Biache-Saint-Vaast (Pas-de-Calais) *Bull. Ass. Fr. Et. Quatern.* 15: 65-67.

DER ÜBERGANG VOM ALT- ZUM MITTELPALÄOLITHIKUM IN SÜDDEUTSCHLAND

Gisela Freund

Wie immer man auch die Grenze zwischen Alt- und Mittelpaläolithikum zieht – geochronologisch mit dem Beginn des Riss-Würm-Interglazials oder technisch mit dem Einsetzen der Levalloistechnik –, Süddeutschland kann zu der Frage nach dem Übergang kaum entscheidende Beiträge leisten. Dies liegt daran, dass es in diesem Raum insgesamt nur sehr wenige altpaläolithische Kulturniederschläge gibt und damit auch sehr wenige, die – bei geochronologischer Grenzziehung – in die Riss-Eiszeit oder das Riss-Würm-Interglazial gehören oder die – bei der zweitgenannten Grenzziehung – das Einsetzen der Levalloistechnik belegen.

Auf der Suche nach der Existenz eines echten Altpaläolithikums in Süddeutschland scheiden an dieser Stelle ohnedies alle unter dem Begriff der "Artefakte des *Homo heidelbergensis*" so vielfach diskutierten Komplexe aus. Sollte es sich, bei vielleicht vereinzelten Stücken, überhaupt um intentionell hergestellte Werkzeuge handeln, dann gehören sie, von welchen Fundorten auch immer, einer weit älteren Zeitstufe an und erreichen nicht die Endphase des Altpaläolithikums. Gleichfalls scheidet für den genannten Raum die in umgebenden Landschaften (Böhmen/ČSSR, Oberhessen/nördlich des Mains) sehr wohl vertretene Fazies der Geröllgeräte aus. Für die Frage des Übergangs Alt-/Mittelpaläolithikum leistet allenfalls das Gebiet um K r o n a c h in Oberfranken einen Beitrag, wo, in stark altpaläolithischer Tradition stehend, echte Choppers und Chopping-tools innerhalb eines Mittelpaläolithikums, freilich als Oberflächenfunde auf Terrassen, auftreten (Fig. 1). Der gesamte Komplex, der eine Sonderstellung im mitteleuropäischen Mittelpaläolithikum einnimmt, dürfte aber nicht älter als Ende riss-würm-interglazial sein (Zotz u. Freund 1973).

Da in diesem Symposium zugleich der Ursprung des *Homo sapiens* angesprochen ist, muss aus dem süddeutschen Raum der Schädel des *Homo steinheimensis* mit seinen Praesapiensmerkmalen genannt werden. Seine Zeitstellung im Mindel-Riss (= Holstein) -Interglazial ist nicht umstritten, sehr wohl indes einige Artefakte, vor allem der mit dem Homofund in Verbindung gebrachte Faustkeil (Adam 1974).

Auf der weiteren Suche nach risseiszeitlichen Kulturesten, die an der Schwelle zum Mittelpaläolithikum stünden, bieten sich in Süddeutschland ausser Einzelstücken wie dem Faustkeil von Biburg/Ldkr. Kelheim (Freund 1969) und möglicherweise einigen weiteren Faustkeilfunden von Acheuléencharakter, jedoch alle ohne gesicherte Stratigraphie, nur drei weit auseinanderliegende Fundbestände an. Die Lössfundstelle von M u r g (östlich Basel) am Hochrhein (Ldkr. Säckingen) bleibt trotz der zahlreichen Literatur, die über sie vorliegt, sowie der genaueren Materialvorlage (vgl. zuletzt Gersbach

1968) sowohl vom geochronologischen Aspekt (Interstadial im Riss oder Riss-Würm-Interglazial oder Frühwürm) wie vom kulturellen (Anzahl der Begehungen, autochthon oder verlagert) her, schwer beurteilbar. Zu einigen Faustkeilen oder Vorformen von solchen gesellen sich in der Mehrzahl Abschläge und Geräte in Levalloistechnik, sowie deren Kerne, jedoch auch Choppers und Chopping-tools einerseits, wie Schaber eines eher fortgeschrittenen Mittelpaläolithikums andererseits. Ob primär ein zeitlich einheitliches oder ein gemischtes Inventar verschiedener Zeit- und Kulturstellung vorlag, lässt sich wohl nicht mehr entscheiden. Ferner gehört, wenngleich auch jünger, ein grosser Teil des Oberflächenpaläolithikums aus dem Donaubogen südlich Regensburg (Freund 1977) mit seinen Faustkeilen von Jungacheuléentyp hierher. Einer Übergangsphase vom Riss-Würm-Interglazial zum frühesten Würm zugesprochen, geht die Belegung dieser Freilandfundstellen der reichen Höhlenbesiedlung des nahen Altmühltales voraus. Mit den Faustkeilen vergesellschaftete Kerne und Geräte in Levalloistechnik stellten sie, nach der Gliederung von G. Bosinski, allerdings auch technisch bereits ein Mittelpaläolithikum, freilich seine ältere Phase dar. Wären die Funde ein stratigraphisch geschlossener Komplex, würde man sie wegen des hohen Faustkeilanteils gerne als eine echte Übergangserscheinung vom Alt- zum Mittelpaläolithikum ansehen.

Besser erfasst dürfte eine solche Übergangsphase aber vor allen Dingen, wenngleich auch mit viel zu kleinen Inventaren, in der Höhle von Hunas, Ldkr. Nürnberg-Land in der mittleren Fränkischen Alb sein. Die in Druck befindliche Monographie dieser paläontologisch ungewöhnlich reichen Fundstelle (Heller u. Mitarbeiter, darunter Brunnacker und Freund, vgl. ferner Freund 1978) beinhaltet in mehreren Schichten ein faustkeilfreies Paläolithikum der späten Riss-Eiszeit. Dominierend sind Schaber, z. T. mit sehr sorgfältiger Retuschierung, von insgesamt mittelpaläolithischem Gepräge (Figs. 2–4). Levalloistechnik spielt indes praktisch keine Rolle. Typologische Parallelen finden sich im mitteleuropäischen Raum für die gleiche Zeitphase nicht, wobei allerdings die sehr geringe Anzahl der Geräte, die am ehesten eine Proto-Quina-Phase repräsentieren, in ihrer Aussagekraft nicht überfordert werden darf. Ein leider ebenfalls nur kleiner Komplex aus der hangenden riss-würm-interglazialen Schicht dokumentiert mit deutlicher Anwendung der Levalloistechnik am ehesten ein Charentien vom Typ Ferrassie (Fig. 5). Die Höhle von Hunas, die zugleich das älteste stratigraphisch gesicherte Paläolithikum Bayerns lieferte, stellt zeitlich wie technisch ein spätes Alt–, typologisch jedoch ein schaberreiches Mittelpaläolithikum dar und zeigt, wie fliessend die Übergänge sind.

Eine für einen Teil der Schaber von Hunas sehr auffällige und wohl charakteristische Bearbeitung ist eine äusserst feinschuppige, mehr oder weniger auf die Oberfläche übergreifende Randretusche. Eben diese taucht auch an einer Reihe von Geräten in den tiefsten Schichten der Sesselfelsgrotte im unteren Altmühltal im Ldkr. Kelheim/ Donau auf und zwar so übereinstimmend, dass man die Stücke gegeneinander vertauschen könnte.

Damit sei Einiges, soweit das Thema des Symposiums berührt ist, zu der gegenwärtig wohl bedeutendsten paläolithischen Fundstelle Mitteleuropas der Sesselfelsgrotte (Zotz 1970; Freund 1968, 1973, 1975), gesagt (Fig. 6). Innerhalb ihrer bis nahezu 7 m mächtigen Schichtfüllung mit etwa 20 Kulturhorizonten und sehr wechselnden Sedimenten liegt ein reiches Mittelpaläolithikum (Schichten G1–G5) oberhalb der Mitte in ca. 2.50 m Tiefe. Typologisch steht es dank seiner starken Bifazialität einem Micoquien in der Definition von Bosinski am nächsten, ohne dass es jedoch mit einem seiner 4 Inventartypen identisch sei. Möglicherweise etwas jünger als der Inventartyp Klausennische, dem Fundplatz genau gegenüber auf dem jenseitigen Talhang der Altmühl, unterscheidet es sich dennoch von dem wiederum als etwas jünger angenommenen

Inventartyp Schambach. Die in ihrem Reichtum nur mit westeuropäischen Verhältnissen vergleichbaren Industrien der G-Schichten sollten dem Inventar der Klausennische näher stehen. Sie gehören damit, entsprechend allgemein herrschender Vorstellungen, in die frühe Würm-Eiszeit.

Unterhalb dieser G-Schichten befinden sich ca. 1.50 m mächtige Schutthorizonte mit reichen Einlagerungen von Nagetierresten, die innerhalb der entsprechenden Zeitspanne zwei unterschiedliche Klimaphasen, in Schicht I eine mehr gemässigte, an der Grenze von K zu L eine kältere mit vereinzelten arktischen Formen, repräsentieren. Erst darunter beginnt eine Folge von 10 weiteren Kulturschichten, deren Formenschatz deutlich von dem der G-Schichten unterschieden ist. Die Bifazialität ist geringer, die Anwendung von Levalloistechnik deutlich ausgeprägt, Faustkeile sowohl vom Acheul-, wie vom Micoquetyp kommen, wenn auch anteilig gering, vor. Echte Moustierspitzen, auch eine Quinsonspitze begleiten ein sehr variantenreiches Schaberensemble, unter dem sich solche mit der angesprochenen feinen schuppigen Retusche wie in Hunas befinden. Die Mächtigkeit dieser unteren Horizonte beträgt wechselnd ca. 2 m, wobei basal auf dem grössten Teil der gegrabenen Fläche der Felsen erreicht ist. Nur in der Westecke auf dem Abri-Vorplatz folgt darunter nochmals ein Schichtpaket von mehr als 1 m Mächtigkeit, gleichfalls mit Funden durchsetzt. In dieser Westecke ist wahrscheinlich der obere und hintere Teil eines tiefer im Hang gelegenen verschütteten Abris angeschnitten, dessen Kulturreste einer älteren Phase als denen in den untersten Schichten der Sesselfelsgrotte angehören sollten.

Damit ist unterhalb des mitteleuropäischen Micoquien und noch unterhalb einer durch Nagetierfaunen belegten klimatisch zweiphasigen Sedimentablagerung ein Paläolithikum mit 10 bzw. 13 Kulturschichten entdeckt, dessen Zeitstellung im Augenblick völlig offen ist. Zahlreiche Mollusken, besonders aus Schicht O, aber auch aus S sowie aus den tiefsten Zonen 2-West und 3-West, zeigen an, dass in diesem gesamten unteren Schichtpaket mehrere Klimaschwankungen enthalten sind, und dass das letzte Interglazial nicht lange vor Bildung der basalen Schichten zu Ende gegangen sein kann oder sich sogar schon in diesen ausdrückt. Die angezeigten Schwankungen werden auch durch die Holzkohlenbestimmungen unterstützt. Pollenanalytische und sedimentanalytische Ergebnisse stehen noch weitgehend aus. Die letzteren weisen jedoch beim gegenwärtigen Untersuchungsstand in die gleiche Richtung.

Fundplätze mit parallelen Erscheinungen gibt es im mittel- und mittelosteuropäischen Raum offensichtlich nicht. Damit erhebt sich die Frage, ob das untere Schichtenpaket bereits der Riss-Eiszeit angehört und ob damit in der Sesselfelsgrotte in jedem erdenklichen Fundreichtum ausgeprägt ist, was in Hunas nur angedeutet erscheint. Die in 14 Jahren mit aller notwendigen Sorgfalt durchgeführten Grabungen haben jedoch keinen Anhaltspunkt für die Existenz interglazialer Ablagerungen und auch nicht für einen entsprechenden Hiatus erbracht. Die spärliche Grossfauna widerspricht beim gegenwärtigen Untersuchungsstand ebenfalls nicht der Annahme, dass die Gesamtablagerungen der Sesselfelsgrotte in die Würm-Eiszeit gehören. Erhärtet sich diese, dann werden sich die Vorstellungen von den Anfangsphasen würmeiszeitlicher Höhlenablagerungen, die in der Sesselfelsgrotte in ungewöhnlicher und vielgliedriger Ausprägung sedimentologisch, faunistisch und floristisch belegt sind, erheblich wandeln müssen. Dies gilt erst recht für die Vorstellung von paläolithischen Kulturniederschlägen dieser Frühphasen, die kein Micoquien, auch nicht das vom ältesten Typ, dem der Bocksteinschmiede, darstellen. Diese genauer anzusprechen, ist erst nach der Aufarbeitung aller Funde aus den Schichten M bis S + 1-West bis 3-West möglich. Offenbar stellen diese kein Jungacheuléen dar, sondern zeitlich und kulturell Phasen, die derzeit ohne Parallelen sind.

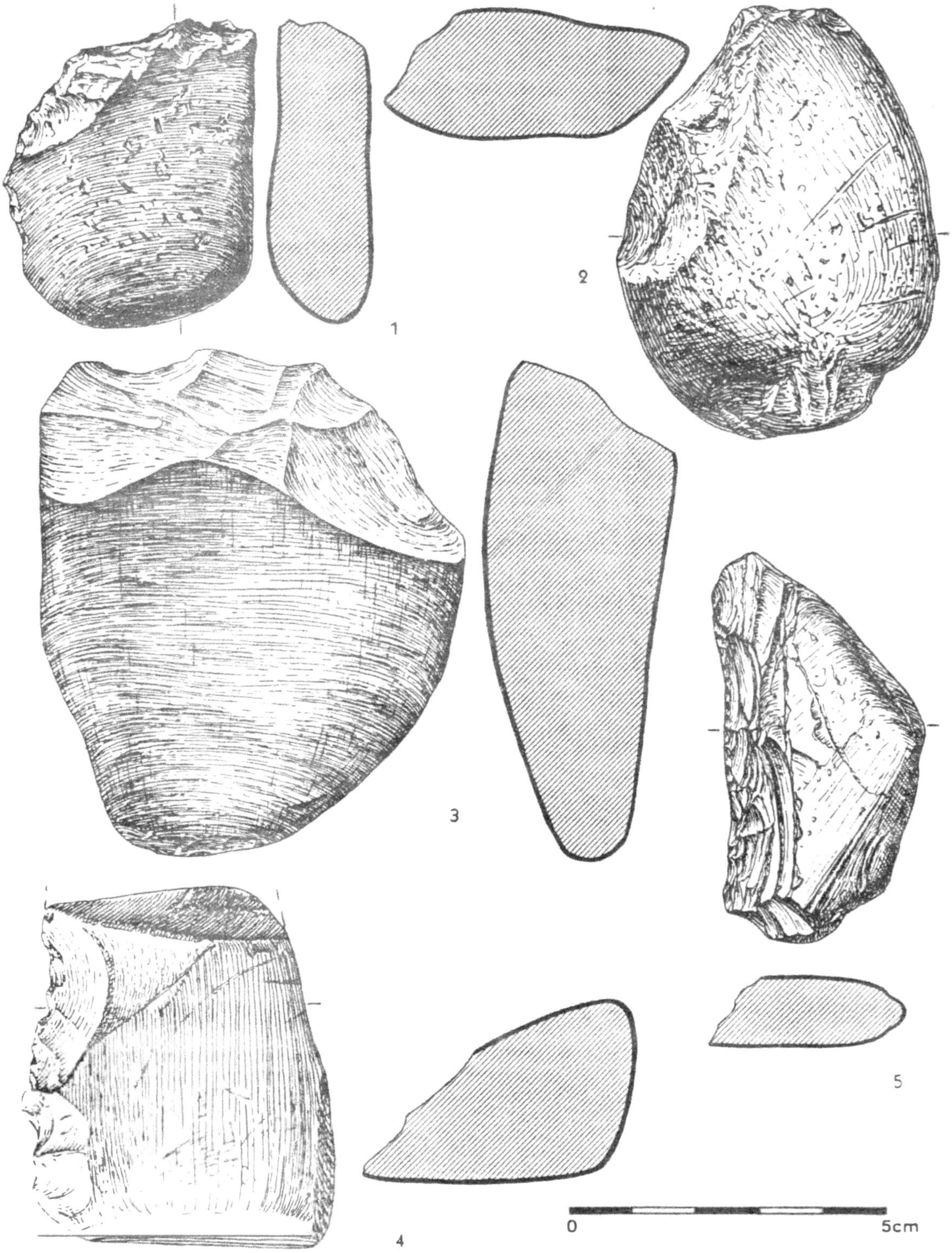

Figure 1 Mittelpaläolithische Geröllgeräte von Kronach. 1-3, Choppers (Typ A, Grundform); 4, 5, Geradschaber (Typ A 1a). 1, 2, 5, Hummendorf; 3, Wachtersmühle; 4, Kreuzberg-Terrasse.

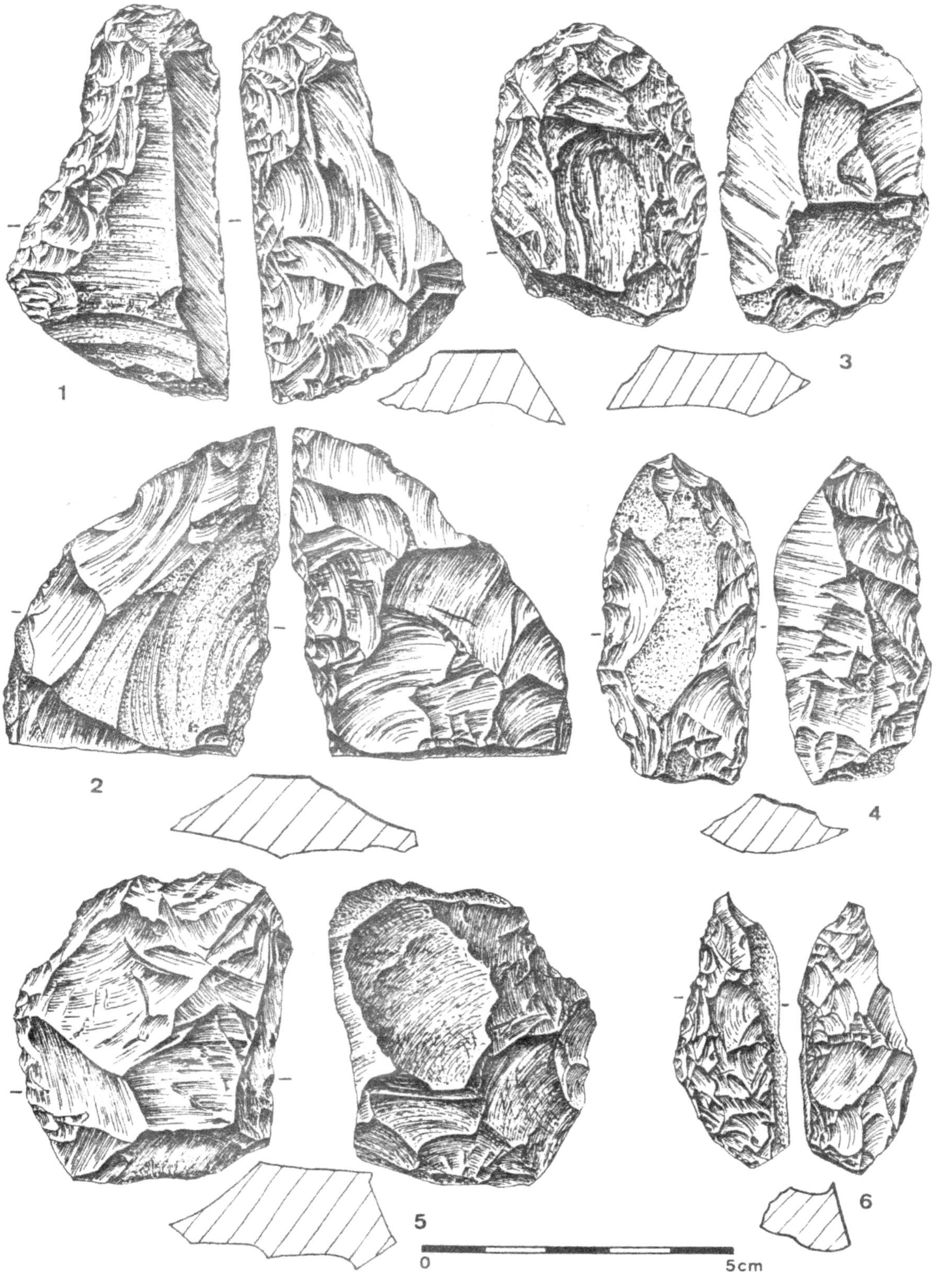

Figure 2 Hunas, Schicht G2.

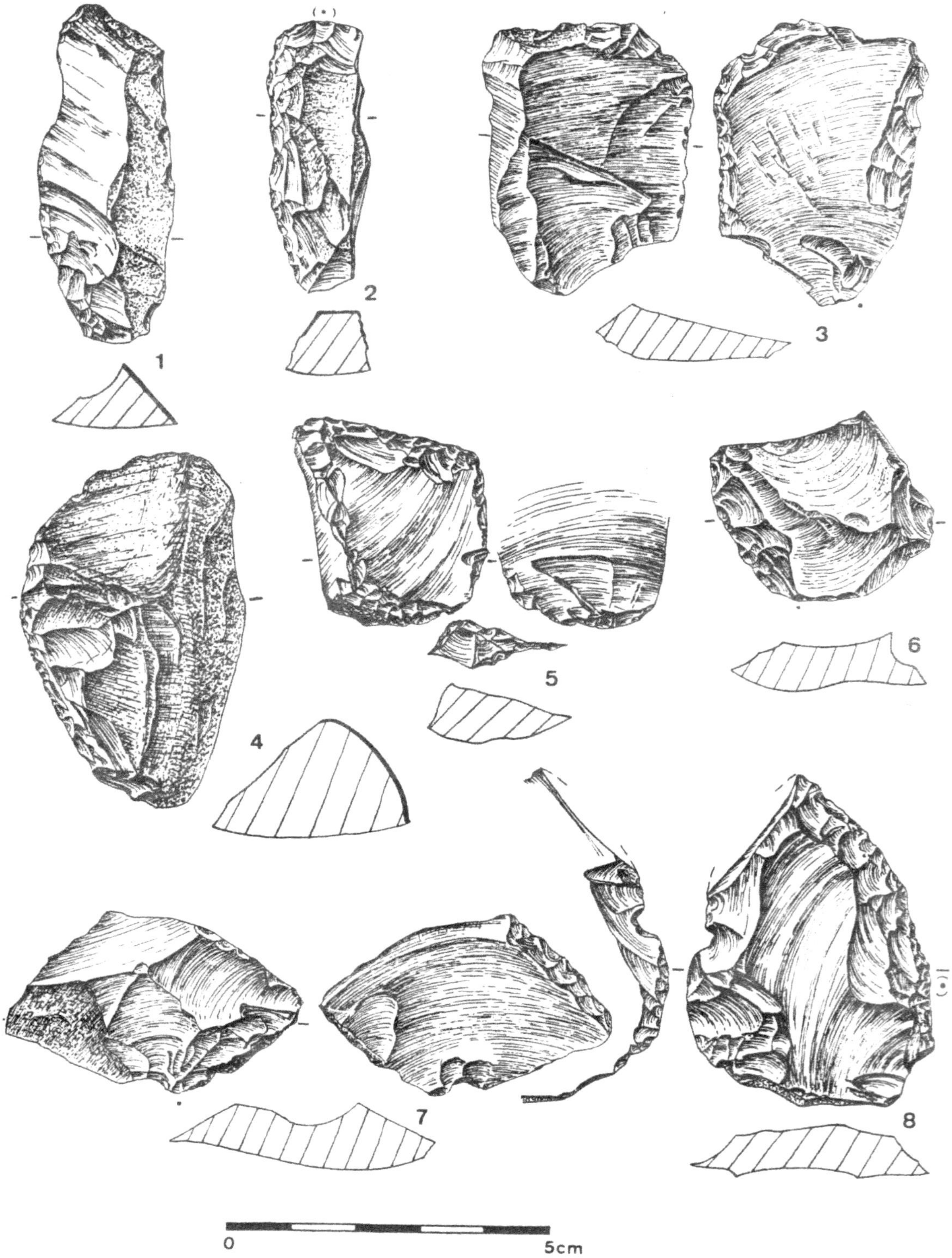

Figure 3 Hunas, Schicht G2.

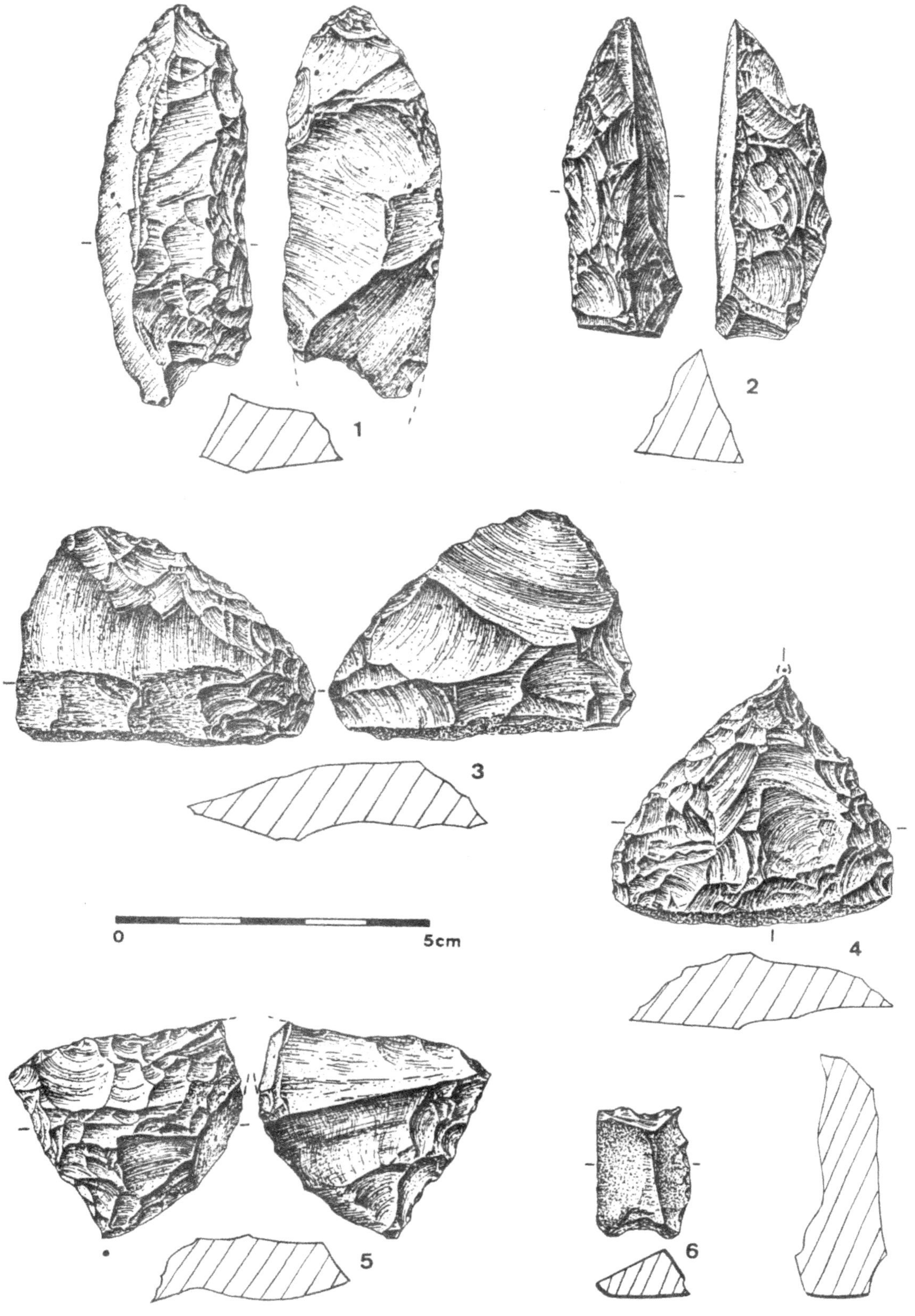

Figure 4 Hunas, Schicht G3.

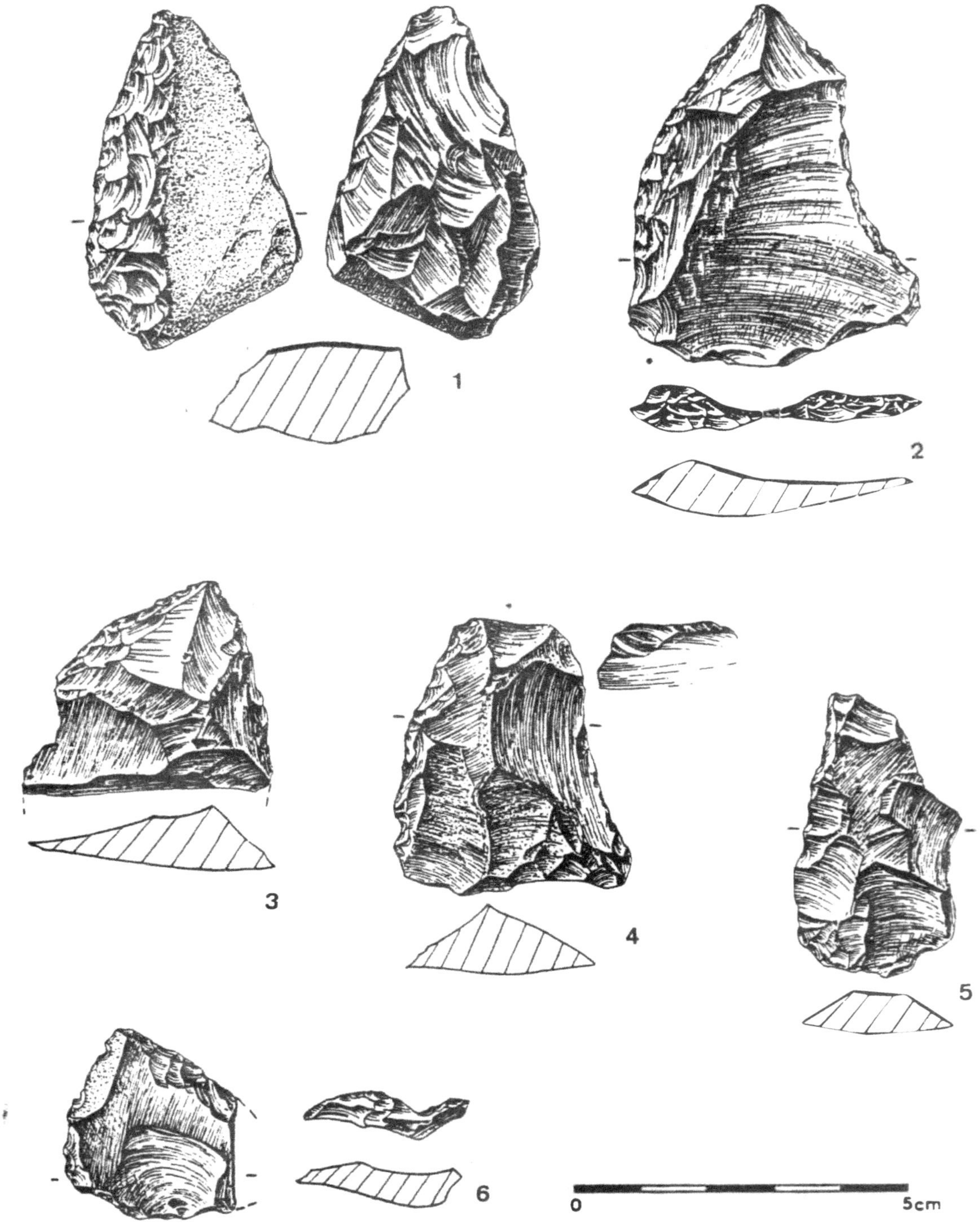

Figure 5 Hunas, Schicht E.

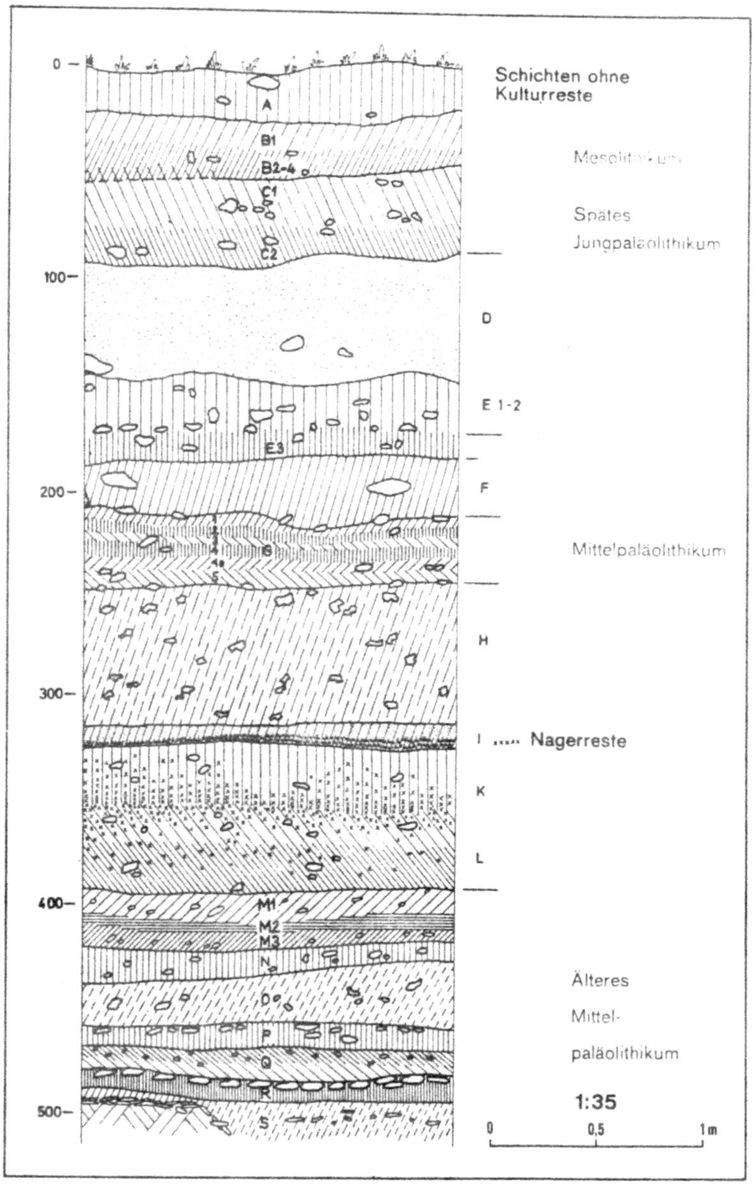

Figure 6 Schichtenfolge der Sesselfelsgrotte ohne die noch 2 m tiefer reichenden Horizonte unterhalb der Schicht S.

Sesselfelsgrotte, section. The two lowest meters, below layer S, are not shown.

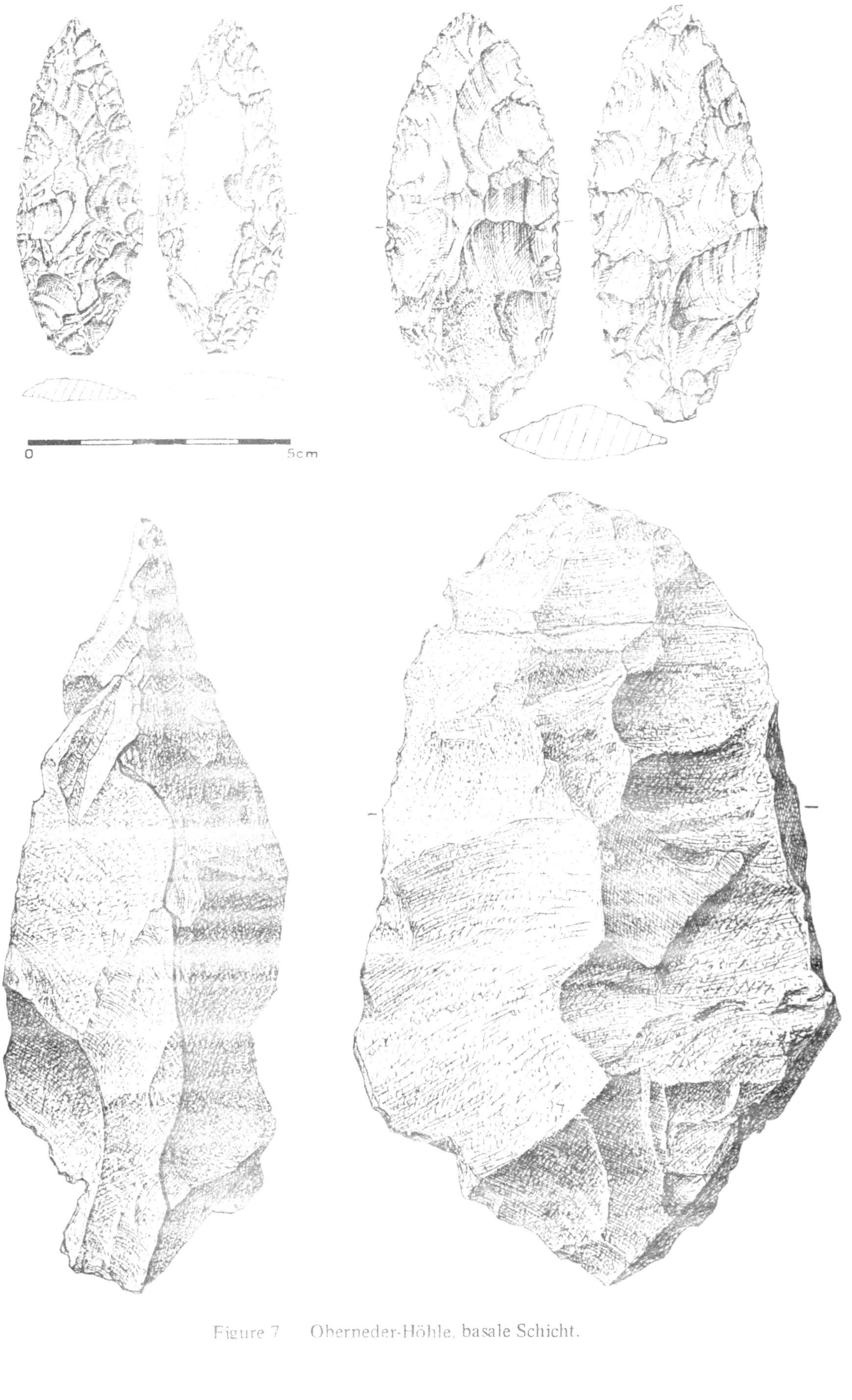

Figure 7   Oberneder-Höhle, basale Schicht.

Die Begehung des oben angezeigten, wohl im Hang verschütteten Abris, der ältere Funde als die ältesten der Sesselfelsgrotte beinhalten sollte, könnte möglicherweise einer Zeitphase entsprechen, die die Freilandfunde im Donaubogen südlich Regensburg, also die eines jüngeren Acheuléen belegen.

Obgleich beim gegenwärtigen Stand der Aufarbeitung das untere Kulturschichtenpaket der Sesselfelsgrotte, 1.5 m tiefer beginnend als der micoquienartige Komplex der G-Schichten, als Mittelpaläolithikum angesprochen wird und die zugehörigen Sedimente für frühest-würm-eiszeitlich gehalten werden, mag die Sesselfelsgrotte einen gewissen Beitrag zu der Frage des Übergangs vom Alt- zum Mittelpaläolithikum leisten. Dies ist so zu verstehen, dass das Mittelpaläolithikum nach derzeitiger Definition auch im mitteleuropäischen Raum eine weit reichere Entwicklung in seinen Anfangsstadien durchlaufen hat, als man dies bis dahin belegt glaubte. Die Übergänge sind im süddeutschen Raum gewiss fliessend.

Etwa gleichzeitig mit den unteren Kulturschichten der Sesselfelsgrotte ist möglicherweise die durch Faustkeile einerseits und durch vollendet gearbeitete Blattspitzen andererseits gekennzeichnete basale Kultur (Fig. 7) der nicht weit entfernt gelegenen Oberneder-Höhle (Freund 1962-63, 1968). Sedimentologisch ist dies angesichts der insgesamt geringen Schichtmächtigkeiten (maximal 1.5 m) allerdings schwer zu erhärten. Faustkeile und Blattspitzen liegen unmittelbar auf dem Felsboden der Höhle. Die bevorstehende Aufarbeitung wird die Frage ihrer Gleichzeitigkeit zu klären haben. Das Inventar stellt jedenfalls einen anderen Typus wohl frühwürmeiszeitlicher Kulturen als in der Sesselfelsgrotte dar und unterstreicht die schon erwähnte Vielfalt in den Übergangsphasen zum Mittelpaläolithikum. Eine Grenzziehung Alt-/Mittelpaläolithikum ist für den süddeutschen Raum, wie immer man sie vornimmt, beim gegenwärtigen Forschungsstand, eine künstliche.

Zitierte Literatur

Adam, K.D. 1974 Die *"Artefakte des Homo steinheimensis"* als Belege urgeschichtlichen Irrens. Stuttgarter Beiträge zur Naturkunde, Serie B, Nr. 6.

Freund, G. 1962-63 Ein neuer Höhlenfundplatz mit mittelpaläolithischen Blattspitzen aus Süddeutschland *Arheološki Vestnik* (Serta Brodariana) XIII–XIV: 71-84.

Freund, G. 1968 Le Paléolithique moyen à pointes foliacées de la Grotte Oberneder sur Altmuhl en Bavière *La Prêhistoire – Problèmes et Tendances* Mélanges Raymond Vaufrey, pp. 175-82, Bordeaux.

Freund, G. 1968 Mikrolithen aus dem Mittelpaläolithikum der Sesselfelsgrotte im unteren Altmühltal, Ldkr. Kelheim *Quartär* 19: 133-54.

Freund, G. 1969 Der Faustkeil von Biburg bei Abensberg, Ldkr. Kelheim/Donau *Quartär* 20: 163-74.

Freund, G. 1973 Die Sesselfelsgrotte im unteren Altmühltal (Landkreis Kelheim, Bayern). Neue paläolithische und mesolithische Ausgrabungen in der Bundesrepublik Deutschland. Zum IX. INQUA-Kongress (Neuseeland, 1973), pp. 19-22, Tübingen.

Freund, G. 1975 Zum Stand der Ausgrabungen in der Sesselfelsgrotte im unteren Altmühltal. *In: Ausgrabungen in Deutschland,* Monographien des Römisch-Germanischen Zentralmuseums 1, Teil I: 25-41, Mainz.

Freund, G. 1977 Das Paläolithikum im Donaubogen südlich Regensburg *Materialhefte zur Bayerischen Vorgeschichte,* Reihe A, Bd. 32. Kallmünz.

Freund, G. 1978 Zum Paläolithikum aus der Höhlenruine von Hunas in der nördlichen Fränkischen Alb. *Archäologisches Korrespondenzblatt* 8: 259-63.

Gersbach, E. 1968/1969 Urgeschichte des Hochrheins (Funde und Fundstellen in den Landkreisen Säckingen und Waldshut) *Badische Fundberichte,* Sonderheft 11, Freiburg.

Heller, Fl. u. Mitarb. Die Höhlenruine Hunas bei Hartmannshof (Ldkr. Nürnberger Land). Eine paläontologische und urgeschichtliche Fundstelle aus dem Spät-Riss (Erscheint als Bd. 4 der *Quartär-Bibliothek* in Druck).

Zotz, L. 1970 Das Paläolithikum der Sesselfelsgrotte *Actes du VIIe Congrès International des Sciences Préhistoriques et Protohistoriques* Prague 1966, pp. 373-76.

Zotz, L. und Freund, G. 1973 Die mittelpaläolithische Geröllgeräteindustrie aus der Umgebung von Kronach in Oberfranken *Materialhefte zur Bayerischen Vorgeschichte,* Heft 27, Kallmünz.

Summary

Whether a geochronological border is drawn at the beginning of the Riss-Würm Interglacial, or a technical one with the appearance of the Levallois technique, southern Germany can hardly contribute to the discussion of the transition from the Lower to the Middle Palaeolithic, due to the scarcity of data.

All the assemblages designated as "tools of the *Homo heidelbergensis*" have to be excluded. If at all they are intentionally made tools, they belong to a much older period. Likewise, the facies of pebble tools, so well represented in surrounding regions (Bohemia/ČSSR, Upper Hessia/north of the Main), has to be eliminated for the same reason.

There are three localities for Riss Glacial cultural remains in southern Germany: the Loess-stations of Murg (east of Basle), the homogeneity of which is difficult to judge; the surface finds from the Danube area south of Regensburg, with Late Acheulian handaxes and Levallois technique; and the Cave of Hunas near Nürnberg. Amidst very rich faunal remains found here (Late Riss), there are small Lower Palaeolithic series with Middle Palaeolithic affinities.

The most important Palaeolithic site known at present in Central Europe is the Sesselfelsgrotte. In a 7 m thick deposit there have been found some 20 cultural horizons. Layer G (at a depth of 2.50 m below the surface) yields a rich Middle Palaeolithic industry of Micoquian affinities; such an assemblage is usually dated to an early phase of the Würm. A thick layer of rubble (1.5 m) separates these layers from the underlying sequence of 10 cultural horizons (2 m thick), with handaxes of both Acheulian and Micoquian types. The lowest horizons, attributed to the very beginning of the last glacial, show marked climatic fluctuations. This indicates that the beginning of the Last Glaciation had a variety of climatic phases.

The basal assemblage of the nearby Oberneder Cave seems to match chronologically the lower cultural layers of the Sesselfelsgrotte. These two assemblages testify to the manifoldness of the Early Würm transition phases to the Middle Palaeolithic. At the present state of research, wherever the borderline from Lower to Middle Palaeolithic is placed in southern Germany, it would be artificial.

Discussion

A. Ronen: Do you think that the industry of Hunas is the exact counterpart of the lower layers of Sesselfelsgrotte, or that one may have developed from the other?

G. Freund: The industry of Hunas is represented by few implements, as you know and saw yourself. They are typologically the same as the lower layers of Sesselfelsgrotte. But the stratigraphy and the fauna are quite different. In Hunas we should be at the end of the Riss, while in the Sesselfelsgrotte we think that we are at the beginning of the Wurm, perhaps at the end of the Interglacial. The snails found there are species from the Interglacial in the opinion of Losek and others. The other fauna is rather very rare.

I have to add that there are some radio carbon dates for layer N - 45,900, but that means nothing. We have a lot of charcoal remains to be analyzed by Prof. Stieber. This is not yet finished, and it is the same concerning the pollen analysis and other radio carbon dates. We are in full work – the excavation has come to an end – and I think that we will have to go on for more than ten years, because the material is extremely abundant.

THE TRANSITION LOWER/MIDDLE PALAEOLITHIC IN NORTHWESTERN GERMANY

Gerhard Bosinski

Definition of 'Lower Palaeolithic" and "Middle Palaeolithic"

In order to discuss the transition Lower/Middle Palaeolithic it is first necessary to define these two periods. "Lower Palaeolithic" is taken as the period before the differentiation of finds into different industries (Formengruppen) of stone-artifacts. It covers the pebble-"culture" and the handaxe-culture (Acheulian) before the existence of different Space-Time-Units (= industries). There are no "cultural" determinated differences in the Lower Palaeolithic, and the material is open to world-wide discussion. The undeniable existing and generally known differences of Lower Palaeolithic assemblages are not based on different traditions but on different situations (raw-material, function).

From this definition of "Lower Palaeolithic" as well results the definition of "Middle Palaeolithic". The spectrums of stone-artifacts are more differentiated, and the assemblages can be grouped to different industries (Formengruppen) which represent Space-Time-Units. An important role is played by the appearance of Levallois technique, which constitutes a marked development of knapping stone resulting in a greater variability of flake types, as well as of retouched tools.

Such an archaeological definition of Lower and Middle Palaeolithic seems more adequate than calling Last Interglacial and early Würm material "Middle Palaeolithic". This geochronological periodization poses problems; e.g. the classification of Riss glacial assemblages without bifaces (Biache, La Chaise, etc.; compare Tuffreau 1979), or the unlikely situation that assemblages taken as "Middle Palaeolithic" may shift to "Lower Palaeolithic" (recently, Fontechevade), or vice versa, by different dating results. The geochronological periodization would, in addition, render impossible the classification of finds of unclear date, or of surface material, as "Lower" or "Middle" Palaeolithic. Finally, any classification of later material (Upper Palaeolithic, Neolithic, Bronze Age, etc.) is based on an archaeological definition.

The differentiation of the industries which marks the beginning of the Middle Palaeolithic, linked with the appearance of Levallois technique, starts during Riss glaciation. Accordingly, the Middle Palaeolithic of our region is subdivided into:

Early Middle Palaeolithic of the Riss (Saale)
Middle Middle Palaeolithic of the Riss/Würm (Eem)
Late Middle Palaeolithic of the Würm (Weichsel)

The Early Middle Palaeolithic of Northwestern Germany

Concerning the transition from Lower to Middle Palaeolithic, the early Middle Palaeolithic finds from the Riss glaciation are of interest. There are two types of assemblages known:

I. The Lebenstedt Group

This group consists of finds with many Levallois flakes, some unifacially retouched flake tools, and rare bifaces (Figures 1–3). The character of these assemblages is highly influenced by Levallois technique. In our region Levallois technique is best developed in this Lebenstedt group. There are Levallois flake-cores, blade-cores, and point-cores. Correspondingly, Levallois flakes – mostly unretouched – are numerous (Figure 1). Regular blades are well represented (Figure 2: 5-7). The Levallois points are elongated and relatively large (Figure 2: 1-4). Retouched tools normally are not very numerous. First of all there are simple side-scrapers (Figure 1: 1,2. Figure 3: 2,4), more seldom double side-scrapers. Linked with the fact that these tools are mostly made on flat Levallois flakes, the retouch is confined to the edges. Thicker working edges with "stepped" retouch do not occur. Also, due to the dominating Levallois technique, there are almost no transversal scrapers. Further, pointed scrapers and rare points occur.

The percentage of bifacially worked tools is low. There are leafpoint-like scrapers (blattförmige Schaber), the most regular specimens of which look like a broad leafpoint (Figure 3: 1,3). The leafpoint-like scrapers have one mostly convex edge, retouched on both faces, which forms the working-edge. The opposite edge is mostly alternately retouched (= forms a zigzag) and/or blunted.

There is a very small percentage of handaxes. Elongated handaxes with continuously thickened bases are important (Figure 3: 5). There are also almost triangular, sometimes heavy handaxes. Cleaver-like tools occur in this material. In any case there is a cleaver in the old Markkleeberg collection (Bosinski 1976, fig. 3: 1), and the Reutersruh cleavers (Luttropp and Bosinski 1971, pl. 122-123) may fit into this connection too.

The most important site of this group is Salzgitter-Lebenstedt (Tode 1953; Bosinski 1967: 103ff.; Grote 1978). Besides stone-artifacts there are also worked bones. The fauna comprises reindeer, mammoth, bison and horse. Palaeontological and palaeobotanical data permit the reconstruction of a cool and "open" biotop. A human occipital from Lebenstedt has not been sufficiently published (Kleinschmidt 1977). It is neither a classical Neanderthal nor a *Homo erectus*. Thus it seems that this skull is comparable to other Riss glacial finds (Biache; La Chaise). The problem of Lebenstedt is its erroneous dating as early Würm (Tode 1953), which unfortunately has been repeated in more recent works (Grote 1978). In our region there is absolutely no early Würm material comparable to Salzgitter-Lebenstedt. Other sites of the Lebenstedt group are Rethen and Hannover-Döhren as well as Herne (Bosinski 1967: 34ff.). Some of the Reutersruh material should also be included in this group (Luttropp and Bosinski 1971). A small series from the base of a Rissian loess at Rheindahlen is important in dating this material (level B 5; Bosinski and Brunnacker 1973). A similar dating is supported by the corresponding finds of Markkleeberg (Grahmann 1955). New excavations at Markkleeberg not only uncovered an identical assemblage, but reconfirmed the dating to the early Saale glaciation (excavations by Baumann and Mania).

On the whole, the Lebenstedt group belongs to a later stage of the Acheulian ("Upper Acheulian"). But the difference between the Lebenstedt group and the Lower

Palaeolithic Acheulian is as great as the difference between Lower Palaeolithic Acheulian and the Mousterian of Acheulian Tradition which is undoubtedly regarded as Middle Palaeolithic. Thus, it should not be too difficult to classify this late Acheulian phase as early Middle Palaeolithic.

II. The Rheindahlen Group

In the middle of Rissian loess (Fleckenlehm) at Rheindahlen there occurs a rich series (level B 3) which is less marked by Levallois technique than those of the Lebenstedt group. Its other characteristics include a well-defined spectrum of unifacially worked flake-tools and the absence of bifaces (Figures 4-5). Levallois cores and flakes are smaller and less regular. Large elongated Levallois points are rare. The percentage of retouched tools is relatively high. First of all there are simple and double side-scrapers (Figure 4), which are also made on thicker flakes and thus have a thick working-edge with a "stepped" retouch (Quina-type). Points are very characteristic, including symmetrical and asymmetrical points with convex edges (Figure 5). In our region the Rheindahlen-type assemblage is only known from Rheindahlen itself (Bosinski 1967: 64; Thieme 1978). The assemblage resembles very much the Ferrassie type of Charentien, and would be classified as "Ferrasie", were it not of the Riss glaciation. The situation is similar to that of Rigabe (Bonifay 1959; Escalon de Fonton and de Lumley 1960) and La Chaise (Debenath 1974, 1977). The same is true in terms of the Quina type: similar finds already occur in Riss context (High Lodge (Breuil 1932 and new finds by Sieveking), Les Tares (Texier 1979)).

Characteristics of the Middle Palaeolithic

The differentiation between the early Middle Palaeolithic of the Riss glaciation and the Lower Palaeolithic can be seen in three points:

1. During the Lower Palaeolithic of Central Europe, man was present only in the temperate periods, mainly the interglacials. The first "ice-age hunters", living in an open landscape and hunting "cold" animals (reindeer, mammoth) arrive in the Middle Palaeolithic. The settlement of the "cold" loess-steppe rich in animals requires a better mastery of environment (habitation, clothing) than life in a "warm" biotop. This more developed mastery of environment (mostly indirectly represented as an improvement of knapping technique) belongs to the definition of the Middle Palaeolithic.
2. A new human type occurs with the Middle Palaeolithic. It is no longer *Homo erectus* but a human being mostly named Pre- or Ante-Neanderthal (Biache, La Chaise, Fontechevade, Lebenstedt). We should ask the anthropologists whether they could name those hominids "Early Neanderthals" instead of "Pre-Neanderthals".
3. The subdivision of our finds into different groups of industries starts with the Middle Palaeolithic. In our region these are the Lebenstedt group and the Rheindahlen-type assemblage. This indicates the development of distinct areas of tradition and communication, which characterizes the human way of life ever since.

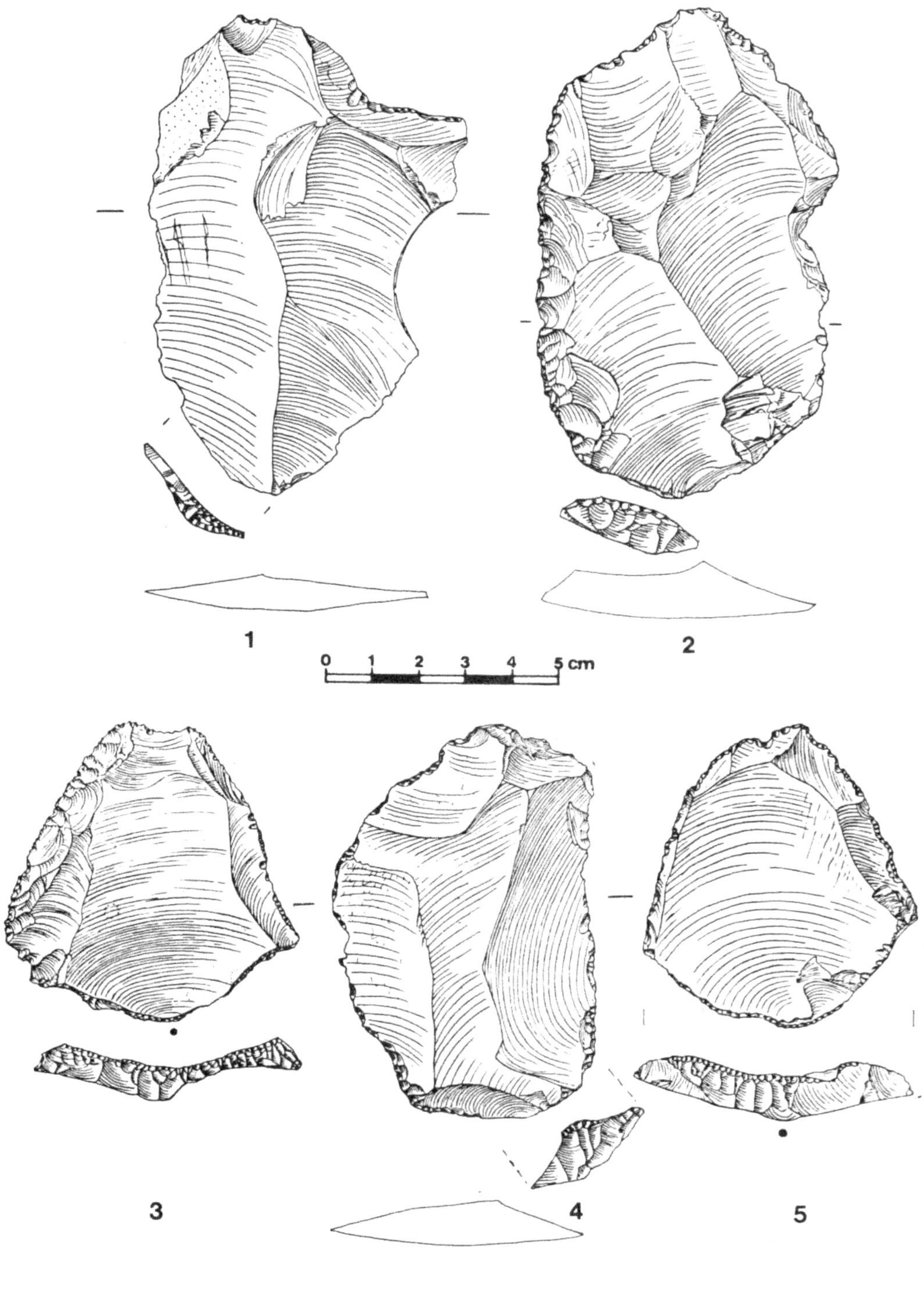

Figure 1 Salzgitter-Lebenstedt.

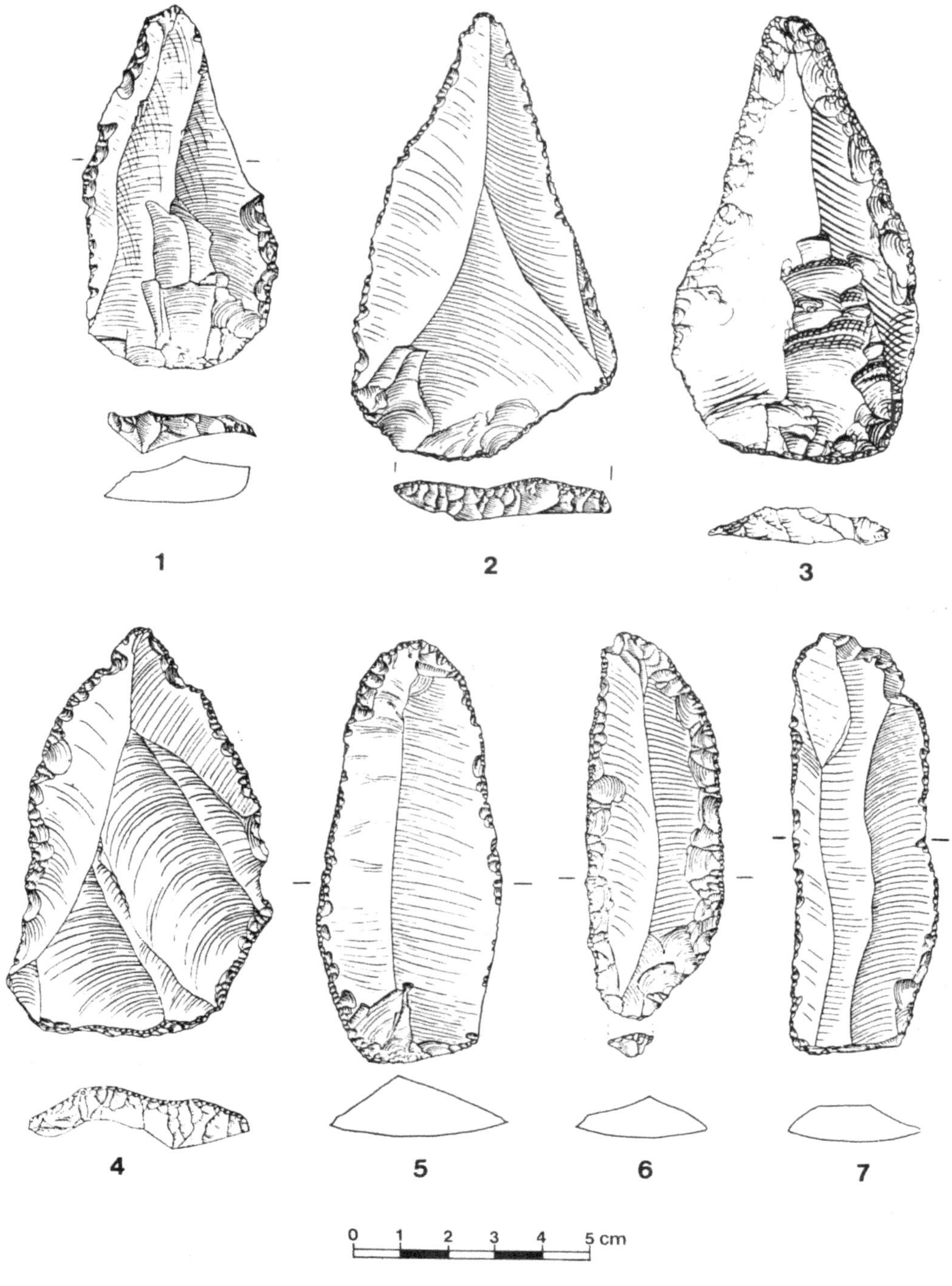

Figure 2 Salzgitter-Lebenstedt.

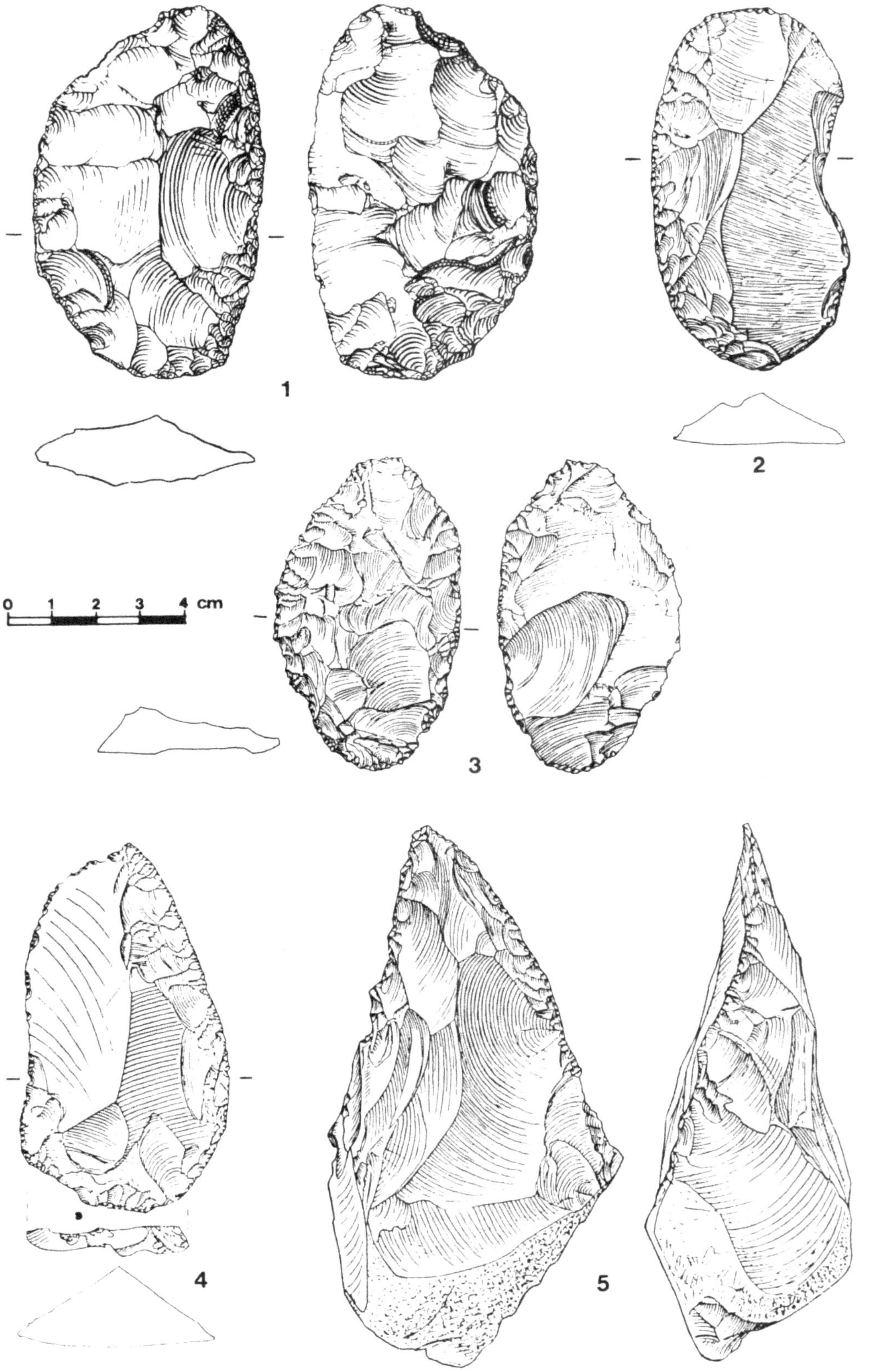

Figure 3 Salzgitter-Lebenstedt. 1, after Grote 1978.

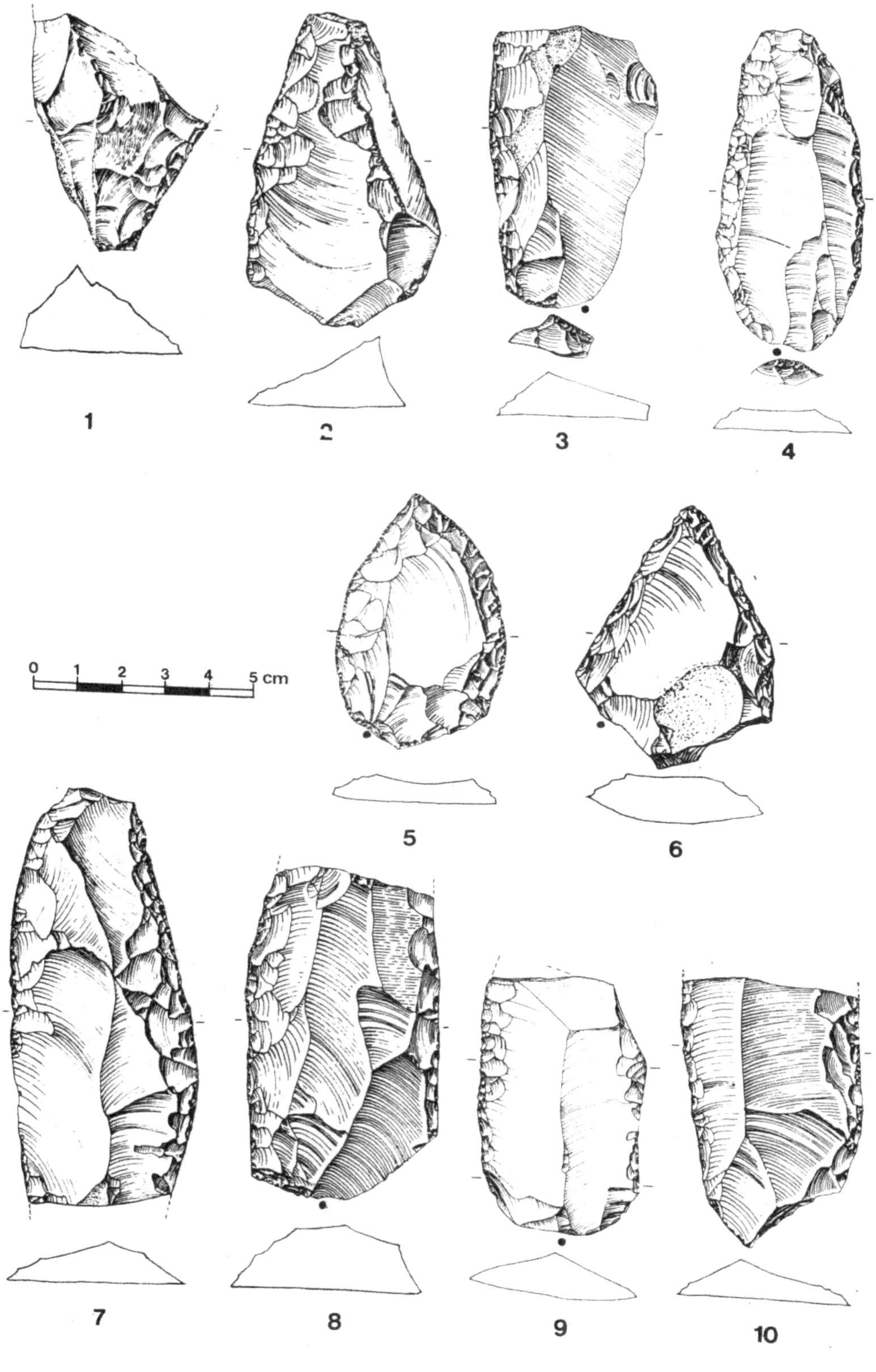

Figure 4 Rheindahlen, level B 3.

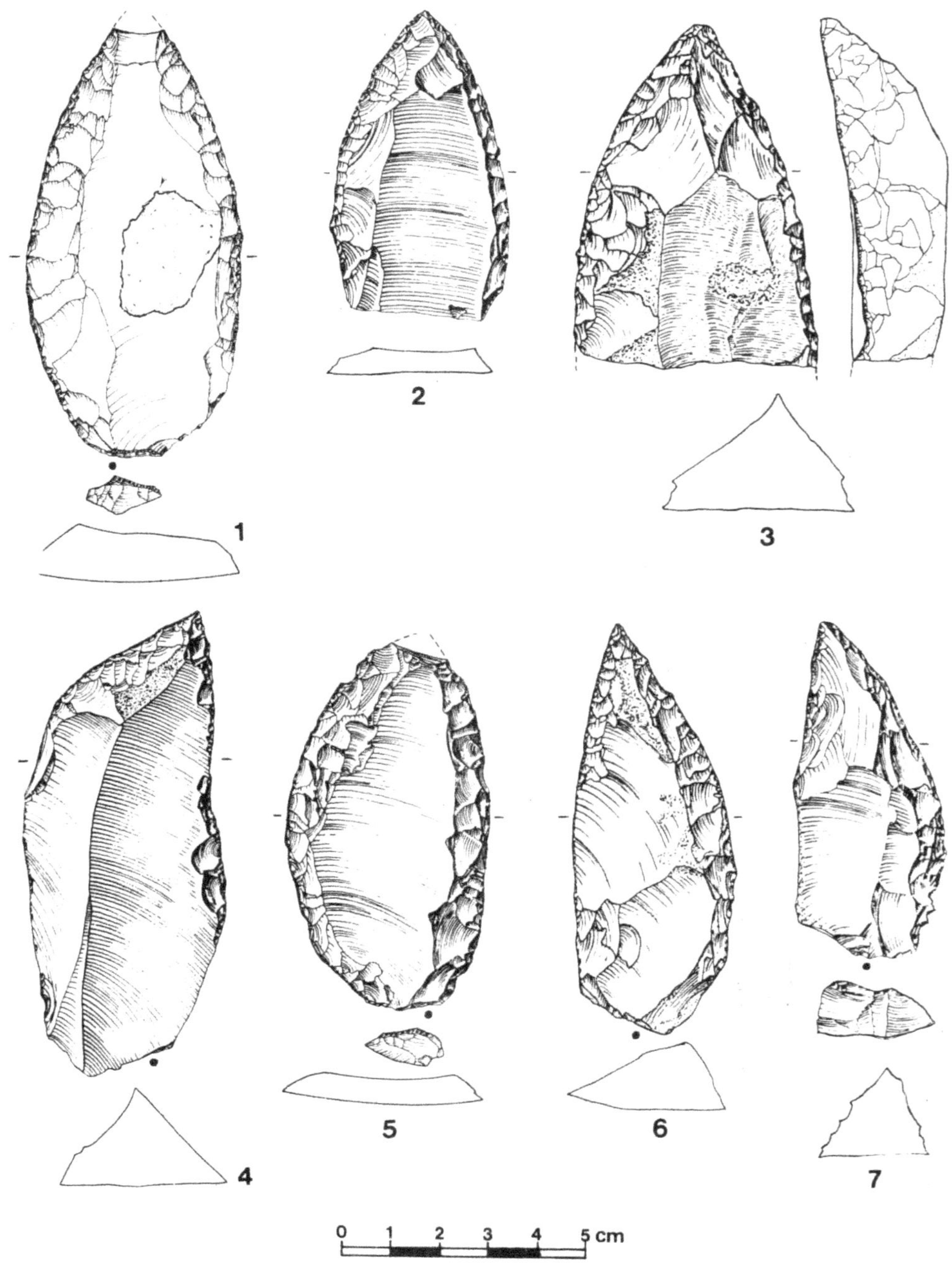

Figure 5 Rheindahlen, level B 3.

References

Bonifay, E. 1965 Etude géologique de la grotte de Rigabe, Artigues (Var) *Congr. Préhist. Monaco* 16, 1959: 265-291.

Bosinski, G. 1967 *Die mittelpaläolitischen Funde im westlichen Mitteleuropa* Fundamenta A/4, Köln and Graz.

Bosinski, G. 1976 L'Acheuléen en Europe centrale du Nord *IX Congr. UISPP, Colloque X,* Nice. pp. 52-64.

Bosinski, G. and K. Brunnacker 1973 Eine neue mittelpaläolithische Fundschicht von Rheindahlen *Arch. Korrbl.* 3: 1-6.

Breuil, H. 1932 Les industries à éclats du Paléolithique ancien. I. Le Clactonian *Préhistoire* 1: 125-190.

Debenath, A. 1974 *Recherche sur les terrains quaternaires charentais et les industries qui leur sont associées* Thèse de Doctorat, Bordeaux.

Debenath, A. 1977 The Latest Finds of Antewurmian Human Remains in Charente (France) *J. Human Evolution* 6: 297-302.

Escalon de Fonton, M. and H. de Lumley 1960 Le Paléolithique moyen de la grotte de Rigabe *Gallia préhist.* 3: 1-46.

Grahmann, R. 1955 The lower palaeolithic site of Markkleeberg and other comparable localities near Leipzig *Transact. Americ. Phil. Soc.* N.S. 45: 509-687.

Grote, K. 1978 Die Grabung 1977 in der mittelpaläolithischen Freilandstation Salzgitter-Lebenstedt *Arch. Korrbl.* 8: 155-162.

Kleinschmidt, A. 1977 Die morphogenetische Einordnung der neandertaloiden Reste von Salzgitter-Lebenstedt. Vortrag auf der 17. Tagung der Hugo-Obermaier-Gesellschaft in Bielefeld. Resümee in: *Quartär* 27/28: 163.

Luttropp, A. and G. Bosinski 1971 *Der altsteinzeitliche Fundplatz Reutersruh bei Ziegenhain in Hessen* Fundamenta A/6, Köln and Wien.

Texier, J.P. 1979 *Recherches sur les formations superficielles du bassin de l'Isle* Thèse de Doctorat, Bordeaux.

Thieme, H. 1978 Rheindahlen. *In:* Alt- und mittelsteinzeitliche Fundplätze des Rheinlandes *Kunst und Altertum am Rhein* 81: 56-69.

Tode, A. *et al.* 1953 Die Untersuchung der paläolithischen Freilandstation von Salzgitter-Lebenstedt *Eiszeitalter und Gegenwart* 3: 144-215.

Tuffreau, A. 1979 Les débuts du Paléolithique moyen dans la France septentrionale *Bull. Soc. préhist. franc.* 76: 140-142.

Discussion

E. Trinkaus: I would like to make one comment concerning the transition from *Homo erectus* to the earliest *Homo sapiens*. From what evidence we have, the transition does not appear to have been a very abrupt one. Most people consider the transition to have been a relatively gradual shift. At any rate, it is very difficult, given the nature of the hominid fossil record, to say that one specimen is entirely *Homo erectus* and that immediately afterwards we have entirely *Homo sapiens*. I think that, from a biological point of view, where one draws the actual species boundary is to a certain extent arbitrary. And so I would question the usefulness of using the evidence of the species transition as a way of corroborating or substantiating a transition in the archaeological record.

G. Bosinski: Of course it was a danger to use archaeological arguments, and for this reason I saved them for last. Bar-Yosef spoke about evolution and revolution. He named evolution "Transition". Of course, the shift from Lower to Middle Palaeolithic was not a revolution but a gradual process. Levallois technique and Middle Palaeolithic industries did not appear suddenly. Nor did Upper Palaeolithic industries. I think that the boundary between Lower and Middle Palaeolithic is of the same range and importance as the transition from Middle to Upper Palaeolithic; and there, the anthropological remarks you have just made are true, too. The distinction between Middle Palaeolithic and Upper Palaeolithic fits well our material, even though the transition between them was not a sudden one.

Do not forget that in the Near East it was always warm, and man was never forced to leave the region due to a rigorous climate. In our region, until the Holsteinian Era, we have finds only from warm periods. It is the growing capacity of mastering the environment which enabled man to inhabit the colder steppe. He could do it only when he was able to build houses and to clothe himself. The occurrence of Levallois technique surely was a very marginal thing, which proves the progress only indirectly. This is something we can not see in the Near East, where it was always warm and housing and clothing were not so necessary – like nowadays.

A. Tuffreau: I agree with Dr. Bosinski on the basis of Northern France, but climate in the Loess countries was not that harsh: Pollen analysis from good sites indicate mild climate. It seems that the principal determinant of habitation periods was to avoid the deposition of Loess.

G. Bosinski: But do not forget, mamouth and reindeer did not like leaf forest.

O. Bar-Yosef: Did I understand correctly that our colleague would like to define the transition from the Lower to the Middle Palaeolithic on the basis of climatic changes? If the situation, as we know it, that bifaces do occur in Middle Palaeolithic industries is true, then how would you differentiate between the Lower and the Middle Palaeolithic on the basis of typology?

Perhaps instead of using continuously such terms as "Lower" and "Middle Palaeolithic" we should try and build a different framework. Such a framework might be constructed from "cultures" (defined on the basis of assemblages in space and time). These will bear names of "type-localities".

G. Bosinski: Concerning the presence of bifaces in the Middle Palaeolithic: it is widely known that good handaxes occur even in the late Middle Palaeolithic, for example in the Mousterian of Acheulian Tradition. This is neither new nor wrong. Another question is whether we should use the old terms. I meant to underline that in the Lower Palaeolithic there are no different industries or "cultures", and I do not know what to call the material other than "Lower Palaeolithic". Take Choukoutien and Bilzingsleben, and I think Ubeidiya too, and there will be no significant differences between them. In the Middle Palaeolithic, on the contrary, you really get different groups. This is first of all an archaeological definition, but in our region it corresponds with a climatic change from the warm Holsteinian to the cold Vaale conditions.

O. Bar-Yosef: One additional remark: If I am not mistaken you have said that the tools imply a certain adaptation to environment. But what I understand from your last comment indicates that the tools were made by man following some concept in his brain and had nothing to do with the environment.

G. Bosinski: I did not mean to say it, and do not believe I said it.

THE TRANSITION FROM LOWER TO MIDDLE PALAEOLITHIC, WITH PARTICULAR REFERENCE TO BRITAIN

Derek A. Roe

Introduction

Before any discussion of particular British Palaeolithic industries that might count as 'transitional' between Lower and Middle Palaeolithic, it seems appropriate briefly to consider the expected nature of the transition in question, to establish just what is being sought. It is also worth asking whether there are any factors peculiar to Britain that need to be taken into account, or whether the situation should be the same as that in, say, France, Central Europe or the Middle East.

Transitional Periods

Transition stages have always had a particular fascination for Palaeolithic archaeologists. It is relatively easy to distinguish in the remote past successive bodies of archaeological evidence, remarkably consistent internally, which represent main stages of human development; what is not so clear is how or why there was a change from one of these generalized stages to the next, and this is what 'transitional industries', if such can be isolated, may help to tell us. Some might regard the ascending path of human progress through the Palaeolithic period as traversing not a simple inclined plane but some form of staircase, whose steps are somewhat unequal and irregularly spaced, in which sudden and often quite rapid transitional stages carry us from plateau to plateau. Our predecessors at various times in the development of our discipline have given the broad stages such titles as 'age', 'period', 'culture' or 'techno-complex', or have used designations like Acheulian, Mousterian or Aurignacian – which may have begun as local names but have usually come to be applied over vast distances, sometimes rather uncritically. Persistence of the broad stages over long periods of time is undeniable, however one may decide to name them. When good evidence exists in any one region, plenty of interesting local diversity and much useful detail can of course be discovered within them, without destroying their integrity. Of this the Mousterian of France is perhaps the classic example. But what were the mechanisms of change? After all, it is changes which carry the archaeological record forward, not long periods of relative stability. No doubt human physical and mental evolution played an important role on the one hand, and so too on the other did ecological change and environmental pressures caused by, *inter alia*, the fluctuations of climate and temperature during the Pleistocene. We might see the former as enabling 'voluntary' change in the sense of social, economic and technological progress, and the latter as causing 'involuntary' change by undermining the foundations of a current state of stability and demanding adaptation to meet a new situation.

Transitional industries are those made by human groups during the periods of relatively rapid progress or adaptation which link one broad stage to the next. Inevitably, archaeologists must see them in terms of the surviving material remains and any information that can be deduced from them. If we believed firmly in the typological and technological evolution of artifacts as a process with its own momentum, we might expect to find a succession of artifact types through any transitional phase that was perfectly logical to our hindsight: does not the process start with the classic tool-kit of the original stage and end when the classic tool-kit of the next one is achieved? In fact, however, a clear succession of that kind is the exception rather than the rule, and the reasons for this are not far to seek. At any one moment in time, in the past just as today, human groups in different parts of the world will be operating at very different levels of technology, some advanced and some retarded. Human groups are not normally static, certainly not those whose economy depends on hunting and gathering, and it is usually those with more advanced technological skills that are most likely to make a success of mobility. Therefore new technological ideas will sometimes appear suddenly in any human territory, by contacts from outside, rather than gradually by evolution from within. Transition is accordingly a process that may have many different forms and rates of occurrence. Certainly 'backwater' areas may exist, by-passed by the currents of human migration at important moments, and here development will be slower.

There is nothing very surprising about any of this, but it is necessary always to recall that 'Lower Palaeolithic' and 'Middle Palaeolithic', not to mention 'Upper Palaeolithic', are terms of modern manufacture: the names rather imply a general and orderly succession from one to the next but, because of the differential rates of progress referred to, no hard and fast general correlations can be made between these stages of the Palaeolithic and successive periods of time; in many areas there may be very substantial periods of overlap and co-existence between advanced groups that have passed what we regard as a technological threshold and retarded ones that have not. There can be little doubt that such a situation prevailed in Europe at the start of the Upper Pleistocene: 'archaic Mousterian' or 'pre-Mousterian' industries certainly go back to the complex of cold and warm events which is still most often known as the Riss glaciation (Bordes 1968; de Lumley 1969, 1971), but handaxe industries of the Acheulian group are also present at this time and continue into the subsequent interglacial, while other tool-making traditions may also be represented. It is clearly in this part of the Quaternary time-scale that industries transitional between Lower and Middle Palaeolithic can most profitably be sought, at least so far as Europe is concerned.

What we do not of course see clearly at present is whether different human types belong exclusively to the different archaeological entities: this is one of the ultimate questions towards which so much Quaternary research is aimed, directly or indirectly, but it is hard to solve when hominid remains themselves are still so scarce. The biological processes that created 'transitions' from one human type to another are an entirely different matter from the mechanisms that caused Palaeolithic industries to change. But the two sequences go on side by side against the same Quaternary background and they cannot be wholly separated from each other even if we should wish to disassociate them, because it was man who made the artifacts. To the question of what human types were current during the critical part of the Upper Pleistocene, Britain, alas, has very little evidence to contribute.

Lower and Middle Palaeolithic: General Considerations

If we turn now from generalities to the particular transitional period under consideration, we may note that although much attention has been devoted to *Homo sapiens* and the emergence of the sub-species *Homo sapiens sapiens* (cf. for example Bordes 1972), matched by great archaeological interest in the first appearance of blade-using industries and the spread of the earlier Upper Palaeolithic, less research has been carried out specifically on the emergence of the Middle Palaeolithic and the associated problems of hominid evolution. The present conference is rather overdue. For the Middle Palaeolithic, discussion has been more concerned with the nature of the fully-fledged industries and their variability, and the so-called 'classic Neanderthalers' seem to have proved a more popular attraction than the rather sparse traces of their immediate predecessors.

One can however extract from the literature a fair measure of agreement as to what constitutes 'typical' Lower Palaeolithic and Middle Palaeolithic technology and artifact typology, in a general sense, and it is worth listing some of the main points since the differences are important if we are to consider what may lie between, though space does not permit a detailed analysis.

Lower Palaeolithic	**Middle Palaeolithic**
Prepared core techniques are rare or may be absent. Many flakes for direct use or for retouch were obtained in the early stages of biface manufacture, though simple types of core do occur in some industries as waste products of flake production. The blanks for flake tools are accordingly of very variable shapes.	Prepared core techniques are always important and usually dominant; Levalloisian cores, disc cores, flake-blade cores and even simple prismatic cores are all found, the first two kinds abundantly. Such cores yield predictable and regular flake blanks, whose shapes do not vary greatly except by the maker's intention.
In many industries, notably the widespread Acheulian series, shaped bifacial tools, particularly handaxes and sometimes cleavers, are common: at some sites they occur in thousands. Although individual industries may specialize in particular handaxe types, the overall variety of shapes is considerable. In some Lower Palaeolithic industries, however, for example the Clactonian	Except in a very few instances, handaxes are rare and cleavers rarer still in Middle Palaeolithic industries. In many variants they are wholly absent and they can be called common only in the West European Mousterian of Acheulian Tradition (M.T.A.). Even here, the overall variety of types is not large.

of Britain, handaxes are absent. Such occurrences are not necessarily pre-Acheulian.	
Since retouched flake tools are often made rather informally on flakes selected as suitable rather than specially struck for the purpose, there are few standardized types or closely repeated shapes in many industries. In this, Lower Palaeolithic flake tools contrast notably with Lower Palaeolithic handaxes, where there are clear morphological preferences carefully repeated. The range of types of retouch on the flake tools is also relatively narrow, and the retouch itself is usually robust and purposeful rather than elegant in appearance.	Most of the numerous flake tools are precisely designed and carefully executed, so that numerous clear types exist (cf. Bordes 1961a, etc.) and are accurately repeated. There are many different types of retouch, usually accurately and elegantly applied. The flake tools are normally made on specially struck blanks, to whose size and initial shape careful thought has been given.
At individual sites, the range of different artifact types is often small.	At individual sites, the range of different artifact types is often large, though some industries may be very specialized (e.g. certain 'Quina' or 'Denticulate' Mousterian industries in S.W. France).

The foregoing general comparison should offer us some clues regarding at least the technology of industries transitional between Lower and Middle Palaeolithic. We are looking for tool-kits of appropriate date in which: handaxes, if present at all, are losing ground to flake tools; prepared core techniques have made an important impact; flake tools are becoming more varied and more formalized; and retouch is of several different kinds, carried out with a new degree of accuracy. It should be noted that this is a somewhat hypothetical list of criteria for the transitional industries and does not override the earlier comments relating to the nature of transitions and their different rates of occurrence.

No one would wish to see the transitional period between Lower and Middle Palaeolithic wholly in terms of the typology and technology of stone artifacts. This is not the place to make a proper comparison of Lower and Middle Palaeolithic settlement patterns or social and economic organization, or even the respective geographical distributions of Acheulian and Mousterian in their broadest senses, but many major differences can readily be found. As a few diverse examples one might simply mention the relative rarity of Lower Palaeolithic cave sites and their abundance in the Middle Palaeolithic; the greater variety of settlement locations in the Middle Palaeolithic and the clearer specialization of the industries on many occasions, for whatever reason; the presence of burials in the Middle Palaeolithic and their absence in the Lower Palaeolithic; or the Middle Palaeolithic penetration of large areas – for example within the U.S.S.R. – that were previously unoccupied. These are all very important factors, with major social as well as technological implications, but we cannot discuss them here. As regards the use of raw materials in artifact manufacture, the evidence is not complete enough for a proper comparison. In the case of stone, Lower Palaeolithic man certainly successfully used many different rock types and might on occasion travel substantial distances to outcrops or sources of a favored rock, as in the Olduvai area (Leakey 1971; Hay 1976). One might however tentatively conclude that Middle Palaeolithic man was more skilled in selecting particular rock types for particular elements of his tool-kit and adapting his knapping techniques to the nature and abundance or scarcity of local resources.

The Role of Britain in the Palaeolithic Period

It remains in this preliminary section to consider whether there are any special factors that might make the situation in Britain different from that prevailing elsewhere at the time when the Lower Palaeolithic was ending and the Middle Palaeolithic beginning. The present writer, in a forthcoming book on the British Lower and Middle Palaeolithic (Roe 1981), has stressed that Britain lies at the northwest extreme of Old World Palaeolithic settlement and is not an area that is likely to have been permanently occupied during the Middle and Upper Pleistocene. Though Britain now consists of a group of islands, the major part of its land area would have been at least intermittently connected to Continental Europe across the eastern part of the English Channel and the southern part of the North Sea. Access to southern Britain across this land bridge would accordingly have been available to people coming either from the south via France or from the east via the Low Countries – two rather different source areas, archaeologically speaking. The history of the land bridge is not known in any detail, unfortunately; nor is the question of access to Britain necessarily as simple as it might appear. The land bridge was presumably at its widest during periods of lowered sea level, which we might expect to coincide with glacial maxima, yet on such occasions large parts of Britain were fully glaciated, while evidence for periglacial conditions can be found right down to the present south coast: hardly tempting circumstances for permanent Palaeolithic settlement, though summer hunting expeditions might sometimes have been possible and worthwhile. Conversely, if the land bridge were submerged on occasion during the Pleistocene, putting Britain out of reach to would-be settlers, this could only have happened during interglacial marine transgressions – i.e. at periods when conditions within Britain might otherwise have been ideal for occupation.

Of course, if this line of thought were followed too rigidly, we might even come to wonder how there was ever any Palaeolithic settlement in Britain at all, but fortunately there is no shortage of evidence for its abundant existence, some sites being clearly associated with interglacial faunal and floral evidence and others with signs of much cooler conditions. But the points must remain, first that continuous access and occupation are unlikely and, secondly, that more than one natural route led to Britain from more than one source area, whenever a land connection was available. In these circumstances, we surely cannot necessarily expect to see in Britain a unified and continuous Palaeolithic sequence in which transitional stages are bound to occur locally.

Lower and Middle Palaeolithic in Britain

If we now turn to the British industries themselves, we have first to note two unfortunate difficulties concerning the evidence. First, many of the relevant finds are old, and were made by intermittent collection rather than controlled excavation; stratigraphic information is often sparse, and of poor quality, though a few good sites exist. Secondly, some of the most important British sites from our present point of view are currently being restudied by various workers and the results of their research are not yet available. Some of the suggestions made in the present paper may accordingly need to be reconsidered in a year or two's time, when the new publications are to hand, but it is to be hoped that the writer's information is up to date enough meanwhile to make a viable interim report. We may begin by considering briefly the typical Lower and Middle Palaeolithic industries of Britain and pass to the atypical occurrences which might have some kind of transitional significance.

Acheulian and Clactonian

The British Lower Palaeolithic (Roe 1976a, 1976b, 1981; Wymer 1968, 1974) contains plenty of classic Acheulian material, amongst which earlier and later industries can be recognized; the overall time range apparently extends from an inter-Anglian mild phase, or possibly even from the late Cromerian, until a post-Wolstonian warm period, at present referred to as Ipswichian (for the local Quaternary stage names see Mitchell *et al.* 1973; West 1977). There are also Clactonian industries, lacking traces of handaxe manufacture, which apparently date from the late Anglian and the Hoxnian. In pure form, the Clactonian, with its chopper cores, unretouched flakes and simply retouched flake tools, is rather scarce: the occurrences have been summarized by Wymer (1968, 1974 and in Singer *et al.* 1973) and by Roe (1981), and most recent workers have regarded them as separate from the Acheulian (cf. however Ohel 1979). There are of course numerous find-spots where typical Acheulian bifaces and possibly Clactonian artifacts occur mixed in the same gravel, but on the other hand there are no undisturbed stratified occurrences where the two traditions are clearly blended. In Britain, as elsewhere, handaxes are the traditional hallmark of the Acheulian industries, but where undisturbed sites exist retouched flake tools can be seen to be of common occurrence. Prepared core techniques are absent from the pure Clactonian occurrences, and also from the earlier Acheulian industries. However, in the form of Levalloisian cores and flakes, prepared core technology is associated with some of the later Acheulian industries. In these it usually plays only a minor role, except perhaps in the case of Caddington, Bedfordshire, where Levalloisian artifacts are numerous, though C.G. Sampson (1978: 149) takes the view that they belong to an occurrence wholly separate from that which produced the main body of Acheulian material. Apart from its occasional appearance in typical Acheulian industries, however, Levalloisian technique has a somewhat special role to play in Britain (see below).

Mousterian

Given the extraordinary richness and variety of the Mousterian of France, it might seem surprising that the British Mousterian is sparse and apparently confined to a single facies (Mousterian or Acheulian Tradition), were it not for the palaeogeographical factors already mentioned. It is true that the greatest abundance and diversity of Mousterian industries in France lies in the south and southwest (Bordes 1961b; de Lumley 1969, 1971), but several different variants are represented in the north (regional surveys in de Lumley 1976, vol. I/ii) and the absence of all but one of them in Britain can hardly be accidental. Presumably occupation was only viable in Britain for brief periods and these coincided with the currency of M.T.A. in northern France.

The British Mousterian (Collins and Collins 1970; Mellars 1974; Shackley 1977; Roe 1981) includes a few small-scale occupations at cave sites, of which examples are Kent's Cavern (Devonshire), Coygan Cave (Dyfed, Wales), Wookey Hole Hyaena Den (Somerset) and Robin Hood's Cave (Creswell Crags, Derbyshire). There are also a few substantial open sites, such as Pan Farm, Shide (Isle of Wight), Bramford Road, Ipswich (Suffolk), Little Paxton (Huntingdonshire) and Oldbury (Kent), the latter possibly associated with one or more sandstone rock-shelters since destroyed by erosion. Apart from these, there is a considerable scatter of isolated finds in southern Britain of the typical sub-triangular and *bout coupé* handaxes (Figure 1: 1,2), often occurring as single artifacts but occasionally loosely associated with Levalloisian flakes or cores.

The impoverished nature of the cave occupations is not too surprising when we consider the location of the sites, all at the extreme margins of the settled area of Britain, relatively remote from good sources of flint and in a terrain which must soon have become hostile in cold conditions.

Little clear dating evidence is available for the British Mousterian. There is some slight suggestion from the nature of some containing deposits or accompanying faunal remains that a few of the typical *bout coupé* handaxes may have been made before the end of the Last Interglacial; other of the Mousterian occurrences, on similar grounds, clearly belong within the Devensian (Last Glacial) complex, doubtless to one of the interstadials, of which the full number is not yet clear. These matters are discussed further by the present writer elsewhere (1981).

The foregoing summary of the British Mousterian holds good for every site but one: La Cotte de Saint-Brelade in Jersey, Channel Islands, which is a very prolific site with many occupation levels (McBurney and Callow 1971). The French name, however, is a pointer to the true situation, namely that this site's geographical and archaeological affinities lie clearly with northwest France and not with Britain at all. The material is referred to again below, since it is of direct interest to us here, but the truth would seem to be that only accidents of relatively modern history have made a British site out of a locality which, to the initial Middle Palaeolithic population, was doubtless a rocky outcrop on a coastal plain reached directly from what is now the French mainland.

Levalloisian

'Levalloisian' is properly a name given to a prehistoric knapping technique, a method of obtaining one flake (or at most a few flakes) of predetermined size and shape from a given nodule or large piece of flint or other suitable rock. Unlike the disc core technique, it is very wasteful of raw material and can only be used abundantly when good quality rock is easily available. This is the case in parts of Britain, as in France, and at such localities 'Levalloisian industries' of a specialized kind may occur, i.e. industries in which the production of Levalloisian flakes seems almost the sole concern of the knappers. Baker's Hole, Northfleet, Kent, is the most prolific amongst such sites in Britain: it is of Wolstonian age, and the Levalloisian flakes are typically large, heavy and boldly shaped, usually by radial preparatory flaking, with thick faceted platforms (Figure 1: 3; Smith, R.A. 1911; Wymer 1968: 354-8). A much smaller and less well known, but not dissimilar, site was discovered at Bapchild in north Kent (the Coombe Rock industry of Dines, 1929). There are other essentially Levalloisian sites where flake-blades were manufactured, though not exclusively: these have blade-like proportions, longitudinal dorsal preparatory scars and neat faceted platforms (Figure 1 : 4). Crayford (Kent) and Creffield Road, Acton (Greater London) are among the most striking occurrences of this kind, the former probably of Ipswichian (Last Interglacial) age and the latter undated but not likely to be older than the Ipswichian; for these and similar sites, see Roe (1981), quoting earlier literature.

There are possible associations of a few handaxes with some of the British 'Levalloisian' industries – not, unfortunately, very securely established because the main discoveries took place many decades ago. In the case of Baker's Hole, the handaxes are typically Acheulian, while at Crayford and Creffield Road they are within the range of Mousterian of Acheulian Tradition handaxe morphology, though not necessarily right outside the later Acheulian range in all cases.

It is worth noting that the dating of these 'Levalloisian' sites, where established, falls within that part of the Upper Pleistocene during which we might expect Lower to Middle Palaeolithic transitions to occur. The industries certainly show the use of prepared core techniques on a scale unprecedented during older periods in Britain, though the presence or proximity of abundant supplies of good raw material may be an enabling factor. When we compare these industries with the typological and technological range anticipated for the transitional phase on purely theoretical grounds (see above) we can certainly note that handaxes are rare, if present at all, but so are retouched flake tools. It has to be admitted that these 'Levalloisian' occurrences are not obviously transitional between any one British Lower Palaeolithic stage and any specific Middle Palaeolithic variant, and their actual significance is open to discussion. Baker's Hole was certainly a manufacturing site and the present writer (1981: 275-6) has speculated that it may be a highly specialized activity variant within the Acheulian. At Crayford, the late 19th century finds included animal bones which seem to have been closely associated with the flint artifacts, some of which were conjoinable, in a primary context, and it seems possible that in the 'flake-blade Levalloisian' industries we may be seeing traces of the tool-kit used by Mousterian of Acheulian Tradition groups specially for hunting and butchering activities (Roe 1981: 276).

British Industries of Possibly 'Transitional' Status

There are no long sequences in Britain where we can see a gradual introduction of Middle Palaeolithic technology, as has sometimes been claimed elsewhere, for example at Orgnac III in the Ardèche, southeast France (Combier 1967: 122-7). A few individual

Figure 1 Some artifact types referred to in the text. 1, Sub-triangular handaxe of Mousterian type from Whitehill, Thetford, as illustrated by Evans in 1872; 2, Mousterian handaxe of *bout coupé* type from Castle Lane, Moordown, Bournemouth, after Calkin and Green, 1949; 3, Typical large Levalloisian flake from Baker's Hole, Northfleet, after Wymer, 1968; 4, Levalloisian flake-blade from Creffield Road, Acton, perhaps classifiable as a *pointe Levallois allongée* of Bordes (1961a: 18), after Smith, 1931; 5, 6, Flake tools from High Lodge, Mildenhall, after Marr, Moir and Smith, 1921. (Approx. half size).

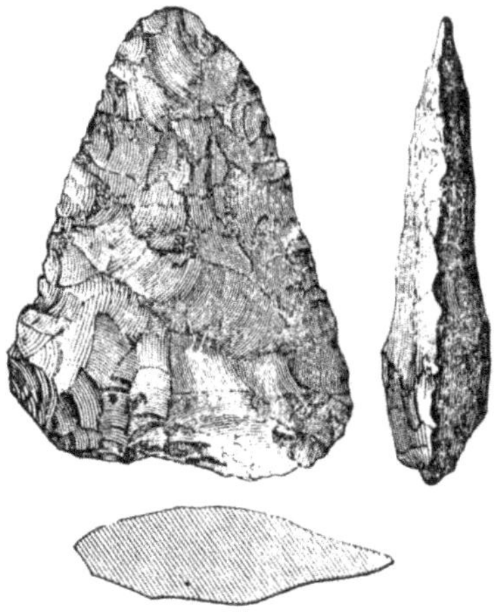

1

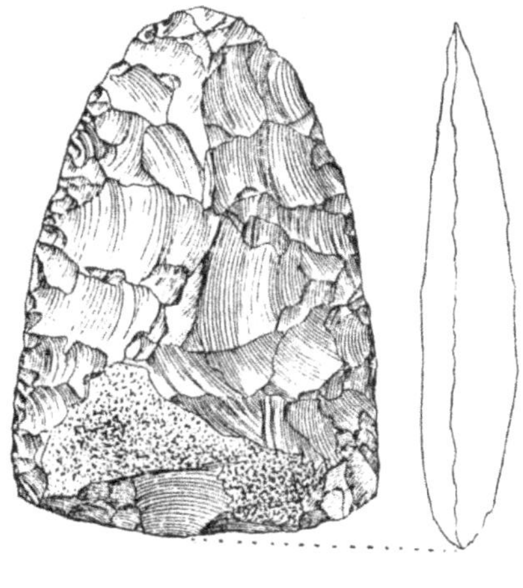

2

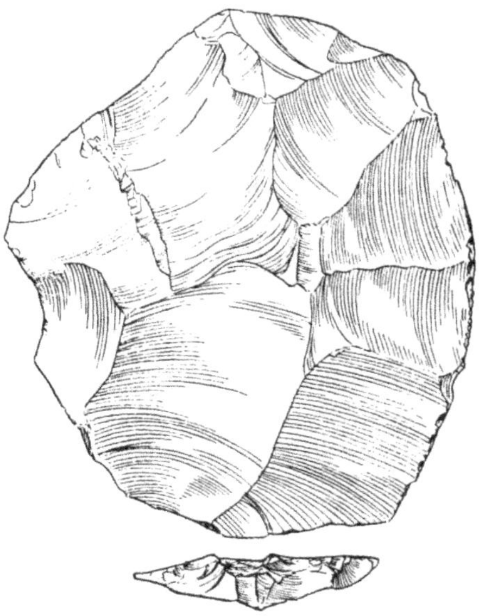

3

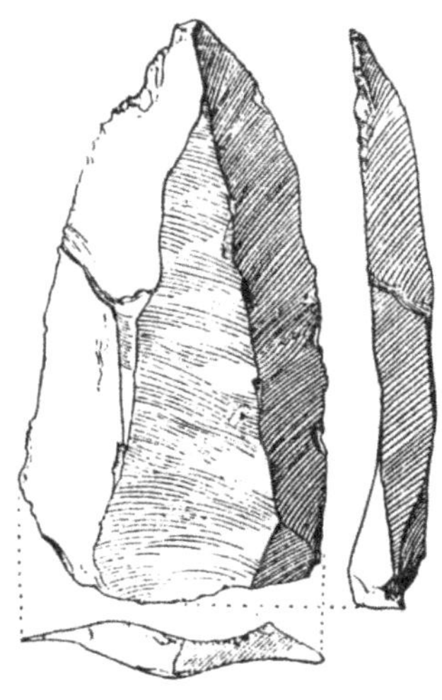

4

5

6

industries may however be mentioned in which technology and perhaps also typology seem to have passed beyond an 'ordinary' Lower Palaeolithic level in one way or another. At Hoxne (Suffolk) in the Upper Industry, which seems likely to date within the Wolstonian complex of colder and less cold phases, we see a handaxe industry with abundant finely made flake tools, including many carefully shaped convex side-scrapers, which surpass the flake tools of British Middle Acheulan industries in both quantity and quality and substantially outnumber the handaxes themselves. There is no report of prepared core flaking in this industry, however (Wymer 1974; Singer and Wymer 1976; monograph in preparation by these authors and their colleagues). To judge from old accounts, a similar site may have existed at Stoke Newington, London (Smith, W.G. 1894): many finely made flake tools accompanied typical Acheulian handaxes and a few cleavers, but there was little or no Levalloisian or disc-core technique. For a review of the likely age of this site and a new assessment of the old evidence, we must await the completion of work in hand and a monograph now being prepared by Dr. J.B. Campbell and Mrs. G.M. Cook. For the present, it might be remarked that these two sites perhaps remind one in general terms of such French industries as l'atelier Commont (Bordes and Fitte 1953), but only the definitive publications will reveal whether such a comparison is justified.

High Lodge Mildenhall (Suffolk) may well prove to be the most important 'transitional' industry located fairly and squarely in Britain. This site has been known and discussed since the late 19th Century, and its problems remain unsolved pending the publication shortly by G. de G. Sieveking and his colleagues of the results of their major excavation campaigns there in the 1960's and early 1970's. From old reports, and from a brief note about the new excavations published in *Nature* in 1969, we may gather that lake sediments representing a mild period within the Wolstonian contain two flake industries, of which the younger is the so-called 'classic flake industry of High Lodge', formerly known as 'Clactonian III'. The lake deposits are overlain by a gravel stratum in which Acheulian ovate handaxes occur. The present writer examined the old High Lodge collection some years ago, and can confirm the opinion of various authors that the typology of many of the flake tools, and the high quality of the retouch used to shape them, closely resemble fine Mousterian work (Figure 1: 5,6); the convex side-scrapers and certain limace-like pieces seem to anticipate industries of Last Glacial age within the Charentian tradition of the French Mousterian. However, subject to what publication of the new results may reveal, it seems that prepared core techniques were not used, the flake implements being made on plain-platform flakes. Nor is the present writer aware that any handaxes or by-products of their manufacture are associated with the flake tools. This industry certainly seems to meet some of the criteria suggested above for incipient Middle Palaeolithic technology, and some authors (e.g. de Lumley 1971: 355) have made general comparisons between it and so-called Tayacian occurrences in France and elsewhere – La Micoque levels 3 and 4, Baume Bonne, and so forth. In due course we shall doubtless hear how appropriate such comparisons may be. To what tradition in Britain (if any) the High Lodge industry belongs is not clear: it may of course be a specialized tool-kit used for some particular activity. It does not mark an obvious stage of some technological transition that took place wholly within Britain, and there is no clear link with the Clactonian industries of Hoxnian age.

Reference has already been made to the prolific site of La Cotte de Saint-Brelade in Jersey. Here again we must await the publication of a monograph that is in preparation, and sadly this conference will not have the benefit of the presence of Professor Charles McBurney, whose tragically early death is such a loss to British and Old World Palaeolithic studies: he of course knew the site and the material better than anyone, after his many seasons of excavation in Jersey. The essential point for our present purposes is that the earlier deposits at La Cotte, older than a Last Interglacial marine transgression, contain some material which is either of Middle Palaeolithic typological and technological status, or else right on the verge of it. There is some Late Acheulian with handaxes and cleavers, but in other levels flake tools are very numerous and various kinds of prepared core techniques are used, with some 'well made discoid cores' present (McBurney and Callow 1971: 197-203) The age of this industry is described as 'Late Rissian'. It may well be, when a full classification and assessment is available, that this site, out of all those mentioned in the present paper, will best satisfy the criteria for transitional status suggested above, and it is likely to prove of considerable importance to Lower and Middle Palaeolithic studies in Western Europe. The present writer remains unconvinced, however, that La Cotte can usefully be described as a British site. There is no particular reason to suppose that the Palaeolithic people who inhabited Jersey during the Late Riss ever set foot in what is now England, though consideration of the raw materials which they used for stone implement manufacture establishes their clear connections with adjacent areas of Continental Europe.

Lastly, it is worth mentioning briefly the distinctive handaxe industry from Wolvercote, Oxfordshire, apparently dating from an interglacial phase younger than the Wolstonian glaciation (details and discussion in Roe 1976, 1981: 118-128). The handaxe element here does not suggest any reduction in importance of that class of tool, though a few well made and elegantly shaped flake tools are also preserved in the old collections. There is no trace of prepared core techniques. The aspects of the artifacts most interesting to us here are the deliberate 'plano-convex' working of the handaxes and their pointed shapes, features which can both be paralleled in industries of the Micoquian series of Central Europe and perhaps in occurrences a little further east (Bosinski 1967; Gábori 1976). In Europe, such industries seem to lie on the threshold between Lower and Middle Palaeolithic and to have some kind of connection with variants of the Eastern Mousterian and with early leaf-point industries, though this is not the place and the present writer is not the person to discuss the Central European evidence in any detail. It is sufficient to observe that the Wolvercote site may represent a visit to Britain by a group of people whose industries in their 'home' area may have a certain transitional significance. If they did indeed come to Britain from Central rather than Atlantic Europe, this illustrates the point previously made about the differing source areas contributing to the British Palaeolithic sequence.

This section, then, has mentioned some of the main sites we should consider when searching for Lower/Middle Palaeolithic transitional industries in Britain. There are other possible candidates. A few sites exist, for example, where it is possible to argue the presence of artifacts in the Clactonian manner but of more regular shape and more evenly applied retouch, though the flake tools never approach the standards of High Lodge or Stoke Newington. Grovelands, at Reading (Berkshire) and Southacre (Norfolk) are examples, but these are mixed assemblages from derived contexts, in which handaxes also occur (and Levalloisian artifacts at Southacre) with no clear evidence as to how much of the material belongs together; the dating is also uncertain, and one cannot really regard such material as useful evidence, however suggestive it might seem.

Conclusion

There are certainly a few industries in Britain which deserve consideration in any discussion in technological or typological terms of the transition from the Lower to the Middle Palaeolithic. The most important names, for one reason or another as indicated above, are Hoxne (Upper Industry), Stoke Newington, High Lodge, La Cotte de Saint-Brelade and perhaps Wolvercote. We can also see in Britain industries in which one or another version of Levalloisian technique may play a very dominant role, whatever the general status of the whole occurrence: Baker's Hole, Bapchild, Crayford and Creffield Road, Acton, were quoted as examples and others exist. While these are not necessarily themselves 'transitional' industries, they may well be relevant to the wide adoption of various prepared core techniques in the Middle Palaeolithic.

We have seen, however, that there is no clear evidence that in Britain *in situ* technological transition took place from any British Lower Palaeolithic industry to any British Middle Palaeolithic industry: no such process can be seen at a single multi-level site or even at a series of related single-level sites within one particular region. To those who have considered the geographical position of Britain and its implications, including the inevitably intermittent nature of British Palaeolithic settlement, this negative outcome to our search need occasion no surprise. The writer envisages that various Lower Palaeolithic groups visited Britain from time to time during the later Middle Pleistocene and earlier Upper Pleistocene, and so did a smaller number of Middle Palaeolithic groups during a shorter period within the Upper Pleistocene (mainly during the first half of the last glacial). Any processes of transition that turned the first of those broad phases into the second had their origins and development within Continental Europe, so far as Britain is concerned. However, it may be that some groups of people, who in their European homeland had reached what we now see as a transitional stage, also made brief visits to Britain and left artifacts for us to find. But that is not at all the same thing as the occurrence of a transitional phase within Britain itself.

References

Bordes, F. 1961a *Typologie du Paléolithique Ancien et Moyen* Bordeaux: Delmas (Publ. de l'Inst. de Préh. de l'Univ. de Bordeaux, Mém. 1).

Bordes, F. 1961b Mousterian Cultures in France *Science* 134: 803-10.

Bordes, F. 1968 *The Old Stone Age* London: Weidenfeld and Nicolson.

Bordes, F. 1972 (ed.) *The Origin of Homo sapiens* Paris: Unesco.

Bordes, F. and P. Fitte 1953 L'atelier Commont *L'Anthropologie* 57: 1-45.

Bosinski, G. 1967 Die mittelpaläolithischen Funde im westlichen Mitteleuropa *Fundamenta* Reihe A, Band 4, H. Schwabedissen (ed.). Köln: Böhlau.

Calkin, J.B. and J.F.N. Green 1949 Palaeoliths and terraces near Bournemouth *Proc. Prehist. Soc.* 15: 21-37.

Collins, D.M. and A. Collins 1970 Excavations at Oldbury in Kent: cultural evidence for Last Glacial occupation in Britain *Bulletin of the Institute of Archaeology* (University of London) 8-9: 151-76.

Combier, J. 1967 *Le paléolithique de l'Ardèche dans son cadre paléoclimatique* Bordeaux: Delmas (Publications de l'Inst. de Préhist. de l'Univ. de Bordeaux, Mém. 4)

Dines, H.G. 1929 The flint industries of Bapchild *Proc. Prehist. Soc. of East Anglia* 6:1: 12-26.

Evans, (Sir) J. 1872 *The Ancient Stone Implements, Weapons and Ornaments of Great Britain* First Edition, London: Longmans, Green and Co.

Gábori, M. 1976 *Les civilisations du Paléolithique moyen entre les Alpes et l'Oural* Budapest: Akadémiai Kiadó.

Hay, R.L. 1976 *Geology of the Olduvai Gorge: a study of sedimentation in a semi-arid basin* Berkeley, Los Angeles and London: University of California Press.

Leakey, M.D. 1971 *Olduvai Gorge, vol. 3: Excavations in Beds I and II, 1960-63* Cambridge: Cambridge University Press.

De Lumley, H. 1969 *Le paléolithique inférieur et moyen du Midi méditerranéen dans son cadre géologique. Tome I: Ligurie-Provence* Paris: C.N.R.S.

De Lumley, H. 1971 *Le paléolithique inférieur et moyen du Midi méditerranéen dans son cadre géologique. Tome II: Bas-Languedoc - Roussillon - Catalogne* Paris: C N.R.S.

De Lumley, H. 1976 (ed.) *La Préhistoire française* 2 vols. Paris: C.N.R.S.

Marr, J.E., J.R. Moir and R.A. Smith 1921 Excavations at High Lodge, Mildenhall, in 1920 A.D. *Proc. Prehist. Soc. of East Anglia* 3: 353-379.

McBurney, C.B.M. and P. Callow 1971 The Cambridge excavations at La Cotte de Saint-Brelade, Jersey – a preliminary report *Proc. Prehist. Soc.* 37:2: 167-207.

Mellars, P.A. 1974 The Palaeolithic and Mesolithic. *In: British Prehistory, a new outline* A.C. Renfrew (ed.) London, Duckworth. pp. 41-99, 268-79.

Mitchell, G.F., L.F. Penny, F.W. Shotton and R.G. West 1973 *A correlation of Quaternary deposits in the British Isles* Geol. Soc. Lond., Special Report no. 4.

Ohel, M.Y. 1979 The Clactonian: an Independent Complex or an Integral Part of the Acheulian? *Current Anthropology* 20:4: 685-726.

Roe, D.A. 1976a The earliest industries in Britain. *In: Les premières industries de l Europe* K. Valoch (ed.) IX U.I S.P.P. Congrès, Nice, France, Colloque VIII, *prétirage.* pp. 76-95.

Roe, D A. 1976b The evolution of the Acheulian in Britain. *In: L'évolution de l'Acheuléen en Europe* J. Combier (ed.) IX U.I.S.P.P. Congrès, Nice, France, Colloque X, *prétirage.* pp. 31-46.

Roe, D.A. 1981 *The Lower and Middle Palaeolithic periods in Britain* London: Routledge & Kegan Paul.

Sampson, C.G. 1978 (ed.) *Paleoecology and Archaeology of an Acheulian site at Caddington, England* Dallas: Southern Methodist University (Dept. of Anthropology)

Shackley, M.L. 1977 The *bout coupé* handaxe as a typological marker for the British Mousterian industries. *In: Stone Tools as Cultural Markers: change, evolution and complexity* R.V.S Wright (ed.). pp. 332-39.

Singer, R. and J.J. Wymer 1976 The sequence of Acheulian industries at Hoxne, Suffolk. *In: L'évolution de l'Acheuléen en Europe* J. Combier (ed.) IX U.I.S.P.P. Congrès, Nice, France, Colloque X, *prétirage.* pp. 14-30.

Singer, R., J.J. Wymer, B.G. Gladfelter and R.G. Wolff 1973 Excavation of the Clactonian Industry at the Golf Course, Clacton-on-Sea, Essex *Proc. Prehist. Soc.* 39: 6-74.

Smith, R.A. 1911 A palaeolithic industry at Northfleet, Kent *Archaeologia* 62: 515-32.

Smith, R.A. 1931 *The Sturge Collection, an illustrated selection of flints from Britain bequeathed by William Allen Sturge, M.V.O., M.D., F.R.C.P.* London: British Museum.

Smith, W G. 1894 *Man the Primeval Savage: his haunts and relics from the hill-tops of Bedfordshire to Blackwall* London: Stanford.

West, R.G. 1977 *Pleistocene Geology and Biology, with especial reference to the British Isles* (second edition, revised) London: Longmans.

Wymer, J.J. 1968 *Lower Palaeolithic Archaeology in Britain as represented by the Thames Valley* London: John Baker.

Wymer, J.J. 1974 Clactonian and Acheulian industries in Britain: their chronology and significance *Proc. Geol. Ass. Lond.* 85: 391-421.

Discussion

A. Ronen: Could you please clarify the relation between the industry and the 8 m. seashore at La Cotte Saint-Brelade: Did the cutting by the sea postdate the industries?

D.A. Roe: Yes.

A. Ronen: Was the industry embedded in the beach deposits, or on top of them?

D.A. Roe: The cave stands some way above the modern sea. The Upper Pleistocene sea rose enough to truncate the deposits that come down from the cave, thereby creating a marine cliff. At the bottom of that cliff lies the 8 meter beach, so it perhaps marks a regression stage from the full height of the sea that cut that cliff. In any case, the cave deposits and their archaeological contents are older than that marine transgression. There are other reasons for thinking that it is a "Rissian" site, for example, the supplies of raw material: there are no sources of flint on the Island of Jersey, and those that were available were on the sea bed that was exposed only during the glaciation. Jersey was then a rocky outcrop on a coastal plain. It must have been reached and occupied within the glacial age.

G. Bosinski: I think that we have Markkleeberg as a parallel to Baker's Hole. This also seems to be a workshop – otherwise it is difficult to explain the hundreds and hundreds of cores and flakes found there. Among the tons of material in Markkleeberg there are a few handaxes, and the same is true for the open-air site Reutersruh near Kassel.

G. Freund: You compare the last series by their Micoquian bifaces and you talk of flatness comparable to clumsiness. At Markkleeberg this flatness is due to the raw material, which is extremely thin plaques of flint.

D.A. Roe: Yes. That is certainly so – it is a tabular raw material, is it not? But another comparison that I made was in the number of flake scars on the face that has fewest of them. In the case of the British example, the Wolvercote industry, this is sometimes demonstrably a main flake surface. In other words, those handaxes in some cases were made from a large flake, but not in all cases. And even so, when you look at their sections, it is not only that each implement is flat on one face, but it is also that the other face has been made domed, deliberately, apparently – this can be done when the implement is made from a pebble that starts life with equal curvature, or it can be done on one made from a tabular piece.

In the case of Wolvercote, anyhow, the handaxes were deliberately made to have a plano-convex section, and this was associated with preference for a pointed shape and was achieved by a particular technique that left fewer scars on the flatter surface.

I think there are potentially some very important connections there with material which goes right across Eastern Europe, under the heading of what has been called, by Charles McBurney among others, East Mousterian; and I think one could go further afield even than that, into Poland and European Russia, and find particular types which are made by a very similar plano-convex technique. Certainly as far as Kiik-Koba in the Crimea, and maybe further. It is also true, of course, that some of the German Micoquian industries, as I understand it, have some kind of connection with the leaf point series. But at Wolvercote there are no leaf points and no sign of them, or anywhere else in Britain until the beginning of our Upper Palaeolithic. That is something that perhaps did not come so far West.

THE LOWER/MIDDLE PALAEOLITHIC TRANSITION IN CZECHOSLOVAKIA

Karel Valoch

The dividing line between the Lower and Middle Palaeolithic is a matter of convention based on certain concrete characters. Many authors limit the notion of the Middle Palaeolithic to the Mousterian assemblages from the period of Early Würm. In their view the entire Middle Palaeolithic represents, in the development of human civilization, an episode of some thirty thousand years. This however, does not correspond to the real duration of all phenomena that can be related to the Middle Palaeolithic.

The Middle Palaeolithic can be represented by the stone tool assemblages of the so-called Mousterian complex. In contrast to the previous Lower Palaeolithic, the technique of stone flaking developed greatly and the shapes of the tools were more differentiated. The beginning of this process can be set approximately at the beginning of the Penultimate Glacial Period or Riss. Roughly in this period begins the proper "évolution buissonante" of the Middle Palaeolithic techno-complexes (Valoch 1968), appears the so-called Levalloisian technique of the preparation of tortoise cores (Bosinski 1967). In this period the real Mousteroid industries arise (Tuffreau 1979), based on the purposeful manufacture of flakes (Fridrich 1980). In this sense, the Middle Palaeolithic comprises a period ranging from the beginning of the Penultimate Glacial Period to Middle Würm (Pleniglacial A), spanning through at least two very warm and three cold periods.

The changes we want to follow here have to do in principle with stone tool assemblages. The limit we have set for the beginning of the Middle Palaeolithic is a conventional one, i.e. it is an artificial dividing line, not representing any revolutionary breach in the steady development of the material culture of the human society. There is no doubt that the first indications of the above-mentioned criteria had existed already in the course of the previous Holstein Interglacial, but they became more marked with the onset of the Glacial Period. We can say, however, that the process of development and changes depended only slightly on the changing ecological conditions, and that Man with his existing material inventory always managed to adapt himself to these conditions. Changes in the inventory of tools reflect, more likely, the growing manual and psychical capabilities of Man, his developing intellect and making the firm cultural traditions in the course of the development of the *Homo erectus* population into a very differentiated *Homo sapiens (sensu lato)* population. The population of Neanderthal Man is morphologically very differentiated. His material culture is also very diversified from both the technological and typological viewpoints. The different techno-complexes are characterized by certain technological features and sets of types, whose individual elements are in certain mutual relations. They develop parallelly, side by side, in the course of warm and cold periods, and it seems that independently of the activities performed (for the Siberian Upper Palaeolithic see Abramova 1979). Presumably the stone industries represent that part of the material culture that is most closely connected with traditions, and thus it better reflects the ethnic differences of their manufacturers (Lyubin 1977).

The Lower Palaeolithic on the territory of Czechoslovakia and in the adjacent regions is representative of the problems of the period in its entire breadth. Central Europe is among those regions of the world that were relatively early visited by the hunting groups of *Homo erectus*. Together with Upper Biharian fauna of the Cromerian period appeared also tools made of fragments of local rocks in Prezletice near Prague (Fridrich 1976a) and on Stránská skála near Brno (Musil and Valoch 1968). These assemblages are so atypical, although without any doubt intentional, that it is completely impossible to attach them to any known techno-complex. They have therefore been called Prezletician (Fridrich 1976a) and their further development is unknown. Pebble-tool assemblages with a large number of heavy-duty tools, namely of choppers, appear in central Bohemia and in south Moravia ("épannelé" from fossil soil PK VI in Sedlesovice (Valoch 1981; Valoch *et al.* 1978) perhaps as early as the Cromerian (Suchdol near Prague (Fridrich 1976a), Pouzdrany-Strachotín in south Moravia (Valoch 1982)), but without any doubt at the beginning of the Middle Pleistocene (Mindel Glacial). No genuine bifaces have been found in this context. In view of the fact that most of these finds come from surface collections, heavily ground by wind, or from fluvial gravels, their light-duty component is still little-known. These types of industries probably still existed in the Holsteinian.

In addition to this group, there appears already during the Mindelian in Central Europe a "Microchopper complex" (the Buda industry from Vértesszöllös in Hungary (Kretzoi and Vértes 1965)), forming the beginning of a long tradition. The tools are made of small pebbles of various rocks, their débitage is really microlithic and the retouched tools are unusually polymorphous. They comprise real side-scrapers, borers, end-scrapers and burins. In the course of the Holsteinian this tradition is represented by Bilzingsleben in the G.D.R. (Mania 1974), a significant sample with a large number of stone tools and also interesting bone tools (Mania and Dietzel 1980). Both sites, Vértesszöllös and Bilzingsleben, also yielded skeletal fragments ascribed to *Homo erectus* (Vlcek and Mania 1977).

From the Rissian we have so far no finds documenting the existence of the Buda industry, however, from the Eem Interglacial we know several sites in the G.D.R. (Taubach, Rabutz) and also in Czechoslovakia (the Kůlna Cave, layer 11, Gánovce, Bojnice III, and others), which without the slightest doubt continue in the tradition of microlithic tools made of small pebbles, the assemblage I call Taubachian (Valoch 1976). It is quite possible that this techno-complex lived to see not only the beginning of Würm (Tata in Hungary), but also the end of the Middle Palaeolithic towards the end of Early Würm. This phenomenon has interesting ecological features: most localities are linked to the travertines of mineral water wells (Vértesszöllös, Bilzingsleben, Taubach, Gánovce, Bojnice III, Tata). The interesting thing is that the travertine endocranium from Gánovce has certain erectoid characters (J. Jelinek, pers. comm.).

The Acheulian, relatively quickly following the Oldowan in East Africa (Leakey 1971), soon appeared in the Levant as well (Copeland and Hours 1979), but came relatively late to Central Europe. If we neglect the isolated, and perhaps even problematic find of a handaxe from Steinheim (Valoch 1968, fig. 2), supposed to belong to Holsteinian, early and middle Acheulian have not been represented at all. The rich and well stratified industry from Markkleeberg (G.D.R., Mania and Baumann 1980) is

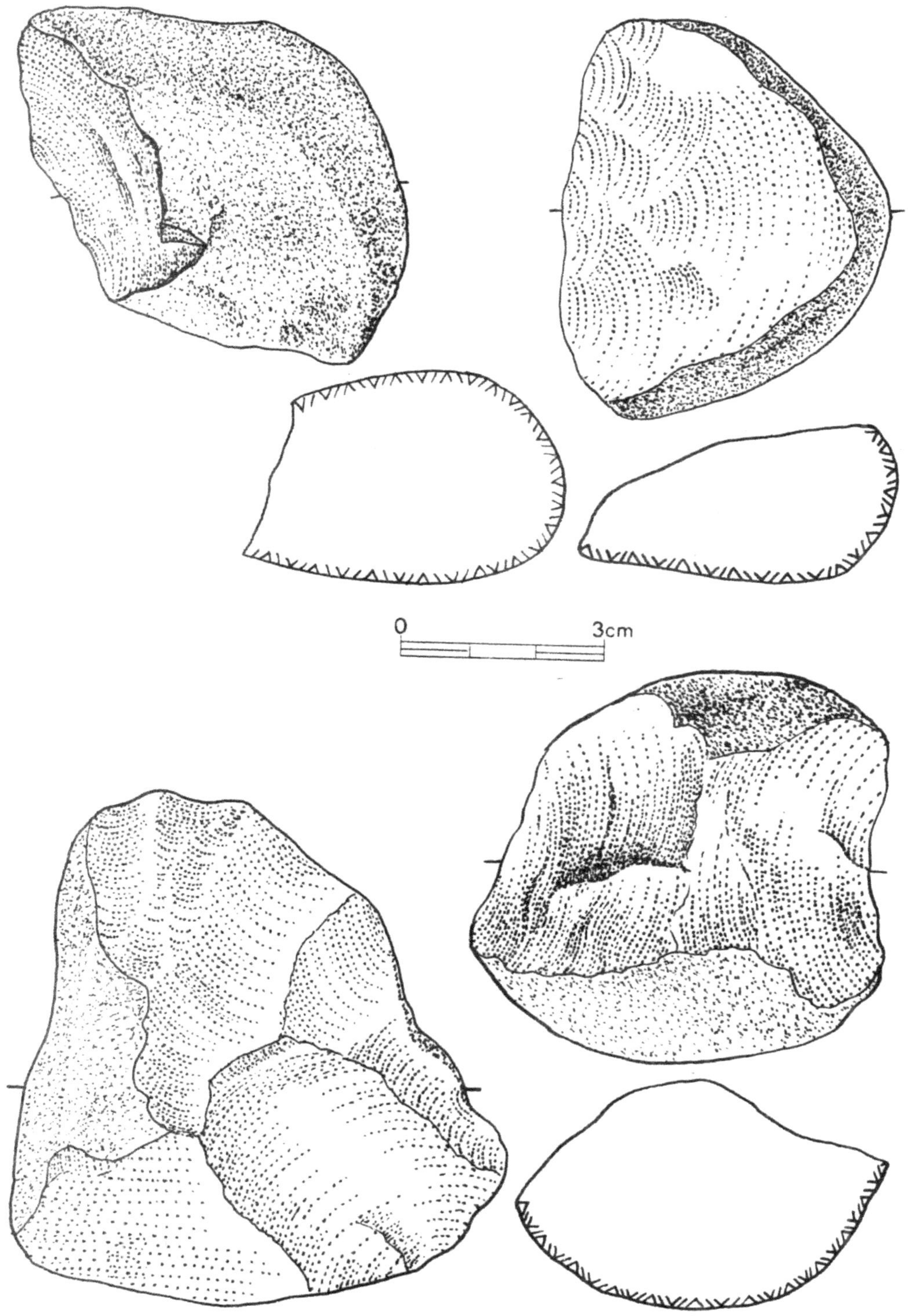

Figure 1 Pebble-tools. Site: Přibice I.

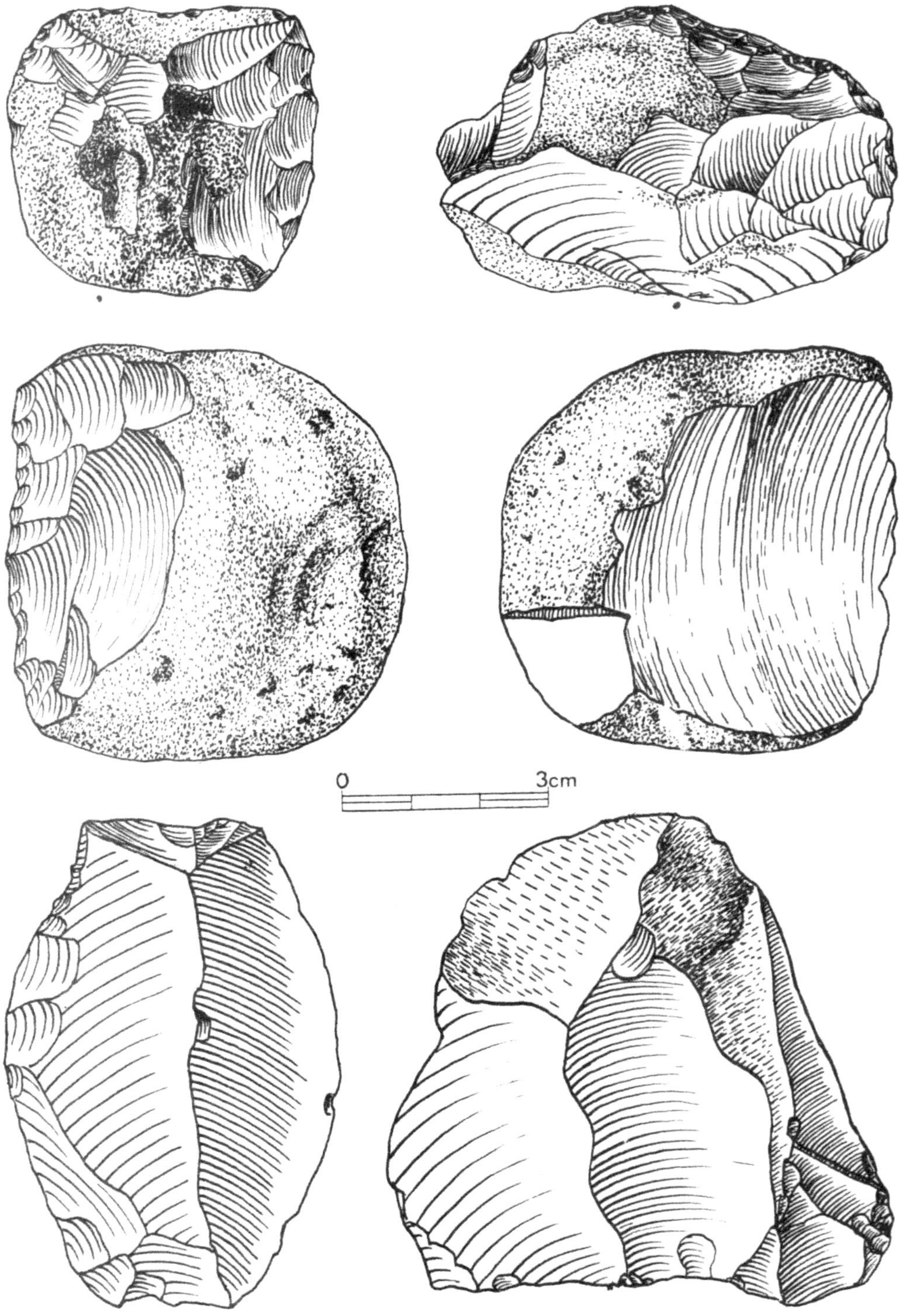

Figure 2 Taubachian. Site: Kůlna Cave, layer 11.

dated as late as the transitory phase from Holsteinian to the Saale Glacial. Acheulian extended over the territory of the G.D.R. as late as the Penultimate Glacial (Toepfer 1981), and it reached also to Bohemia (Fridrich 1980) and to Moravia (Oliva 1981). In view of the fact that, besides two isolated bifaces from south-western Poland, Acheulian is not known in the surrounding countries, the relatively rich collections from Zhitomir and Khotylevo in the Ukraine (Mesyats 1962; Zaviernyayev 1978) and from Punikve in northern Croatia (Malez 1979) are extremely interesting.

We can presume that Acheulian had created in Central Europe the technological basis for the Micoquian, an important Middle Palaeolithic techno-complex, appearing towards the end of the Eem and in Early Würm in south and western Germany, in south Poland and also in Moravia.

From the territory of the G.D.R. and from the Holsteinian come the finds called Clactonian (Toepfer 1968; Collins 1968). The problem of the Clactonian has been recently revised and the conclusion is that the so-called Clactonian sites are only Acheulian preparatory areas (Ohel 1979). In my view this study is unable to resolve questions in the relations between Clactonian and Acheulian and I personally am inclined to side with the authors expressing certain doubts with regards to Ohel's ideas (Bordes; Cahen; Collins; Newcomer; Roe; Wymer *in* Ohel 1979). There can be no doubt that the Lower Palaeolithic in Europe is not represented by a single technological tradition – by Acheulian – to which *a priori* all stratigraphically old assemblages belong. We must take into account that in the course of Mindel and Holstein there was in Europe, besides the Acheulian, another important techno-complex of pebble-tool industries, and the Clactonian offers itself for possible connection with it. I would like to emphasize also that the occurrence of the genuine Clactonian is limited to the regions of Europe rich in high-quality flint (England, northern France, Belgium, and the G.D.R.); on the contrary, pebble-tools appear in the regions poor in high-quality raw materials.

We must take into account the possibility that Clactonian is only a variant of the pebble-tool industries, based on different raw material. We can imagine that flint nodules easy to flake, but of irregular shapes, could be more easily made into biconical cores, from which it was possible to obtain a number of flakes suitable for use as tools. On the other hand, although the pebbles are of more or less regular and reoccurring shapes, they are of unsuitable materials (prevailingly quartz, quartzite, etc., in southern France, western Germany, in Bohemia and Moravia). It was very difficult to obtain regular-shaped flakes from them. However, they were suitable for making tools of almost standard shape (choppers), by removing a minimum number of cortex flakes. The flakes found in small numbers in these pebble-tool industries all have Clactonian characters.

On the other hand, a real Acheulian workshop on outcrops of quartzites (similarly as in Reutersruh bei Ziegenhain (Luttropp and Bosinski 1971)) exists also in Bohemia (Becov near Most (Fridrich 1980)). It is accompanied by Levallois technique and has no Clactonian characters.

This means that the Lower Palaeolithic is represented in our region by two basic traditions: by the pebble-tool complex and by the bifaces complex. Clactonian forms part of the first and its specific branch is formed by the microlithic tradition (Buda industry, Taubachian). Although there is no doubt that the Acheulian had developed from Oldowan (in Africa), the two complexes came to our region independently of each other, one of them declining and the other already fully developed. It would be preposterous to consider any genetic relations between them in this part of Europe. In Middle Palaeolithic the steady development of both the Acheulian and Taubachian continues, analogous to the Tayacian in southern France (de Lumley 1971). Completely

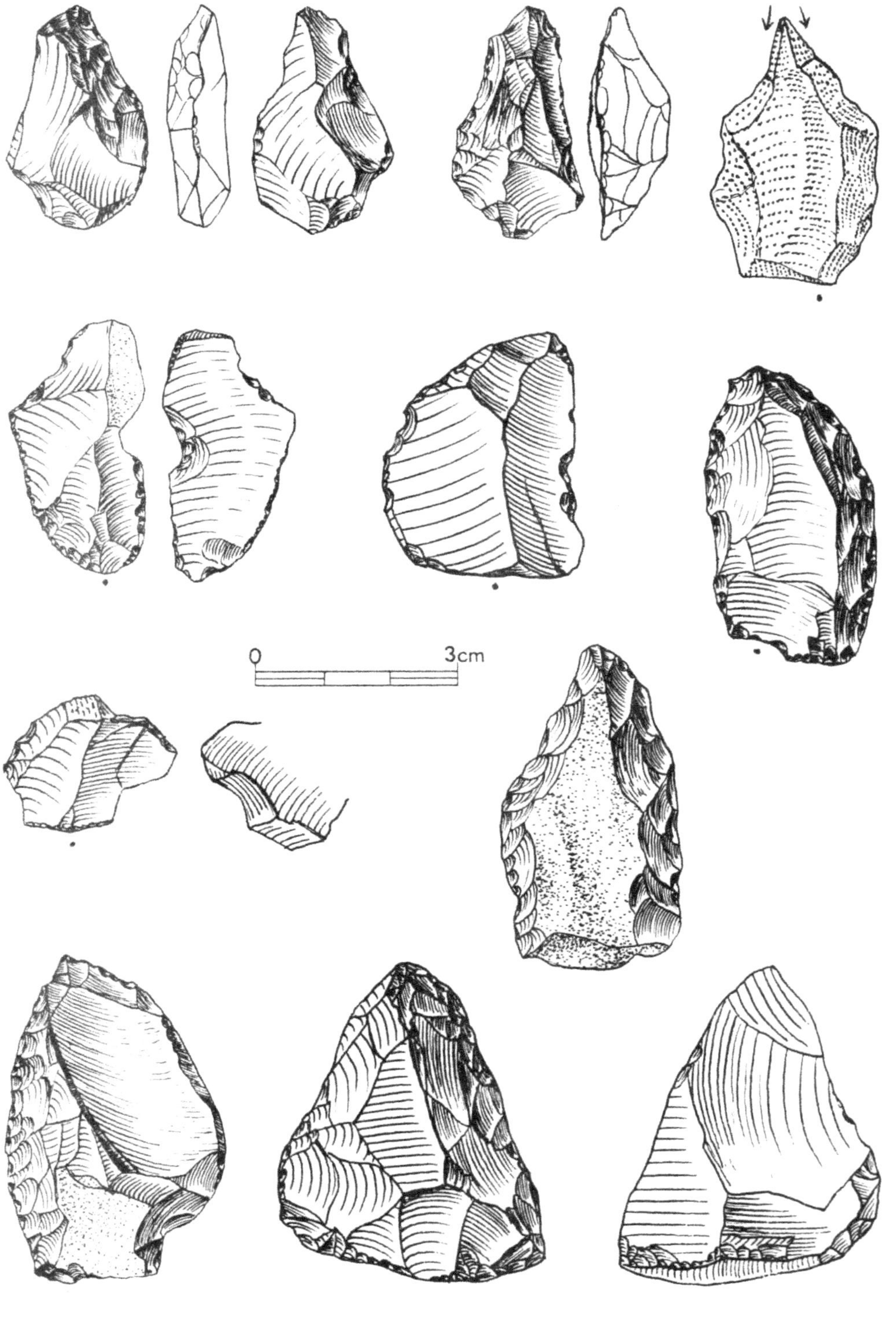

Figure 3 Krumlovian. Sites: Maršovice IV, Vedrovice VII.

new industries arise as well. So, for example in Becov, in the oldest layer of site I, appears a very rich flake industry of Mousteroid type, classified as Protocharentian. The layer on the basis of Palaeopedological characters is put into the temperate Riss interphase (Riss I/II = PK IV). The industry comes from a dug-in dwelling lined with stones (Fridrich 1976b). Towards the end of Riss we have in south Moravia an industry processing large cobbles from chopping-tools and unidirectional prismatic cores are made. They are accompanied with atypical light-duty tools (side-scrapers, denticolates; the so-called Krumlovian (Valoch 1971)).

The initial phases of Middle Palaeolithic appears very rarely in our territory and in the whole of Central Europe we can find only very few stratified industries from the course of the Riss. However, they also document the qualitative changes occurring in this period in the composition of stone-tool assemblages. The process of their development had entered into a new stage of strong differentiation, culminating in Early Würm through the existence of a large number of technologically and typologically parallel groups of the Mousterian complex (Bordes 1961). It is quite probable that the process is mutually dependent with the development and differentiation of the human population entering the *Homo sapiens neanderthalensis* stage.

In the morphologically heterogeneous population of the Neanderthal Man, and with the mosaic-shaped development, appeared the first groups with characters of the Modern Man (Qafzeh); in a similar way in the material culture in some assemblages (Mousterian of the Acheulian Tradition in France) the trend towards the formation of pure Upper Palaeolithic prevailed.

On our territory the Man from the Mladec Cave can be regarded as the oldest representative of *Homo sapiens sapiens*. He is a typical representative of the Aurignacian. We must presume that Modern Man already appeared in Moravia before the end of Early Würm, where the typical Aurignacian is for the first time stratigraphically documented (Vedrovice II (Valoch and Oliva, in press)). Vedrovice II, together with the Bacho Kiro Cave in Bulgaria and the Istállóskö Cave in Hungary are the earliest documents of the existence of real Upper Palaeolithic in Europe.

Our region in Central Europe was not only settled in the very ancient Cromerian period and had contributed to the development of the Middle Palaeolithic, but formed also one of the local points of the origin of both the Upper Palaeolithic and of Modern Man.

References

Abramova, Z.A. 1979 *Paleolit Jeniseya. Kokorevskaya kultura* Novosibirsk.

Bordes, F. 1961 Mousterian Cultures in France *Science* 134: 803-810.

Bosinski, G. 1967 *Die mittelpaläolithischen Funde im westlichen Mitteleuropa* Fundamenta A/4, Köln.

Collins, D.M. 1968 Metrischer und typologischer Beweis für die Selbständigkeit der Kulturtraditionen des Acheuléens und Clactoniens in England und Deutschland *Jahresschrift für mitteldeutsche Vorgeschichte* 52: 27-38, Halle.

Copeland, L. and F. Hours 1979 *Le Paléolithique du Nahr el-Kébir. Quaternaire et Préhistoire de Nahr el-Kébir septentrional*, 29-119, Paris.

Fridrich, J. 1976a The first industries from Eastern and South-Eastern Central Europe. Les premières industries en Europe 8–23; *Colloque VIII, IXe Congrès Int. UISPP; Nice.*

Fridrich, J. 1976b Ein Beitrag zur Frage nach den Anfängen des künstlerischen und ästhetischen Sinns des Urmenschen *Památky archeologické LXVII:* 5-27, Praha.

Fridrich, J. 1980 Becov IV (distr. Most) An Acheulian Site in Bohemia *Anthropologie* XVIII/2-3: 291-298 Brno.

Kretzoi, M. and L. Vértes 1965 Upper Biharian (Intermindel) Pebble-Industry Occupation Site in Western Hungary *Current Anthropology* 6: 74-87.

Leakey, M.D. 1971 *Olduvai Gorge. Excavations in Beds I & II 1960–1963* Cambridge.

Lyubin, V.P. 1977 *Must'erskie kultury Kavkaza* Leningrad.

Lumley-Woodyear, H. de 1971 *Le Paléolithique inférieur et moyen du Midi méditerranéen dans son cadre géologique* Tome II. Gallia Préhistoire, Suppl. V, Paris.

Luttrop, A. and G. Bosinski 1971 *Der altsteinzeitliche Fundplatz Reutersruh bei Ziegenhain in Hessen* Fundamenta A/6, Köln.

Malez, M. 1979 Paleolitsko i mezolitsko doba u Hrvatskoj *Praistorija Jugoslavenskih Zemalja, Paleolit i Mezolit* 195-296, Sarajevo.

Mania, D. 1974 Bilzingsleben, Kr. Artern. Eine altpaläolithische Travertinfundstelle im nördlichen Mitteleuropa *Zeitschrift für Archäologie* 8: 157-173.

Mania, D. and A. Dietzel 1980 *Begegnung mit dem Urmenschen. Die Funde von Bilzingsleben* Leipzig.

Mania, D. and W. Baumann 1980 Neufunde des Acheuléen von Markkleeberg bei Leipzig (DDR) *Anthropologie* XVIII/2-3: 237-248 Brno.

Mesyac, V.A. 1962 Nachodki drevnepaleolititcheskich orudii v rayone Zhitomira *Kratkie soobchtchenia Instituta Archeologii AN SSSR* 92: 4-55.

Musil, R. and K. Valoch 1968 Stránská skála: Its Meaning for Pleistocene Studies *Current Anthropology* 9: 534-539.

Ohel, M.Y. 1979 The Clactonian: An Independent Complex or an Integral Part of the Acheulian? *Current Anthropology* 20: 685-726.

Oliva, M. 1981 The Acheulian finds from Karolín in Moravia *Anthropologie* XIX/1: 27-32 Brno.

Toepfer, V. 1968 Das Clactonian im Saale-Mittelelbegebiet *Jahresschrift für mitteldeutsche Vorgeschichte* 57: 1-26, Halle

Toepfer, V. 1981 Das Acheuléen auf dem Boden der Deutschen Demokratischen Republik *Anthropologie* XIX/1: 55-77 Brno.

Tuffreau, A. 1979 Les débuts du Paléolithique moyen dans la France septentrionale *Bulletin de la Société Préhistorique Française* 76: 140-142.

Valoch, K. 1968 Evolution of the Palaeolithic in Central and Eastern Europe *Current Anthropology* 9: 351-391.

Valoch, K. 1971 Eine mittelpaläolithische Industrie von Maršovice in Südmähren *Anthropologie* IX: 29-47, Brno.

Valoch, K. 1976 Un groupe spécifique du Paléolithique ancien et moyen d'Europe Centrale. L'Evolution de l'Acheuléen en Europe, pp. 86-91; *Colloque X, IXe Congr. Int. UISPP,* Nice.

Valoch, K. 1981 Stratifikovaný valounový nástroj ze Sedlesovic u Znojma *Archeologické rozhledy*,XXXIII: 92-94 Praha.

Valoch, K. 1982 Altpaläolithische Geröllgeräte in Südmähren *Anthropozoikum* 14, Praha.

Valoch, K. and M. Oliva In press Das Frühaurignacien von Vedrovice II und Kuparovice I in Südmähren *Anthropozoikum* Praha.

Valoch, K., L. Smolíková and A. Zeman 1978 The Middle Pleistocene Site Přibice I in South Moravia *Anthropologie* XVI: 229-241, Brno.

Vlcek, E. and D. Mania 1977 Ein neuer Fund von *Homo erectus* in Europa: Bildingsleben (DDR) *Anthropologie* XV: 159-169, Brno.

Zaviernyayev, G.M. 1978 *Chotylevskoe paleolititcheskoe mestonachozhdenie* Leningrad.

THE TRANSITION FROM LOWER TO MIDDLE PALAEOLITHIC IN CENTRAL ITALY: AN EXAMPLE FROM LATIUM

Marcello Piperno and Aldo G. Segre

Some Comments on the Concept of Transition in Prehistory (M.P.)

The concept of transition in Palaeolithic prehistory is largely the result of the nomenclature conventionally used in the schematic subdivision of cultural traditions as reflected in stone tools throughout the Pleistocene.

However, the complexity of available data makes it increasingly difficult to include them in the "classic" framework hitherto proposed by current terminology for the cultural history of the Quaternary.

In a dynamic conception of stone technology, the total number of tool associations of the whole Pleistocene may be said to represent nothing but a continuous transition in which a number of artificial divisions, e.g. Acheulian, Mousterian, etc. are made *a posteriori*. These terms are often used in a chronological sense, sometimes in reference to lithic technology, and other occasions as an indication of cultural facies.

The impression obtained from a diachronic analysis of the Pleistocene industries is actually one of substantial overall unity with no apparent gaps, except perhaps in the documentation, characterized by a relationship between different facies which is by no means simple and straightforward but rather due to the combination of the various parameters interacting to form each cultural whole, and undergoing changes occurring at different rates in a way fairly closely resembling the mosaic pattern development so often observed in human palaeontology (Chavaillon *et al.* 1978, 1979).

The transformation of the Pleistocene stone industries does not seem to correlate significantly with the fluctuations in climate and consequent environmental changes occurring throughout the Quaternary. Nor has any (probably non-existent) correlation been found between cultural traditions and specific human species. Suffice it to say that the distribution area of the Acheulian industries takes in such a wide range of differentiated ecological niches and such a long time span that it can be concluded that the technological development of tools is virtually independent of external factors.

It is inside a common tool background that the transformations occurring in the stone industries took place as both a reaction and stimulus to probably changes in way of life during the Palaeolithic. It is thus against this very background that the technological innovations and improvements gradually occurred. The latter occasionally took the form of actual discoveries, e.g. the Levallois technique, certain kinds of retouch, etc. in an increasignly better fit between form and function and an ever-widening range of adaptation between tools and differentiated activities.

The result of neglecting the existence of a large basis common to the whole of prehistoric technology leads to the hierarchization of certain characteristics, such as the presence or absence of certain types and their percentage frequency, as a result of which false distinctions are made between technological stages that can actually be equated.

The difficulties facing an exact interpretation of the transition stages is immediately apparent even in the Oldowan/Acheulian transition, a problem that is still being debated and consists of two diametrically opposite conceptions. The first of these, based substantially on the interpretation of the Olduvai data, in which the upper limits of the Oldowan are chronologically superimposed on the first appearance in the Olduvai Gorge of industries defined as Acheulian (Leakey 1971, 1976). The second is taken from Melka Kunture, according to which the Oldowan/Acheulian transition does not consist in a superimposition of different, but simultaneous, traditions, but occurs as a continuous process in which the innovatory Acheulian factors coexist in equilibrium with typical Oldowan features in the early stages of the transition, until they prevail at the end of the transformation process.

At the opposite end of the sequence, also the transition from Acheulian to Middle Stone Age at Garba III shows the persistence of Acheulian features and the rise of the Levalloisian industry and other innovations typical of the more recent periods (Chavaillon *et al.* 1978, 1979).

Generally speaking, the relationship between Acheulian and earlier and later stages is likely often to be misjudged owing to the undue importance attributed to the presence of bifaces, with the consequent exclusion from this facies of industries in which this type is either absent or present in only very small quantities.

A broader definition of the Acheulian, in which the presence of bifaces is once again ranked on a par with other tool types, and also involving the other features characterizing this cultural tradition, thus seems to be a necessary condition for a better understanding of the transition towards the Middle Stone Age. During the transition, this tradition, which certainly did not disappear without a trace at the end of the Riss or the last interglacial, definitely played an essential role, comparable to that played by the pre-Wurmian industries lacking bifaces. Furthermore, the latter industries could, at least in some cases, have had a much closer relationship with the Acheulian than is generally supposed.

This is the context in which, for instance, the preliminary data obtained from the new excavations in the Venosa basin by the Istituto Italiano di Paleontologia Umana are to be considered (Piperno and Segre 1980). At the lower limit of the sequence known in this area lies the "Tayacian" living-floor, assigned to the Mindel and already known at time of the IIPU excavations (Chiappella, 1964), which have recently been continued by the Musée d'Anthropologie Préhistorique de Monaco (Barral *et al.* 1978).

The upper part of the sequence in the same basin contains an exceptionally abundant documentation of Acheulian industry in the area of Zanzaniello, Castelluccio, etc. Owing to their typological characteristics, the latter may be largely assigned to upper and later facies of this culture, although the abundant industry gathered in the various sites has so far lacked any definite stratigraphic references.

The gap between "Tayacian" level and the abundant evidence of upper and/or final Acheulian could apparently by filled, at least partially, by recent finds in the Notarchirico site in the Venosa basin. By means of sondages it was possible to identify a sequence consisting of six levels (top to bottom I–VI), some of which corresponding to occupation periods on the shores of lacustrine basins or the gravel banks of watercourses. In them, the industry and palaeontological remains are associated with pebble beaches, while in others all the stone material may be considered as having being deliberately transported. A preliminary attempt at making a cultural classification of these industries, which consist mainly of tools on limestone pebbles and chips, scrapers, denticulated flakes, etc., mostly on flint, can be based on the presence at level IV of a few bifaces that are morphologically different from those usually representative of the upper and final Acheulian at Venosa. At this stage of the excavation, the presence of bifaces seems to be the only feature distinguishing level IV, which can almost certainly be assigned to the Middle Acheulian, and the other levels that, although lacking bifaces, must be considered as part of a context definable as Acheulian *sensu lato* because of their stratigraphic position and the sum total of other factors, either typological in nature or connected with living-floor organization.

The Lower/Middle Palaeolithic Transition in Latium

The problem of the transition between Lower Palaeolithic and Middle Palaeolithic in Italy must be tackled taking account of the country's peculiar geographic position with respect to the large areas of techno-complex development in central-western Europe, as well as of the topography of the peninsula, charcterized by the backbone of the Apennines (Piperno 1980).

On the one hand, it must be considered that the peninsula may have formed a kind of cultural 'cul de sac' vis-a-vis continental Europe, or at the same time represented an area of early transit of influences and technologies to and from numerous other areas. On the other hand, the clear-cut longitudinal separation of the eastern side from the western slopes of the Apennines may have been an important factor of diversification in the independent development of the two areas (Radmilli 1976).

It follows that the processes of transformation of stone assemblages during the Quaternary do not necessarily follow patterns that can be extended uniformly to the whole peninsula and that the transition stages observed in certain areas can display peculiar characteristics according to the substrate they derive from. These characteristics cannot thus be generalized to other, even geographically neighboring, areas.

In other words, although the stone technology throughout the Quaternary appears to follow a single consistent guideline, for the above reasons, in a region like Italy, it may be the result of different influences and the various facies can have different outcomes which are not synchronic with those of non-Italian techno-complexes. In view of the aforegoing, we prefer to follow an example of Lower/Middle Palaeolithic transition in a smaller area, i.e. restricted to the coastal region of Latium lying between the Circeo promontory to the south and Torre in Pietra to the north. Reference will be made only in passing to other Italian sites in which, for different reasons, the level of our present knowledge still allows no definite link-up between the events marking the period between an advanced Riss phase and the early Wurm.

In the area considered, this same period appears to be fairly well documented by a series of sites whose mutual relative chronological position may be deemed sufficiently well established, mainly by virtue of the numerous geological and stratigraphic contributions made concerning the region.

In addition to this favorable situation, there is no doubt that, perhaps more than in other Italian regions where an intense Mousterian occupation is also superposed on the

Acheulian population, there is a certain 'family resemblance' between the stone assemblages of the sites examined in the present report, quite apart from their respective chronological positions. This resemblance is not only the result of having chosen a certain type of raw material, but is due above all to the numerous shared techno-typological points and the very structure of the stone assemblages.

The area considered is also characterized by a fairly high degree of geographic isolation in the sense that, as will be seen, it shows no relationship with the technical and typological features of the Rissian and early Wurmian assemblages of neighboring areas.

For the purpose of introducing a schematic subdivision that will be useful in understanding the pattern of the transformations undergone by the assemblages to be examined in the following, together with their interrelations, our presentation is divided up into three arbitrary parts: The premises, consisting mainly of the industry at level m of Torre in Pietra, but which can be extended, as an Acheulian facies of Torre in Pietra, to other recently discovered sites in the same area; the transition, which is represented by the Aniene valley sites Monte delle Gioie and Sedia del Diavolo, by Saccopastore and by level d at Torre in Pietra; the outcome, i.e. the abundant evidence in the Pontinian characterizing, by means of a homogeneous spread during the early Wurm, the whole area of the Circeo promontory, the Pontine plain and at least part of the central-northern hinterland of Latium.

Although these sites have already been described in broad outline, only comparatively recently has a series of exhaustive studies been carried out to specify the typological characteristics in detail (Taschini 1967, 1970, 1972, 1979; Tozzi 1970; Piperno and Biddittu 1978). In one work a phylogenetic approach has been made to the problem of the roots of the Pontinian and thus of the Lower/Middle Stone Age transition (Taschini 1967).

The actual presentation of the data differs somewhat from the interpretation given by Taschini (1967) as regards what she calls the "Rissian proto-Pontinian" of Sedia del Diavolo and Monte delle Gioie. This interpretation was proposed before the stone assemblage of the levels m and d of Torre in Pietra had been studied (Piperno and Biddittu 1978).

The Premises. The Acheulian Facies of Torre in Pietra Level m

The level m industry of Torre in Pietra assigned to the Riss comprises 53 bifaces and more than 300 artifacts, the typological classification of which is shown in Table 1, together with a few typological and technical indices.

Thirty-nine of the bifaces are made of limestone, 5 of siliceous limestone and 9 of flint. Most of them have a cortex base and bilateral symmetry; a secondary retouch is sometimes present on the edges. Also present are several lanceolated types, ficrons and amygdaloids and others resembling the cordiforms. However, the majority of the bifaces is characterized by the complete absence of any morphological standardization.

The main technical/typological characteristics of the complex are as follows:

- a certain degree of differentiation in the choice of raw material: 62% flint; 25% siliceous limestone; 13% limestone, most of the retouched tools being of flint.

Table 1 Torre in Pietra, levels *m* and *d*. Typological list and major indices (after Piperno and Biddittu 1978)

	Level m		Level d	
	N	%	N	%
Typical Levallois flakes	–	–	19	5.9
Atypical Levallois flakes	1	1.01	2	0.62
Levallois points	–	–	3	0.93
Mousterian points	–	–	4	1.24
"Limaces"	–	–	1	0.31
Single straight side scrapers	7	7.07	22	6.83
Single convex side scrapers	12	12.12	32	9.93
Single concave side scrapers	5	5.05	8	2.48
Double side scrapers, straight/convex	–	–	2	0.62
Double side scrapers, straight/concave	–	–	1	0.31
Double side scrapers, biconvex	–	–	11	3.41
Double side scrapers, conc./conv.	1	1.01	3	0.93
Convergent side scrapers, straight	–	–	1	0.31
Convergent side scrapers, convex	2	2.02	3	0.93
"Déjetés" side scrapers	3	3.03	17	5.27
Transverse side scrapers, straight	–	–	5	1.55
Transverse side scrapers, convex	5	5.05	12	3.72
Transverse side scrapers, concave	1	1.01	1	0.31
Inverse side scrapers	1	1.01	5	1.55
Steep side scrapers	–	–	2	0.62
Bifacial side scrapers	–	–	1	0.31
Alternate side scrapers	1	1.01	3	0.93
End scrapers	3	3.03	6	1.86
Burins	–	–	3	0.93
Borers	2	2.02	4	1.24
Naturally backed knives	2	2.02	12	3.72
Truncations	2	2.02	6	1.86
Notches	12	12.12	24	7.45
Denticulates	10	10.10	50	15.52
Alternate thick/thin retouch	–	–	13	4.03
"Tayac" points	–	–	1	0.31
Distally notched flakes	–	–	2	0.62
Choppers	24	24.24	34	10.55
Miscellaneous	5	5.05	9	2.79
Total	99		322	
Flakes	71		157	
Handaxe trimming flakes	6		–	
Utilized flakes	12		–	
Retouched flakes and fragments	23		16	
Debitage and fragments	88		146	
Cores	17		92	
Hammerstones	–		1	
Broken pebbles	11		10	

Real Typological Groups

	m	d
I	1.01	7.45
II	38.38	41.56
III	5.05	4.03
IV	10.10	15.52
+enc	22.22	22.97

Essential Typological Groups

	d
I	–
II	46.95
III	4.55
IV	17.54
+enc	25.96

Real Technical Indices

	m	d
IL	0.5	6.72
ILstr	–	2.3
IF	2.89	25.14
IFstr	2.38	34.17
IFs	2.17	15.25
IFs str	2.38	20.10

Real Typological Indices

	m	d
ILty	1.01	7.45
IR	38.38	40.01
IC	18.18	15.82

Essential Typological Indices

	d
ILty	–
IR	45.2
IC	17.88

- the striking platforms are mostly plain, only in three cases facetted and in one dihedral.
- retouch is usually rather irregular, often only partial, discontinuous and denticulated. The marginal type is the most frequently attested while there is virtually no evidence of embricated retouch.
- about half the unretouched flakes and more than 66% of the tools have cortex.
- there is already a fairly high degree of typological diversification, with 20 types from the Bordes list (1961) present, 10 of which are in the scraper group, which accounts for 38% of tools. There are no points or limaces. The Upper Palaeolithic group is represented by comparatively few, although typical, tools (end-scrapers and percoirs; burins are absent). Together, denticulated and notched flakes account for about 22% of the tools. The most frequent are LD choppers, whether retouched or not (24%), while there are only two HD choppers. The distinction between LD (light duty) choppers and HD (heavy duty) choppers and, among the LD choppers, between retouched choppers and scrapers on pebble is the one followed in the typological classification of the m and d levels of Torre in Pietra (Piperno and Biddittu 1978).

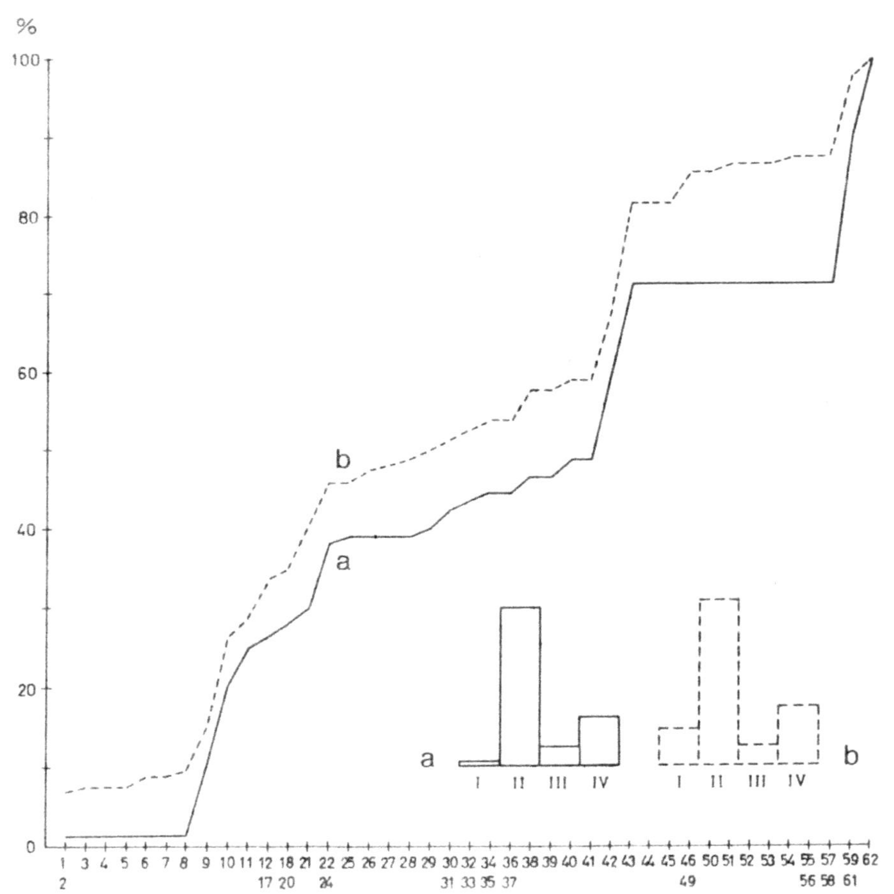

Figure 1 Torre in Pietra. Diagram of tools and typological groups. a, Level *m*; b, Level *d* (after Piperno and Biddittu 1978).

As far as the K/Ar datings are concerned (Evernden and Curtis 1965), it must be pointed out once again that the age of the samples subjected to dating (KA 1185 and KA304 = 431,000; KA334 = 434,000; KA345 = 438,000) merely represents a *post quem* term for the chronological ordering of level m in that the so-called "black pumice tuff" from which these dates were obtained is stratigraphically "older than the palaeo-surface and than the subsequent fluvial formation containing the Torre in Pietra Acheulian, in which pebbles from these same tuffs are found reworked" (Blanc 1956).

Researches carried out over the past decade in the Latium hinterland has led to the identification of a large concentration of pre-Acheulian and Acheulian sites, particularly along the Sacco-Liri river valley (Biddittu 1971, 1974a, 1974b; Biddittu and Cassoli 1966; Biddittu *et al.* 1979). The latter have been assigned to the final Riss or the beginning of the R/W interglacial (Biddittu and Segre 1976).

As far as can be judged from the bifaces (the flake industry being rather poorly represented in all these sites), the Acheulian facies of central-south Latium does not correlate with the Torre in Pietra facies, compared with which it displays basically more highly evolved characteristics. In our opinion, this observation does not necessarily imply an actual more recent dating for the Liri valley assemblages, but can also be explained in terms of the presence of human groups in unconnected niches and in which, albeit contemporary, technological traditions can manifest themselves with partly dissimilar characteristics reflecting a probable ethnic differentiation between non-communicating groups. The Torre in Pietra facies, on the other hand, actually corresponds to other neighboring sites along the Via Aurelia, at Castel Di Guido and Malagrotta (Longo *et al.* 1981a, 1981b; Mallegni and Fornaciari 1981; Longo and Radmilli 1972).

The Transition

This transition phase must be subdivided into two stages. The first of these is represented by the two Aniene valley sites, which still belong to the Riss, although to an advanced stage of it. The second, comprising also level d of Torre in Pietra and the Saccopastore industry, represents as it were the final stage of this transition period, which now abounds in clear-cut tendencies towards the subsequent Pontinian results. At the same time, however, and this is particularly true for level d of Torre in Pietra, it is still in some ways connected with the technical/typological heritage dating back not only to the Monte delle Gioie and Sedia del Diavolo assemblages characterized by the absence of bifaces, but also to the Acheulian facies itself at level m, which we consider, in the framework of an Acheulian/Mousterian continuity, as the remote stimulus, the very premise, of the subsequent structural transformations undergone by the tools during the period and in the area considered.

Monte delle Gioie and Sedia del Diavolo

On the whole, the stone industry found in these two sites comprises just under 90 artifacts (about 30 from Sedia del Diavolo and 62 from Monte delle Gioie). Also in this case they have been made from small siliceous pebbles.

According to Taschini (1967), at Sedia del Diavolo there are some ten reject flakes and about fifteen tools, most of which consist of convex side-scrapers. Two small LD choppers (45 x 30 x 14 mm and 25 x 20 x 15 mm) have also been reported. This is a peculiar type that we have already found in level m at TP and that will again be found unchanged in level d and throughout the Pontinian Wurmian. Quina retouch is documented together with the presence of a few facetted striking platforms.

The Monte delle Gioie industry is somewhat more fully documented. The following is a list taken from Taschini (1967).

Side-scrapers, simple straight	13
Side-scrapers, convex	19
Side-scrapers, concave	5
Side-scrapers, double	2
Side-scrapers, convergent	3
Side-scrapers, déjeté	1
Side-scrapers, transv. convex	1
Side-scrapers, inverse	1
Side-scrapers, with bifacial retouch	1
Side-scrapers, with alternating retouch	1
End-scrapers	1
Natural backed knives	1
Raclettes	1
Notches	2
Denticulates	7
Chopping tools	3
Cores	8
Flakes and waste	ca. 30

Torre in Pietra d and Saccopastore

The typological list of the 744 items from level d at Torre in Pietra is shown in Table 1.

- The first novelty to be stressed is the presence, however slight, of the Levallois technique.
- Flint is now in general use.
- About 70% of the striking platforms are plain, although more than 15% are already facetted and just under 10% dihedral. The platform preparation is more frequent in the case of tools than for non-retouched flakes.
- Embricated and sub-embricated retouch predominates, and is mostly of the Quina type.

- The typological diversification of the assemblage seems already to have almost attained a clear-cut Mousterian character, and some 37 types from the Bordes list (1961), mostly side-scrapers, are present. The Upper Palaeolithic types are quite frequent. As far as pebble tools are concerned, there are a considerable number of side-scrapers on small pebbles and LD choppers; the discoidal cores are standardized.

The industry collected from the Saccopastore gravel quarry unfortunately comprises only eleven artifacts (Blanc 1948). Although it is obviously impossible to make a statistical evaluation of the individual characteristics it displays, this assemblage must clearly be included among those industries of the Last Interglacial which resemble the level d assemblage at Torre in Pietra; there is evidence of Levallois technique and the preparation of striking platforms; the side-scrapers are already typologically well diversified; the Mousterian points and discoidal cores are present in individual specimens; there is no evidence of Quina retouch. All the artifacts are made of flint.

The Results. The "Pontinian" Mousterian

The Pontinian Mousterian is indubitably the most widespread industry of the early Wurmian phase (Wurm I until the Wurm II–III or Gottweig interstage) in the area considered, represented by stratified sites in the numerous caves at Monte Circeo, Sperlonga (Grotta dei Moscerini), Gaeta (Grotta di S. Agostino) and on the Pontine Plain (Canale della Acque Alte), where even surface finds are frequently made.

Typologically speaking the "Pontinian" has been defined as "a Mousterian of the Charentian type in the main, closer to the Quina type, although certain features . . . resemble the assemblages of the eastern Charentian of the Mediterranean Midi in France" (Taschini 1979). Although rather rare, Levallois debitage is always present; the facetting index is quite high; side-scrapers are present in large amounts (57.5 ess. in level 2 of Gr. Guattari); there is little evidence of the Upper Palaeolithic group, denticulated flakes and naturally backed knives; the Quina retouch attains moderately high values (11 in level 2 of Gr. Guattari) (Taschini 1979).

Although in Italy some other Mousterian assemblages have used small siliceous pebbles as raw material, e.g. Grotta delle Mura (Cornaggia Castiglioni and Palma di Cesnola 1967), Grotta dell'Alto (Borzatti von Lowenstern and Magaldi 1967), Santa Croce (Cardini 1956), the Pontinian, according to the typological definition which has been given, is geographically limited to the Tyrrhenian coast of Latium.

As previously observed for the area affected during the Riss by the Acheulian facies of Torre in Pietra, also in the Mousterian period there seems to be a repetition of this characteristic of impermeability of the Latium seaboard to external influences. Not far away, in fact, in central and southern Latium, there is evidence of other Mousterian facies in which any typological relationship with the Pontinian is completely absent. At Sora Valle Radice and Carnello (Biddittu *et al.* 1967) there is a Mousterian of Levallois facies with abundant points; at Selva dei Muli, near Frosinone, there is a kind of micro-Mousterian, which is still comparatively unknown (Biddittu and Segre 1976); the Mousterian found by Costa (1864) near Cassino and illustrated by Mochi (1912) resembles that of Valle Radice.

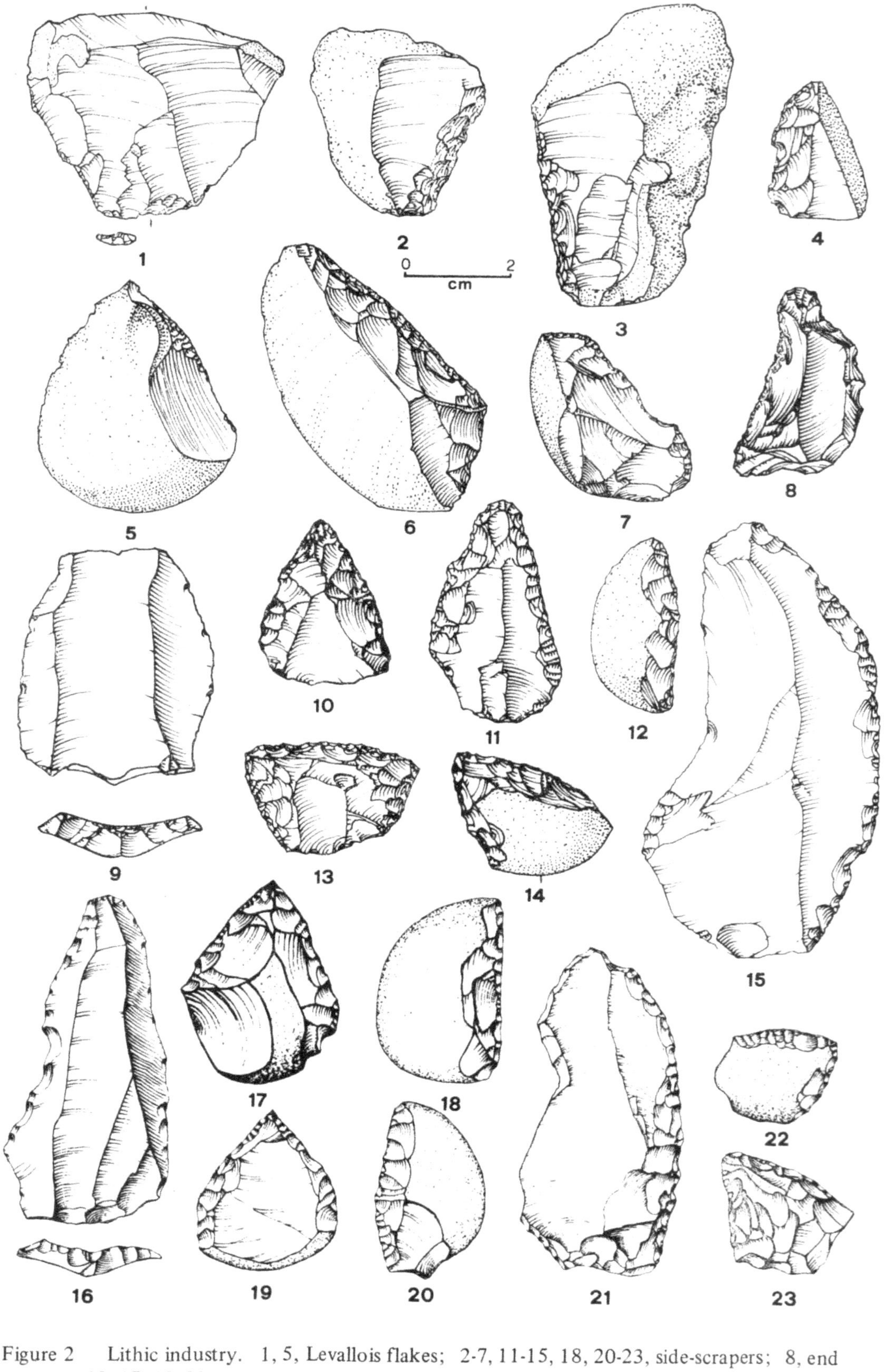

Figure 2 Lithic industry. 1, 5, Levallois flakes; 2-7, 11-15, 18, 20-23, side-scrapers; 8, end scraper; 10, 17, 19, Mousterian points; 16, Levallois point. 1-8, from Torre in Pietra level *m*; 9-15, from Torre in Pietra level *d* (after Piperno and Biddittu 1978); 16-23, from Guattari Cave (after Taschini 1979).

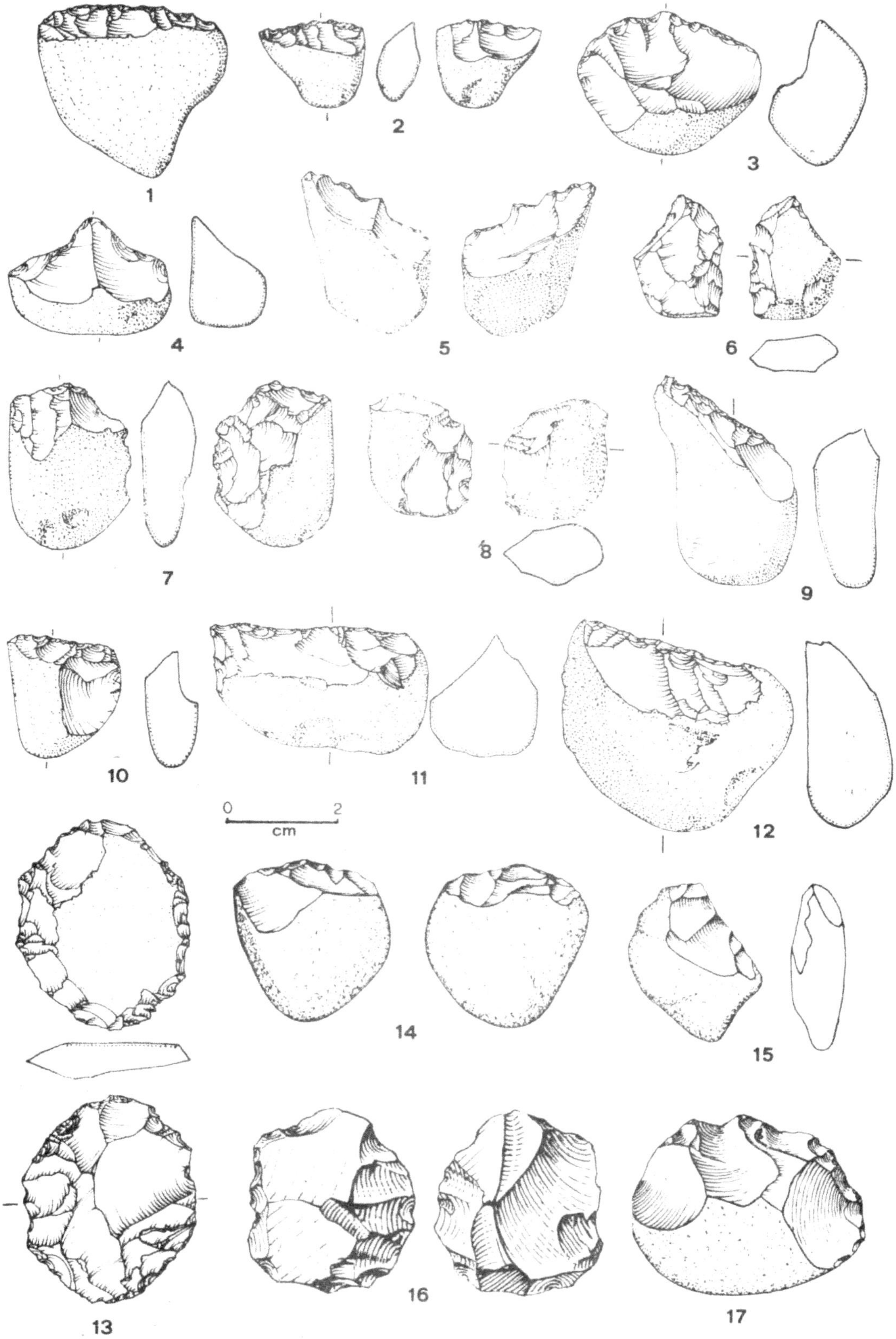

Figure 3 Lithic industry. 1-12, 14, 15, 17, Choppers LD and retouched choppers; 13, 16, discoidal cores. 1-8, from Torre in Pietra level *m*; 9-13, from Torre in Pietra level *d* (after Piperno and Biddittu 1978); 14-17, from Guattari Cave (after Taschini 1979).

In the transition example we have chosen, it seems to be possible to detect a set of tendential features. Already present in level m of Torre in Pietra, these features persist throughout level d of the same site, until the early Wurm. They consist of an enrichment of the group of side-scrapers, the presence of side-scrapers on small pebbles and on choppers, the frequency of LD choppers and, as far as the choice of raw material is concerned, a preference for using small siliceous pebbles for flake tools, despite the fact that the availability of other types of raw material is clearly documented by the hand-axes associated with this industry.

Monte delle Gioie, Sedia del Diavolo and Torre in Pietra d display a series of converging characteristics (recurrent embricated retouch, standardization of side-scrapers, presence of tools on small pebbles, LD choppers and discoidal cores), although the two sites of the Aniene Valley differ from level d at Torre in Pietra in the absence of the Levallois technique, scanty evidence of the preparation of striking platforms, the sparse presence of déjeté side-scrapers, denticulated flakes and notches, and the absence of points.

Most of the characteristics of the Torre in Pietra d level industry are subsequently found in a more mature form in the Pontinian: embricated retouch, high percentage of tools on cortical flakes, high facetting index, standardization of side-scrapers, use of Levallois technique, high frequency of LD choppers and discoidal cores (Figures 2 and 3).

Conclusions

As mentioned above, our concept of Lower/Middle Palaeolithic transition does not include the dichotomy between the Acheulian on the one hand, especially certain aspects of the final Acheulian, and the Mousterian on the other. Within the framework of the Lower Palaeolithic traditions, it also excludes any clear-cut distinction between the phylum of the bifacial industries and the assemblages in which this tool is not represented. We still know too little to be able to identify relationships that nevertheless must have existed between these two assemblages. Nor, on the other hand, are we actually in a position to decide which of these two phyla formed the substrate for the spread of the Mousterian, or whether they were both involved.

It is also clear, and not only from the assemblages we have proposed to clarify an example of transition in Latium, that the onset of the "Mousterianization" of flake industries dates back further than the conventionally accepted Wurmian limits for the various Mousterian facies. The industries of Monte delle Gioie, Sedia del Diavolo, Torre in Pietra d and Saccopastore are already typologically "Mousterian", even though they occasionally display certain differences vis-a-vis the "classic" Mousterian of the Wurmian period. The beginning of the Lower/Middle Palaeolithic transition must probably thus be situated in a period preceding the last interglacial and the various terms "pre-Mousterian", "proto-Pontinian", etc. must be assigned a purely chronological meaning, since these assemblages were often already "Mousterian" in their typological structure and flaking techniques right from the Riss or Riss/Wurm period.

If one is to speak of a transition towards the Mousterian, this phase needs must be solidly rooted in those Acheulian assemblages in which a part of those tendencies that, as a whole, will definitely be standardized at the beginning of the Wurm, are to be found "in nuce".

This obviously does not mean that all the Acheulian assemblages have represented the beginning of that transformation process that was ultimately to lead to the Mousterian. However, the variability of the various Mousterian facies can probably be explained, at least in part, as the result of diversified premises rooted in the Acheulian period, in the flake industry associated with the bifacials in these assemblages.

At this stage of tool evolution, which appears to be characterized by a technical/typological uniformity dominated by the regular production of bifacials, the links with the past and innovative tendencies must have balanced out for a long time and have both been very strong. In certain assemblages, the flake tools seem to break this equilibrium between memories of the past and innovative stimuli during the final phase of the penultimate glacial. Some final Acheulian sites[1] show that, side by side with typical Acheulian features, the innovations are now so extensive and so standardized that the assemblage no longer appears to be in equilibrium, as it were, and the transition process tends towards irreversible results.

Stratigraphical Correlations (A.G.S.)

In peninsular Italy, there are numerous Pleistocene basins with sediments from various facies ranging from fluvio-continental to lacustrine and brackish, with numerous intercalations of volcaniclastics in the middle and upper portions of the series (Segre 1978; Biddittu *et al.* 1979; Segre and Biddittu1981; Piperno and Segre 1980). The Middle and Lower Palaeolithic industries found *in situ* can occasionally be adequately localized from the stratigraphic point of view, particularly in Latium and Basilicata. North-west of Rome, the Torre in Pietra sequence (Figure 4) consists of a terraced residue on the outskirts of the centers of Sabatium (i.e. Bracciano) vulcanism. It is superposed on an older sequence (I), composed, on the bottom (1), of Calabrian marine clay, brackish marls and beach sands; on the top (2), of basically volcaniclastic and limnotufaceous layers characterized by the presence of an extensive ignimbrite, often of considerable thickness, known as "black-scoriae tuff", K/Ar dated as 430,000 (Evernden and Curtis 1965). The terrace II sequence (Figure 4) can be divided into a lower part (3) consisting of epilitoral marine clays and beach sands, with a lagoonal event at Cerastoderma (3'). On top of it, a palaeosurface with a ferriferous duricrust presents a sparse Acheulian industry (Piperno and Biddittu 1978) accompanied by the fragmentary remains of large mammals (level m). The same industry, together with mammal bones, are found scattered over a river gravel bed immediately above it. The rest of the sequence (4), ca. 12 m, consists of a limno-marshy continental facies (palaeosols with solifluctions) with final travertine and with no industry and virtually no macrofauna. In unconformity, based on uneven surface shaped by a preceding erosion cycle there comes an alluvio-detritic sequence (d) with Fagus flora and flake industry. Assemblage (4) is

[1] This is, for instance, the case of the final Acheulian site of Rosaneto (Cosenza), where the flake tools are characterized by a clear-cut predominance of the Levallois technique and the tool range appears to be highly differentiated. Some types have now been definitely standardized, although there are still a large number of pebble tools (unifacial and bifacial choppers, retouched choppers, rabots, etc.) accounting for a good proportion of the assemblage, side by side with bifacials occasionally on flakes and even rather highly developed cleavers (Piperno 1974; Biddittu and Piperno,forthcoming).

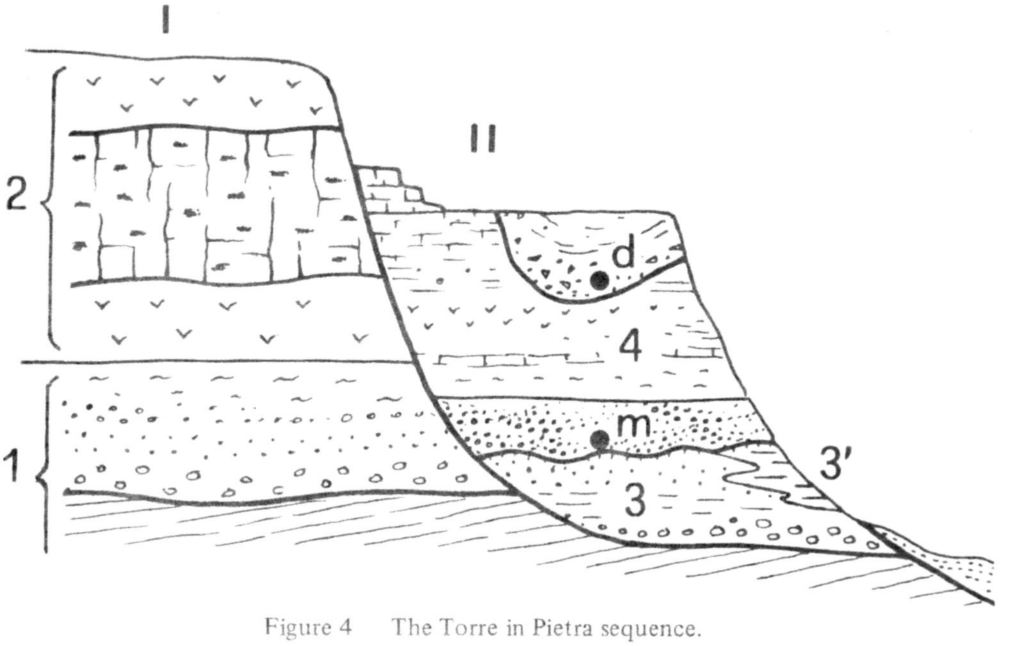

Figure 4 The Torre in Pietra sequence.

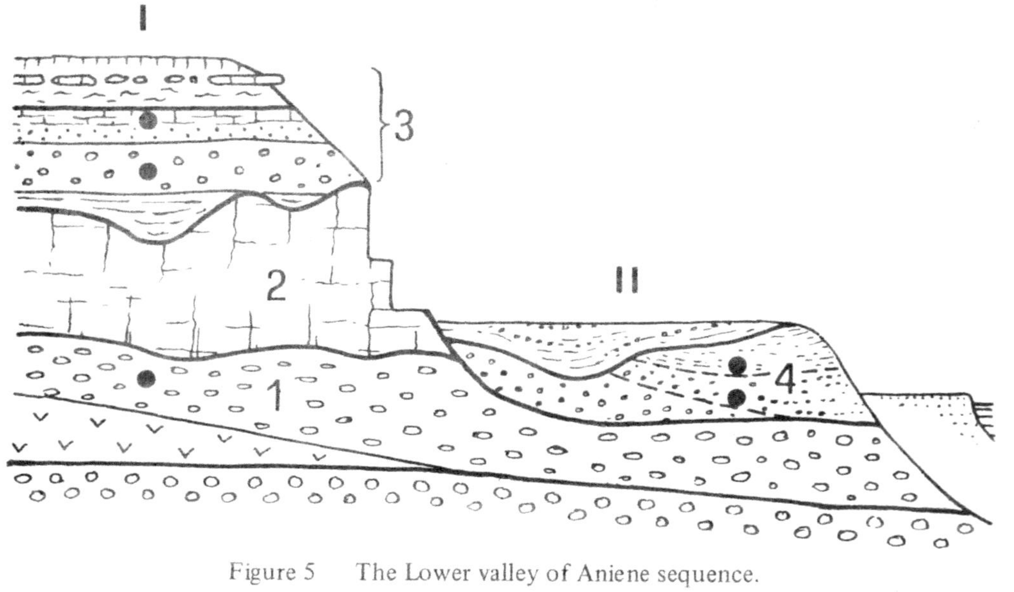

Figure 5 The Lower valley of Aniene sequence.

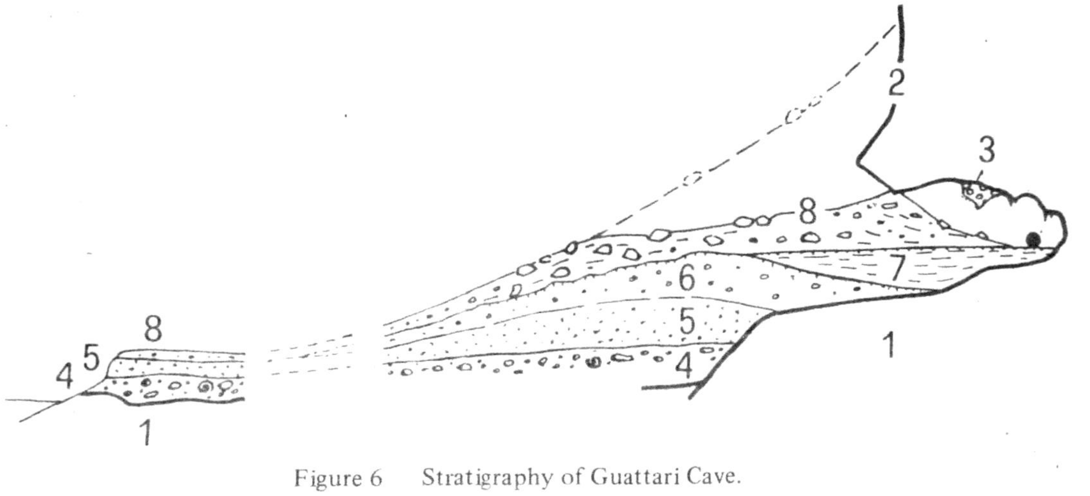

Figure 6 Stratigraphy of Guattari Cave.

generically assigned to an inter-Riss or the cataglacial Riss 1 (Malatesta 1978), while the same author assigns (d) to the beginning of the Riss-Wurm interglacial, although it may conceivably belong to the end of this interglacial.

An important role in the depositing of the earlier part of the Middle-Lower Palaeolithic was played by the terraces of the lower valley of the Aniene river north of Rome (Figure 5), which is now completely built over. The residue of a lower terrace (II), i.e. that of Saccopastore, lay on a middle terrace (I) (Segre1948). The Saccopastore terrace is well known in the literature because of the finding of two pre-Neanderthal skull remains (Sergi 1948) accompanied by a flake industry in the alluvial sand and gravel (4). The flora and fauna (Tongiorgi 1939; Blanc 1948) assign these levels to the Riss/Wurm interglacial, i.e. ca. 130,000 years ago. The intermediate terrace (I) consists of pyroclastic layers mostly from the Latial volcano, as well as by limno-fluvial silt, sand and gravels. The terrace, situated on both the right (Monte delle Gioie) and left (Sedia del Diavolo) of the valley, appears as equivalent sequences with certain facies variations. It must be remarked that the lower part with the gravels (1) underlying the "lithoid tuff" layer (2) that crops at Tor di Quinto on the right bank of the Tiber, has produced a polished bifacial *in situ* (Blanc 1934). In discordance on the "lithoid" lies a fluvio-lacustrine sequence (3) with consolidated sands and travertine at Monte delle Gioie, containing flake industry immediately above a cold avifauna horizon (Blanc 1955, 1956). On the Sedia del Diavolo terrace on the opposite side of the valley, the industry lies in a fine gravel level immediately below the above horizon, which appears here with a clear lacustrine facies. Here the industry is associated with mammals of at least a temperate-warm environment (presence of hippopotamus). However these two levels with industry, which Blanc (1955, 1956) held to be contemporary, are at worst chronologically close together. It must be recalled that even at the time, Blanc correlated the higher level (d) of Torre in Pietra with those of the lower Aniene valley. This correlation could seem satisfactory in the absence of any radiometric dating. However, it must also be remembered that the upper levels of the middle terrace of the Aniene valley (Monte delle Gioie, Sedia del Diavolo) are considered still to be Rissian (Blanc 1948; Taschini 1967) owing to the presence of arctic avifauna at Monte delle Gioie immediately below the industry layer (Blanc 1955). There would thus be a close succession rather than an equivalence between the industries of the middle terrace (I) of the Aniene and that of the upper level (d) of Torre in Pietra.

At Monte Circeo, south of Rome, Grotta Guattari displays a stratigraphy (Figure 6) which facilitates the assignment of the Neanderthal skull discovered there on the surface of a palaeosol inside the cave (Blanc 1939; Sergi and Ascenzi 1974), the remains of jaws, as well as the associated mammal fauna (*Bos primigenius, Cervus elaphus, Crocuta spelaea,* etc.); they are included between a complex of underlying detrital-pedogenic horizons (5–6–7) and a successive detrital cone (8) blocking the cave by closing it completely. This cone was formed by repeated earthslides during the dry lower pleniglacial (W-3) which ended during the xerothermic phase in about 13,000 y.b.p. (Escalon 1972; Cassoli and Segre 1977). The entire lower complex contains but scanty Mousterian industry with a "Pontinian" facies, obtained from pebbles and lying on the "Strombus" Tyrrhenian beach (4).

In southern Italy, the fluvio-lacustrine basin of Venosa displays a very interesting stratigraphic sequence (Figure 7) (Barral *et al.* 1978; Segre 1978). It comprises upper layers ranging from the more highly evolved Acheulian (A) with fauna such as *Palaeoloxodon antiquus, Bos primigenius*, etc.) to a lower one with flake and pebble industry (C) associated with inter-Mindel fauna considerably older than that of the higher levels

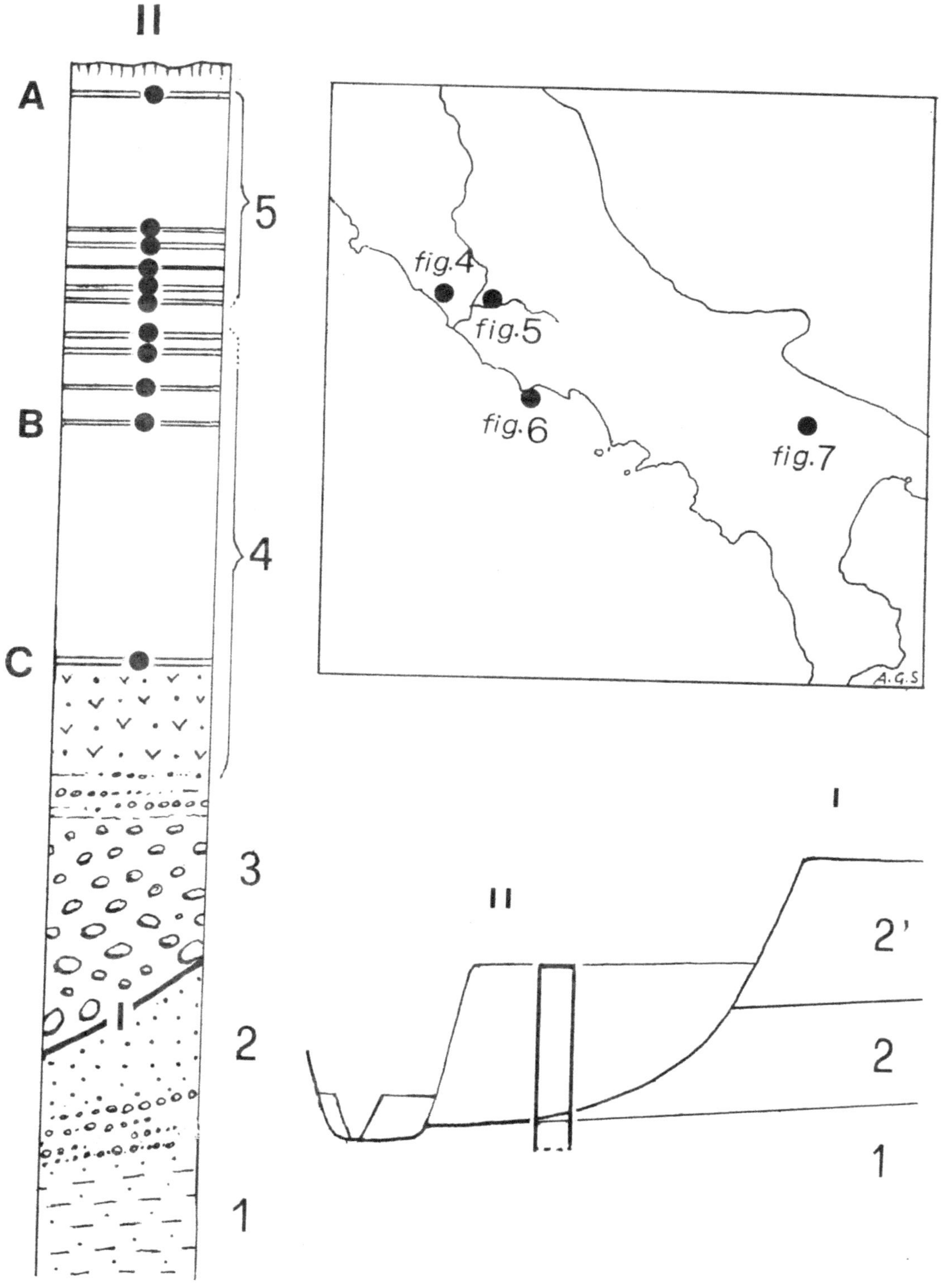

Figure 7 Stratigraphic sequence of Venosa basin.

as it contains, among others, *Dicerorhinus etruscus, Equus stenonis, Ursus deningeri, Homoterium* (Bonifay 1977; Angelelli *et al.* 1978). It may correlate with the similar horizon at Anagni, with K/Ar dating of 458,000 years (Biddittu *et al.* 1979) or be slightly older. The Venosa sequence, which is still being explored (Piperno and Segre 1980), has several archaeological levels in a continuous stratigraphic series extending over about 30 m, the upper levels of which (A-B) contain a flake and bifacial industry, that is more continental and limno-marshy facies at the top (5), and more fluvio-lacustrine facies towards the bottom (4) with fresh-water limestone layers, the whole of which is characterized by numerous, occasionally pedogenized, pyroclastic-tufitic intercalations originating in the next Vulture Pleistocene volcano. The whole sequence ends with coarse gravel (3) and rests on a large high terrace of Villafranchian gravels with *Elephas meridonalis* (2), which in their turn overlie epilitoral marine Calabrian sands and clays (1).

References

Angelelli, F., L. Caloi, A. Malatesta and M.R. Palombo 1978 Fauna quaternaria di Venosa, cenni preliminari *Atti 20° Riun. Scient. I.I.P.P.* Firenze, pp. 133-140.

Barral, L., V. Heinichen-Chiappella and S. Simone 1978 Datazione relativa del giacimento di Loreto, Venosa *Atti 20° Riun. Scient. I.I.P.P.*, pp. 126-132.

Blanc, A.C. 1934 Amigdala chelleana delle ghiaie quaternarie del Tevere presso Ponte Milvio, Roma *Riv. Antropol.* 30: 3-6.

Blanc, A.C. 1939 L'uomo fossile del Monte Circeo: un cranio neandertaliano nella Grotta Guattari a San Felice Circeo *Rend. R. Accad. Naz. Lincei* Roma, s.6, 29, fasc. 5: 205-210.

Blanc, A.C. 1948 Notizie sui trovamenti e sul giacimento di Saccopastore e sulla sua posizione nel Pleistocene laziale *Paleontogr. Italica* 42: 3-23.

Blanc, A.C. 1955 Ricerche sul Quaternario laziale 3°: avifauna artica, crioturbazioni e testimonianze di soliflussi nel Pleistocene medio-superiore di Roma e di Torre in Pietra *Quaternaria* 2: 187-200.

Blanc, A.C. 1956 Il più antico reperto fossile umano del Lazio rinvenuto a Roma: un secondo metatarsale destro umano e industria paleolitica nelle ghiaie superiori della Sedia del Diavolo, Roma *Quaternaria* 3: 259-260.

Blanc, A.C. 1957 On the Pleistocene sequence of Rome. Palaeoecology and Archaeologic correlations *Quaternaria* 4: 95-109.

Blanc, A.C. and A.G. Segre 1953 Le Quaternaire du Mont Circé *INQUA 4e Congr. Intern.*, Rome-Pise, Guides, pp. 23-107.

Blanc, A.C., G. Cova, P. Franceschi, F. Lona and F. Settepassi 1955 Ricerche sul Quaternaria laziale 1° *Quaternaria* 2: 159.

Biddittu, I. 1971 Il Paleolitico inferiore di Arce e Fontana Liri, Frosinone *Arch. Antrop. Etnol.* 101: 251-252.

Biddittu, I. 1974a Giacimenti pre-Acheuleani di Castro dei Volsci, Frosinone *Mem. Ist. It. Paleont. Umana* 2: 51-60.

Biddittu, I. 1974b Giacimento pleistocenico ad amigdale acheuleane nel territorio di Ceprano *Mem. Ist. It. Paleont. Umana* 2: 61-67.

Bidditu, I. and P.F. Cassoli 1966 Una stazione del paleolitico inferiore a Pontecorvo in prov. di Frosinone *Quaternaria* 10: 167-197.

Bidditu, I., P.F. Cassoli and L. Malpieri 1967 Stazione musteriana in Valle Radice nel comune di Sora, Frosinone *Quaternaria* 9: 321-348.

Bidditu, I. and A.G. Segre 1976 Giacimenti preistorici e Quaternarion della provincia di Frosinone *Boll. Ist. Storia e Arte d. Lazio Merid.* 9(1-2): 21-44.

Bidditu, I., P.F. Cassoli, F. Radicati di Brozolo, A.G. Segre, E. Segre Naldini and I. Villa 1979 Anagni a K-Ar dated lower and middle Pleistocene Site, Central Italy: a preliminary report *Quaternaria* 21: 53-71.

Bidditu, I. and M. Piperno In preparation L'Acheuleano di Rosaneto, Cosenza.

Bonifay, M.F. 1977 Liste préliminaire de la grande faune du gisement préhistorique de Venosa (Basilicate, Italie). Fouilles 1974–1976 *Bull. Mus. Anthrop. Préhist. Monaco* 21: 115-125.

Bordes, F. 1961 *Typologie du Paléolithique ancien et moyen* Pubbl. Inst. Préhist. Univ. de Bordeaux, Mém. I. Delmas.

Borzatti von Löwenstern, E. and D. Magaldi 1967 Ultime ricerche nella Grotta dell'-Alto, S. Caterina Nardò *Riv. Sc. Preist.* 22(2): 205-250.

Cardini, L. 1956 Bisceglie, Bari *Riv. Sc. Preist.* 11: 243-244.

Cassoli, P.F., A.G. Segre and E. Segre 1977 Evolution morphologique et écologique de la côte de Castro, Pouilles, dans le Pléistocène final *Colloques Intérnat. C.N.R.S.* n° 271, pp. 325-332, Bordeaux.

Chavaillon, J., N. Chavaillon, F. Hours and M. Piperno 1978 Le début et la fin de l'Acheuléen à Melka Kunturé, méthodologie pour l'étude des changements de civilisations *Bull. Soc. Préhist. Fr.* 75: 105-115.

Chavaillon, J., N. Chavaillon, F. Hours and M. Piperno 1979 From the Oldowan to the Middle Stone Age at Melka Kunturé, Ethiopia. Understanding Cultural Changes *Quaternaria* 21: 87-114.

Chiappella, V. 1964 Il Paleolitico inferiore di Venosa *Bull. Paletnol. It.* 73: 5-23.

Cornaggia Castiglioni, O. and A. Palma di Cesnola 1967 Grotta delle Mura, Monopoli, Bari: l'industria musteriana del livello H *Riv. Sc. Preist.* 22(1): 3-21.

Costa, O.G. 1864 Relazione intorno alle ossa di mammiferi fossili di Cassino e della Melfa *Rc. R. Acc. Sc. Fis. Mat.* Napoli, 3(6): 155-161.

Escalon de Fonton, M. 1972 L'Epipaléolithique et le Mésolithique dans le midi de la France *Coll. Intérnat. s. l'Epipaléol. Méditerranéen* Aix en Prov., 35-59.

Evernden, J.F. and G.H. Curtis 1965 The Potassium-Argon dating of late Cenozoic Rocks in East Africa and Italy *Current Anthropology* 6(4): 343-385.

Leakey, M.D. 1971 *Olduvai Gorge Excavations in Beds I and II, 1960-1963* Cambridge Univ. Press.

Leakey, M.D. 1976 The early stone industries of Olduvai Gorge *IX Congr. U.I.S.P.P., Colloque 5* Nice, pp. 24-41.

Longo, E., F. Mallegni, R. Mariani, A.M. Radmilli and A.G. Segre 1981 Giacimento ad amigdale e resti umani di Castel di Guido a Roma *Atti 23° Riun. Scient. I.I.P.P.* (in press).

Longo, E., C. Pitti and A.M. Radmilli 1981 Prima campagna di scavi nella stazione del Paleolitico inferiore a Castel di Guido a Roma *Atti Soc. Tosc. Sc. Nat.* Pisa (in press).

Longo, E. and A.M. Radmilli 1972 Nuovo giacimento con amigdale a Roma *Riv. Sc. Preist.* 27(2): 403-409.

Malatesta, A. 1978 La serie di Torre in Pietra nel quadro del Pleistocene romano *Quaternaria* 20: 537-577.

Mallegni, F. and G. Fornaciari 1981 Nuovo reperto umano proveniente dal giacimento del Paleolitico inferiore di Castel di Guido, Roma *Arch. Antrop. Etnol.* Firenze, 111: 290-292.

Mochi, A. 1912 *La succèssion des industries paléolithiques et les changements de la faune du Pleistocene en Italie.*

Mochi, A. 1912 A proposito della cronologia del Paleolitico italiano, risposta ad alcune critiche *Archiv. Antropol. Etnol.* Firenze 42(1): 144-151.

Piperno, M. 1974 Presenza di Hachereaux nel Paleolitico inferiore italiano *Mem. Ist. It. Paleont. Umana* 2: 43-50.

Piperno, M. 1980 Considerazioni e problemi a proposito del Paleolitico inferiore italiano *Atti 23º Riun. Scient. I.I.P.P.* Firenze 1982.

Piperno, M. and I. Biddittu 1978 Studio tipologico e interpretazione della industria acheuleana e premusteriana dei livelli *m* e *d* di Torre in Pietra *Quaternaria* 20: 441-536.

Piperno, M. and A.G. Segre 1980 Pleistocene e Paleolitico inferiore di Venosa; nuove ricerche *Atti 23º Riun. Scient. I.I.P.P.* Firenze 1981.

Piperno, M. and A.G. Segre 1980 Il Paleolitico inferiore nel Lazio, Basilicata e Sicilia *Atti 23º Riun. Scient. I.I.P.P.* Firenze 1982.

Radmilli, A.M. 1976 The first industries of Italy *IX Congr. U.I.S.P.P., Colloque 8º* Nice, pp. 35-74.

Segre, A.G. 1948 Sulla stratigrafia dell'antica cava di Saccopastore presso Roma *Atti Acc. Lincei* 345: 743-751.

Segre, A.G. 1978 Il Pleistocene e il Paleolitico della Basilicata *Atti 20º Riun. Scient. I.I.P.P. (1976)* Firenze, pp. 15-39.

Segre, A.G. and I. Biddittu 1981 Paleolitico inferiore: le piú antiche tracce dell'attività umana nella Valle del Sacco *Quaderni d. Centro di Studio p. l'Archeologia Etrusco-Italica* 5: 26-34.

Sergi, S. 1948 Il cranio del secondo paleantropo di Saccopastore *Paleontogr. Italica* 42: 25-164 & *in: Riv. Antropol.* Roma 36: 1-95.

Sergi, S. and A. Ascenzi 1974 *Il cranio neandertaliano del Monte Circeo* Acc. Naz. Lincei, pp. 65.

Taschini, M. 1967 Il "Protopontiniano" rissiano di Sedia del Diavolo e Monte delle Gioie, Roma *Quaternaria* 9: 301-319.

Taschini, M. 1970 La Grotta Breuil al Monte Circeo; per una impostazione dello studio del Pontiniano *Origini* 4: 45-78.

Taschini, M. 1972 Sur le Paléolithique de la Plaine Pontine, Latium *Quaternaria* 16: 203-223.

Taschini, M. 1979 L'industrie lithique de Grotta Guattari au Mont Circé, Latium: définition culturelle, typologique et chronologique du Pontinien *Quaternaria* 21: 179-247.

Tongiorgi, E. 1939 La flora fossile di Saccopastore *Riv. Antropol.* Roma 32: 237-242.

Tozzi, C. 1970 La grotta di S. Agostino, Gaeta *Riv. Sc. Preist.* 25: 3-87.

THE TRANSITION FROM LOWER TO MIDDLE PALAEOLITHIC IN GREECE AND THE SOUTHERN BALKAN

Ludwig Reisch

At any attempt to discuss special aspects of the Palaeolithic in a European or Old World context the Southern Balkan Peninsula is worthy of careful consideration. Leading from Central Europe to the south-east it extends like a side-track far into the eastern Mediterranean. At no time have there prevailed such extreme environmental conditions as in the northern adjoining regions, and necessarily this Peninsula must have received at least some of the eluding movements caused by the quaternary climatic changes. For that reason it is hard to imagine that precisely in this area traces of early man should be absent, as it was suggested by the state of research for a long time.

The process of the transition from the Lower to the Middle Palaeolithic in this region is still unclear because, in spite of a quickly increasing number of Palaeolithic sites in general, precisely dated Middle Pleistocene complexes in particular are lacking till now. Indeed, it should be accepted that the artifacts reported by Poulianos (1971) from the cave of Petralona, which has become so famous for the hominid skull, are relevant here. But, unfortunately, the publications presented so far do not allow a definite comment. It is also difficult to discuss the archaic pebble tools described by Cubuk (1976a, 1976b) from two elevated marine terraces, 75 and 85 meters above sea level, near Nea Skala on Kefallinia.

Therefore this paper can certainly not offer any clarifying answer to the raised questions. But it can, nevertheless, try to summarize all the available data for the space of time referred to. By that two points of view are of particular interest: the occurrence of handaxes or handaxe-containing industries; and the first precisely dated appearance of industries with a Middle Palaeolithic character.

Handaxes are known from the Southern Balkan Peninsula, up to now, only in a very small number. The oldest one is probably an elongated cordiform biface, made from green trachyte, with an unworked base (Higgs 1964; Dakaris *et al.* 1964: 202). It is a solitary surface find from an undated high level gravel, about 60–90 m above the present valley floor, near Palaeokastron in the mountainous part of Western Macedonia, but it resembles quite well those elsewhere known from the Acheulian.

The broken tip of another handaxe was found together with numerous artifacts which had been collected, for the most part, from the surface and from material washed out of red clay deposits at Kokkinopilos, west of the Pindos range in Epirus (Dakaris *et al.* 1964: 213ff). This assemblage dates according to Higgs, but without convincing arguments, from a middle phase of the Last Glaciation (Würm) and has been described as a Middle Palaeolithic industry with a well represented Levallois technique, many

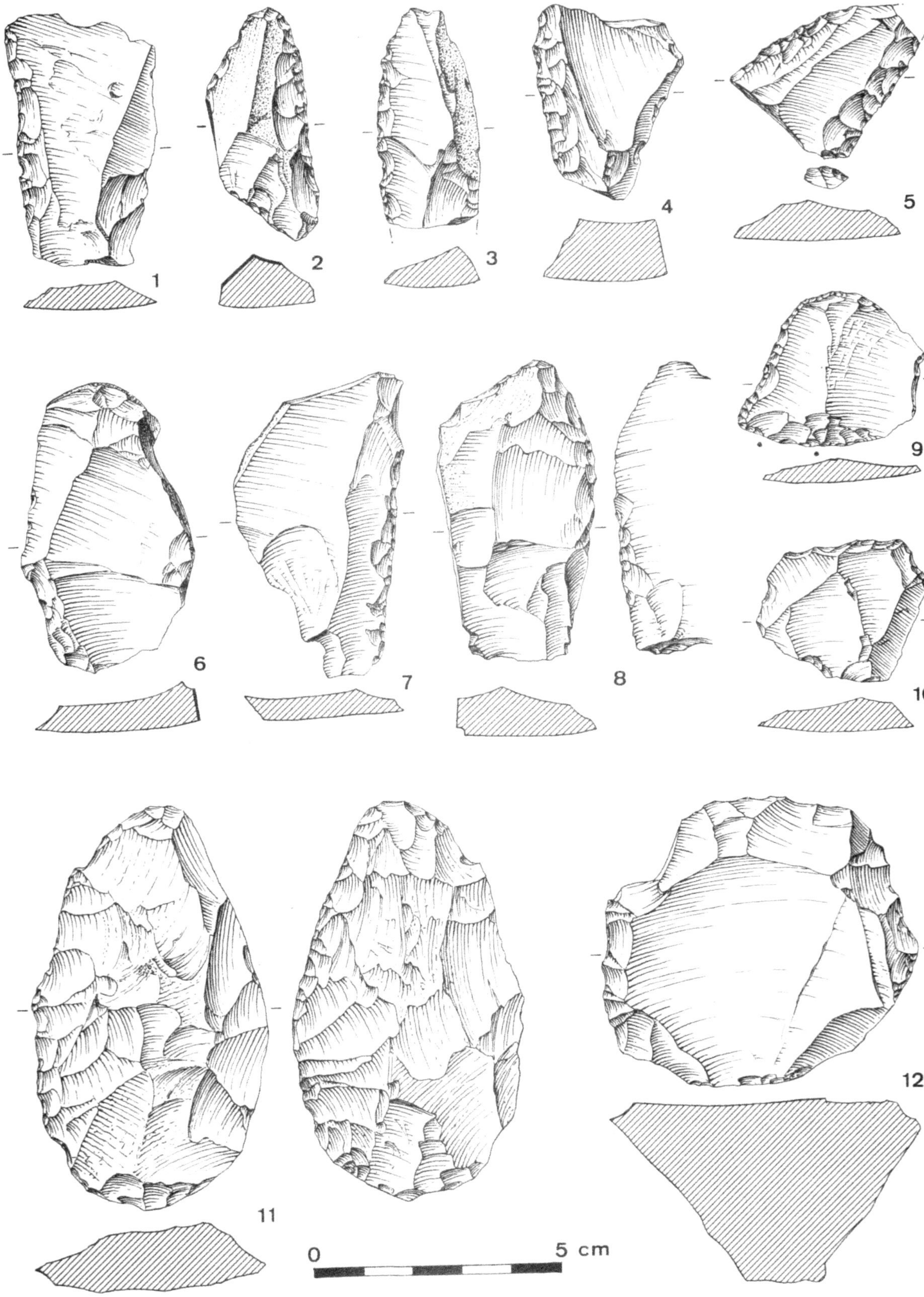

Figure 1 Tripotamia, Nomos Arcadia, Greece. Flint artifacts.

different side-scrapers, Mousterian points and also some leaf-points (Mellars, *in* Dakaris *et al.* 1964: 229ff). If all these pieces really belong to one and the same assemblage, then they demonstrate that handaxes still occur, in Greece too, in an evolved Middle Palaeolithic industry.

Something similar applies to a small assemblage that has been described from a site named "Vasilaki" (Chavaillon and Hours 1964), but which was actually found on the territory of the neighboring village, Tripotamia. The artifacts derive from a thick deposit of red loam, with incorporated gravel horizons, exposed in the upper part of the steep eastern slope of the Erimanthos valley, more than 100 m above the river. There are different side-scrapers (Figure 1: 1-5, 7, 8), a few Levallois flakes, prepared cores (Figure 1: 12) and some other tools, together with a small and flat, almost ovate, limande-like handaxe worth to be noted in particular (Figure 1: 11). Unfortunately this interesting complex, which undoubtedly represents a Middle Palaeolithic, but surely not a late one, can also not be dated precisely: tectonic movements in that area exclude a geochronological dating by means of a sequence of river terraces.

Finally there should be mentioned here a heavily patinated artifact from Kastron on the Killini Peninsula at the western coast of the Peloponnesos (Reisch, *in* Kowalczyk and Winter 1979: 339 and Abb. 10). It was found, together with a few atypical pieces, about 160 m above the sea in allochthonous sands of unknown age, that could be derived from Lower Pleistocene gravels of the Pinios river. The rather small and squat tool, roughly worked on both faces, with a broad base and a rounded tip, resembles some small Middle Palaeolithic handaxes, although it can not be excluded that it might be a prepared core.

Considerable problems exist also with regards to the dating of most of the other Middle Palaeolithic but handaxe-free assemblages, which have meanwhile become known in relatively numerous sites in Greece (Reisch 1980). There are, for example, many objections to the Last Interglacial age of some artifact series from Thessaly, proposed by Milojcić and his collaborators (Milojcić *et al.* 1965). According to the faunistic evidence, with the occurrence of the Saiga Antelope and other inhabitants of an open country, these sites date more likely from an early phase of the Last Glaciation. Particular importance is therefore due to those industries which come from marine beach deposits and immediately connected sediments.

The first, few finds of this kind were mentioned by Keraudren (1971: 115f) from the Killini Peninsula. According to him the oldest finds had a stratigraphic position between the formation of the Eu- and the Neo-Tyrrhenian beaches. This, however, could not be verified during later investigations (Kowalczyk and Winter 1979: 335).

More important is a series of new finds that had been discovered quite casually in the course of geological studies on the southern Peloponnesos. In the area of Elaea and Plitra, on the western coast of the South-East Laconian Peninsula, there occur two different Upper Pleistocene beach levels, in some parts considerably dislocated by tectonics, which had been developed not only by erosion but also by accumulation (Kowalczyk 1977). The most instructive situation within the sequence is found north of Elaea (Figure 2): Above Upper Pliocene sediments cut by gullies which were filled with torrential gravels, there extends a marine coastal terrace which yields shells of *Strombus bubonius* Lmk. among other things. Subsequently, during a glacio-eustatic lowering of the sea level, a channel with fluviatile gravels at the bottom had been eroded. These gravels as well as the older terrace are cut by the cliff of a younger beach, roughly parallel to the modern shore. In this terrace occurs *Strombus bubonius* Lmk. as well. In the course of the next period of sea level lowering, the younger beach was covered

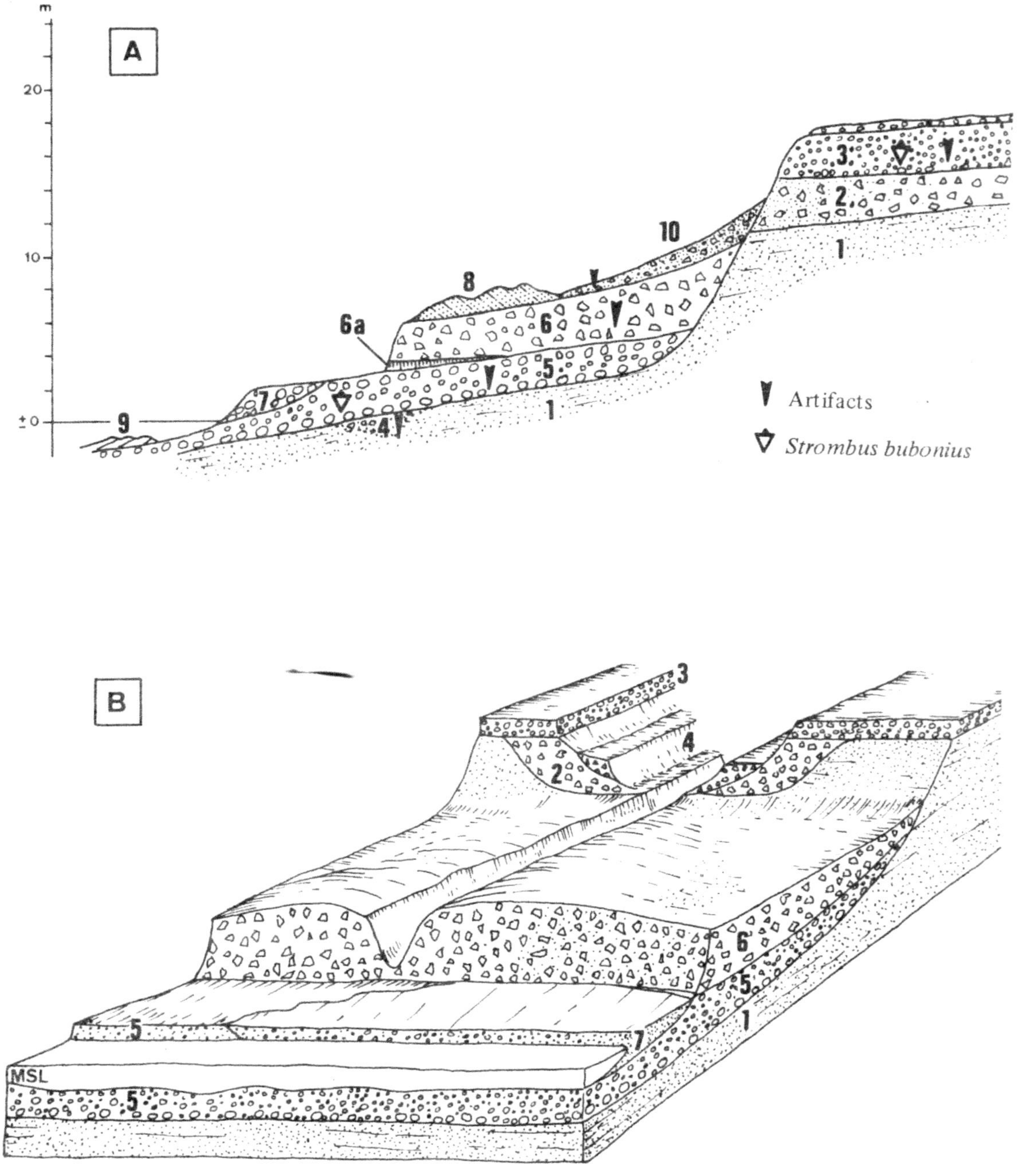

Figure 2 Upper Pleistocene beach terraces and terrestrial deposits near Elaea, Nomos Laconia, Greece. A, compiled section; B, relief-diagram (located north of Elaea).

1, Pliocene; 2, terrestrial debris/torrential gravel of Pre-Eu-Tyrrhenian age; 3, Eu-Tyrrhenian beach terrace; 4, terrestrial debris/torrential gravel between Eu- and Neo-Tyrrhenian; 5, Neo-Tyrrhenian beach terrace; 6, terrestrial deposits/torrential gravel of Post-Neo-Tyrrhenian age; 6a, allochthonous red clay; 7, Holocene beach terrace; 8, subrecent cemented dunes; 9, subrecent beach rock; 10, Holocene scree (After Kowalczyk 1977).

with allochthonous red clays, terrestrial debris from the adjoining slopes, torrential gravels in a fan at the exit of the old reactivated channel and in places even with eolian sands. Because of the presence of *Strombus bubonius* Lmk., the two marine terraces have been attributed to the Eu- and the Neo-Tyrrhenian, respectively, in the sense of the Last Interglacial and early interstadial of the Last Glaciation, whereas the terrestrial sediments would date from the stadial phases between and after them. The youngest beach, adjacent to the Neo-Tyrrhenian terrace and cutting its terrestrial cover, was thus built during the Holocene. Artifacts appear within this sequence at different sites and in different positions. As sources of good flint are absent in the surroundings, the artifacts are primarily made from biabase, in a small part from a dark to black quartzite and very seldom from other stones, altogether a raw material of minor quality.

A quite small assemblage, but so far the oldest of this area, comes from an Eu-Tyrrhenian beach deposit north of Elaea (Figure 3). In addition to some relatively rough flakes with plain striking platforms there should be mentioned in particular a large and flat, almost discoid core with a bifacially worked part of its edge (Figure 3: 1). Even if a cultural determination of such a small collection can hardly be risked, these pieces, nevertheless, give the impression of a different and more archaic industry than the other series.

The archaeological record is much better for the younger Strombus beach, thanks to an increased number of sites, and in spite of the small series found in them (north of Elaea, Elaea, Glifada and Plitra). There are a few simple side-scrapers and fragments of those (Figure 4: 2-4), some more or less regularly prepared cores (Figure 4: 10), typical Levallois flakes (Figure 4: 4-7) and even a Levallois point (Figure 4: 1), the thick platform of which is certainly due to the quality of raw material (diabase) used for its manufacture. These finds show that during the Neo-Tyrrhenian a Middle Palaeolithic industry with Levallois technique prevailed in the southern part of the Balkan Peninsula.

The artifact assemblages found in the terrestrial deposits above the Neo-Tyrrhenian are usually more abundant, and this seems to be the only real difference. They represent also a Middle Palaeolithic with various side-scrapers, points, etc. and with Levallois technique. All the other Middle Palaeolithic industries from Greece are younger, if precisely dated at all, or fit in typologically with the assemblages that can be certainly referred to the time from the Neo-Tyrrhenian.

A different situation seems to be documented by the stratigraphical sequence of the Crvena Stijena in the mountains of Montenegro in southern Yugoslavia (Basler 1967, 1975). In a section more than 20 m deep, there have been distinguished 31 different layers, of which only the lower part needs to be considered here. The basal levels 31–25, that have been dated to the Penultimate Glaciation (Riss) according to sedimentological studies (Brunnacker 1967; Brunnacker *in* Basler 1975: 171ff), yielded several industries which were described as Pre- and Proto-Mousterian. Level 24, of supposed Last Interglacial age (Riss/Würm), contained a Mousterian and above it, in layers 22–11, there was found a sequence of other Middle Palaeolithic assemblages, described as Mousterian and Pontinian. It must be stressed that the cultural determination of the oldest levels relies heavily on the sedimentologically established chronology, to which there can be raised considerable objections. For example, it is peculiar that just those three animal species which, among the whole recorded fauna from this cave, constitute the best indicators of a temperate or interglacial climate – the Fallow Deer, Merck's Rhinoceros and the Greek Tortoise – appear only in the "glacial" levels from 28 downward; on the other hand, the Woolly Rhinoceros, an inhabitant of a more cold and open environment, occurs in the "interglacial" level 24 as well as in the layers

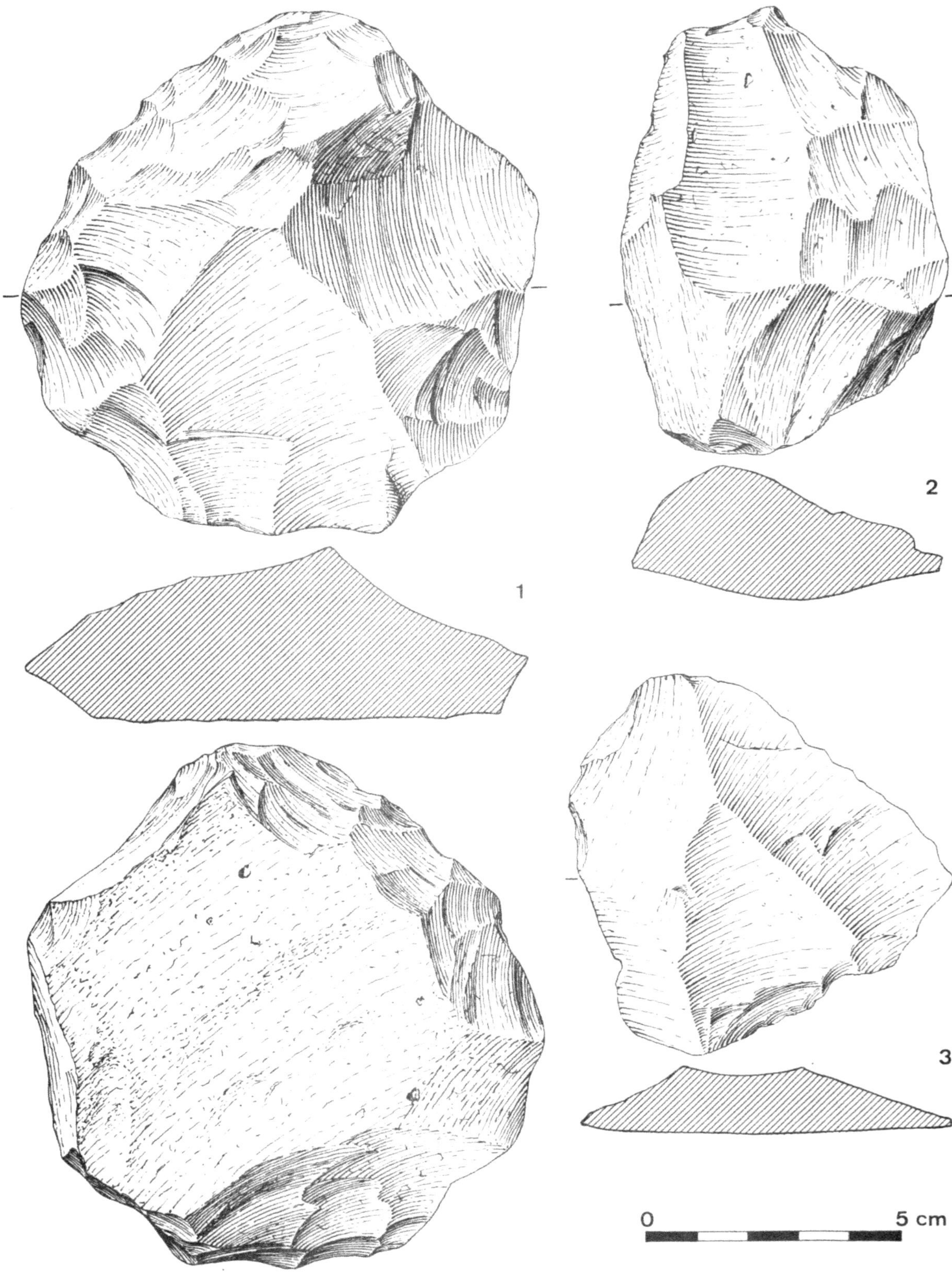

Figure 3 Artifacts from the Eu-Tyrrhenian beach terrace north of Elaea, Nomos Laconia, Greece. Diabase.

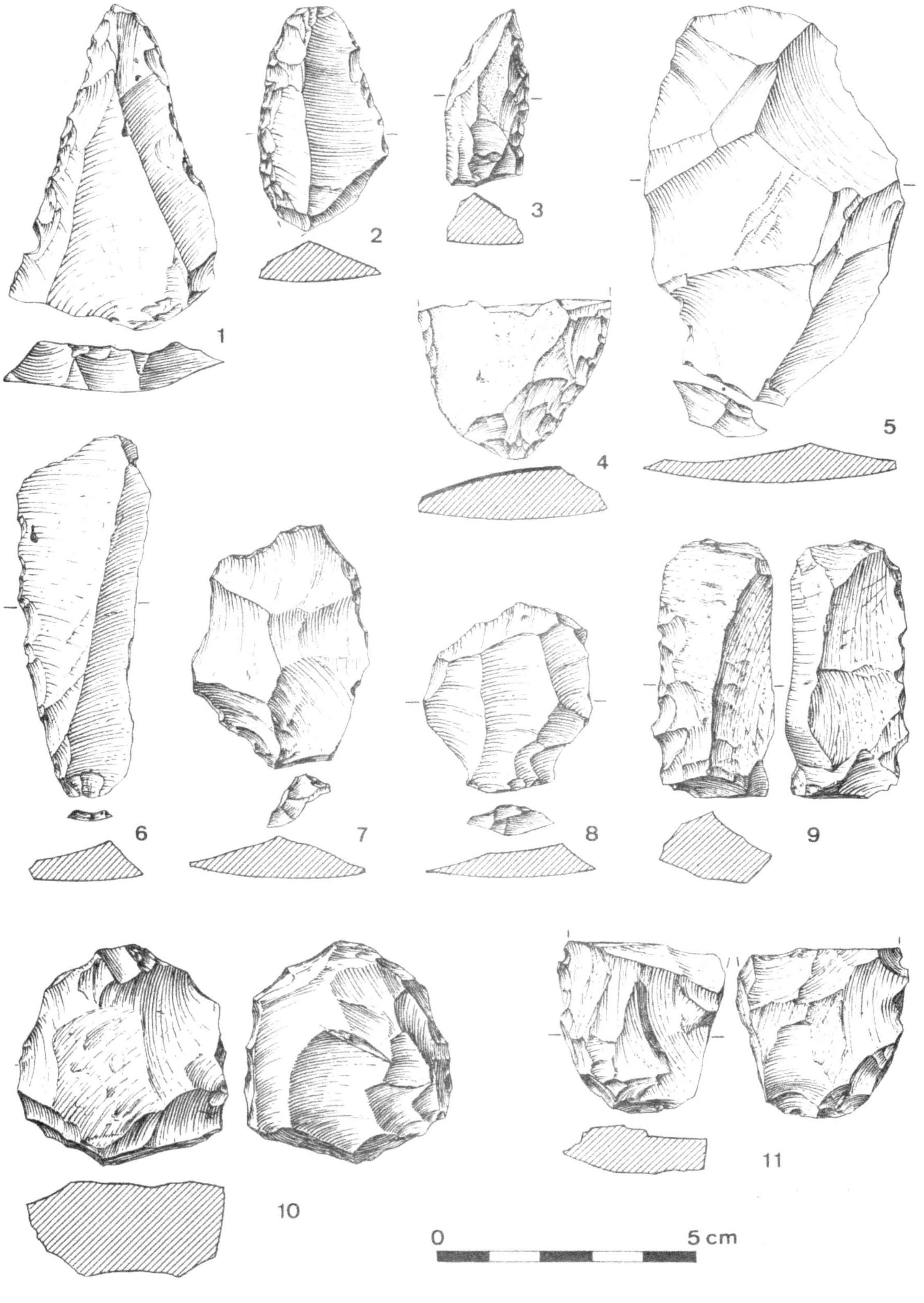

Figure 4 Artifacts from the Neo-Tyrrhenian beach terrace near Elaea, Nomos Laconia, Greece. 1, north of Elaea; 2, 3, Elaea; 4, 5, Plitra; 6-11, Glifada. 1-4 and 6-11, diabase; 5, quartzite.

above. These facts already led Malez (1967) to a divergent interpretation. If one compares therefore, unprejudiced, the sedimentological and the faunistic record, then it seems more probable that all the Middle Palaeolithic industries of the Crvena Stijena date from the Last Glaciation, and that only the lowest part of this sequence approaches the end of the Last Interglacial. According to this interpretation, the cultural remains of the basal horizons lose their particularity, as it is indeed difficult to find in them any essential difference from the Middle Palaeolithic as known from the Neo-Tyrrhenian and the following period in southern Greece.

The picture which has been outlined here for the Southern Balkan Peninsula shows that handaxes, even if not precisely dated, certainly appear during the Lower Palaeolithic. Their small number undoubtedly reflect the present state of research. Without being able to describe a connecting line of evolution, one can state that handaxes occur there also in evolved Middle Palaeolithic industries with Levallois technique, probably during the Last Glaciation. On the other hand, clear Middle Palaeolithic complexes can not be traced back to a time earlier than the Eu-Tyrrhenian, which is thought to correspond to the Last Interglacial, that is the Eemian of North and North-Western Europe. It is not yet possible to describe the process of transition from the Lower to the Middle Palaeolithic in that area. The finds seem, nevertheless, to demonstrate that this transition should have been completed at least during the Last Interglacial.

References

Basler, D. 1967 Arheoloski nalazi u Crvenoj Stijeni. *Glasnik – Arheologija*, N.S. 21/22: 7ff., Sarajevo.

Basler, D. 1975 Crvena Stijena – Sbornik Radova. D. Basler (ed.) Niksić.

Brunnacker, K. 1967 Die Sedimente der Crvena Stijena. *Glasnik – Arheologija*, N.S. 21/22: 31ff., Sarajevo.

Chavaillon, J. and N. and F. Hours 1964 Une industrie paléolithique du Péloponnèse: le Moustérien de Vasilaki *Bulletin de Correspondance Hellénique* 88: 616ff.

Cubuk, G.A. 1976a Altpaläolithische Funde von den Mittelmeerterrassen bei Nea Skala auf Kephallinia (Griechenland) *Archäologisches Korrespondenzblatt* 6: 175ff.

Cubuk, G.A. 1976b Erste altpaläolithische Funde in Griechenland bei Nea Skala, Kephallinia (Ionische Inseln) *IXe Congrès Union Internationale des Sciences Préhistoriques et Protohistoriques* Nice, Colloque VIII, 152ff.

Dakaris, S.I., E.S. Higgs and R.W. Hey 1964 The Climate, Environment and Industries of Stone Age Greece: Part I *Proceedings of the Prehistoric Society* N.S. 30: 199ff.

Higgs, E.S. 1964 A Hand Axe from Greece *Antiquity* 38: 54ff.

Keraudren, B. 1971 Les formations quaternaires marines de la Grèce (Deuxième Partie) *Bulletin du Musée d'Anthropologie Préhistorique de Monaco* 17: 87ff.

Kowalczyk, G. 1977 Jungquartäre Strandterrassen in SE-Lakonien, Peloponnes *Proceedings of the VIth Colloquium on the Geology of the Aegean Region, Volume I* Athens, 435ff.

Kowalczyk, G. and K.-P. Winter 1979 Die geologische Entwicklung der Kyllini-Halbinsel im Neogen und Quartär (West-Peloponnes, Griechenland) *Zeitschrift*

der deutschen geologischen Gesellschaft 130: 323ff.
Malez, M. 1967 Gornjopleistocenska fauna Crvene Stijene. *Glasnik – Arheologija*, N.S. 21/22: 67ff., Sarajevo.
Milojcić, V. *et al.* 1965 Paläolithikum um Larissa in Thessalien. *Beiträge zur Ur- und Frühgeschichtlichen Archäologie des Mittelmeer-Kulturraumes* Bd. 1, Bonn.
Poulianos, A.N. 1971 Petralona – A Middle Pleistocene Cave in Greece *Archaeology* 24: 6ff.
Reisch, L. 1980 Pleistozän und Urgeschichte der Peloponnes (Unpubl.). Habilitationsschrift, Erlangen.

Discussion

A. Ronen: You showed a reddish soil on the Greek coast and one of the typical Levallois flakes which were found on it. This was indistinguishable from what we find in our red sandy soil. Is the Greek red soil similar to that which you have seen on our coastal plain?

L. Reisch: Most of the red sands or red soils are not autochthonous but allochthonous, washed down from the adjoining slopes. They build a thin layer at the base of the terrestrial deposits covering the Neo-Tyrrhenian, as the youngest Strombus beach, in a good sequence. On top of the Eu-Tyrrhenian, however, lies only a very thin sediment cover. I remember nothing from Greece which might be similar to the Kurkar, but the allochthonous red sands may be comparable to the Hamra. At the last site we visited during the excursion, I saw also removed Hamra material redeposited on top of the Hamra.

G. Bosinski: What is your opinion on Petralona, recently much discussed in the literature, and the skull found in it?

L. Reisch: I can say nothing about the skull because I am no anthropologist. I only can say something about the fauna and the tools. The fauna is considered to be of Biharian age, which means from the early Middle Pleistocene. The stone artifacts from this site are rendered only by photographs which are not very good. Poulianos says that he has found choppers, chopping tools and some cleavers. This we have to take on faith, since it is very difficult to judge the tools by the photographs and I have not yet had the chance to see them.

E. Trinkaus: To my knowledge, all of the faunal material and all of the artifacts from Petralona come from a different gallery than the one from which the cranium derives. Although Poulianos correlates the cranium with one of the layers in his excavations, there is no way to confirm that correlation. Yet, a Middle Pleistocene date appears to be reasonable.

K. Brunnacker: Some U/Th dates of Petralona are very young. The problem is that we do not know the exact position of the dated samples.

D.A. Roe: I would like to hear a mention of the Pontinian because, although I do not know the Italian material well myself, is it not so at Torre in Pietra that there is another industry that is very relevant to this transition period, for which a "Rissian" or equivalent date has also been proposed? And it seems to me as I listen to the contributions that the existence of these industries with early Middle Palaeolithic technology is very widely distributed over much of Continental Europe, and over the peninsulas that extend into the Mediterranean; also in the Levant. It is becoming a very frequent phenomenon with many different occurrences – I am sure there must be more in Italy, and I do not know what there may be in the Iberian Peninsula.

G. Bosinski: I agree that there is something in Italy, in Northern Italy, which is dated to the Riss and is in our sense Middle Palaeolithic.

L. Reisch: I think that the Italian region is indeed very important for the questions and the time considered here. Now, some words about Torre in Pietra. There exist two different layers with artifacts: The first one, which yielded many Acheulian handaxes, can surely date from an early Riss. The second one is connected with the first Tyrrhenian complex. Therefore, it may date from the Last Interglacial and not from the Rissian. There are great similarities between the younger industry of Torre in Pietra and the material from the Greek coast, but the collections from the Peloponnesos are still too small to allow an exact determination.

4

AFRICA AND S-E ASIA

THE TRANSITION FROM LOWER TO MIDDLE PALAEOLITHIC IN THE AFRICAN CONTINENT

J. Desmond Clark

Introduction

The Acheulian Industrial Complex appears to have developed in the African tropics about 1.5 m years ago and, by 1.0 m years ago, or thereabouts, had been transmitted to Eurasia. The Acheulian is generally considered to have been replaced in Africa by Middle Palaeolithic traditions sometime between 200,000 and 100,000 years ago, immediately before or during the early part of the Last Interglacial. The associated hominid form – either inferred indirectly from its incorporation in contemporary geological contexts or, more rarely, directly, e.g. at Ternifine (Arambourg 1963), or Melka Kunturé (Chavaillon 1974) – is *Homo erectus* (Pilbeam 1975; Howells 1980). By the time of the late Acheulian genetic modification had brought about a geographical variability and forms possessing some *H. erectus* characteristics but also other, more modern traits, have been classified sometimes as an evolved *H. erectus* (e.g. at Salé (Jaeger 1975: 412)), and sometimes as an early *H. sapiens* (e.g. Kabwe (Broken Hill) *H. rhodesiensis* (Oakley *et al.* 1977b: 192-3)), Ndutu (Clarke, R.J. 1976) and Bodo (Conroy *et al.* 1978).

The Acheulian stone artifact assemblage in Africa is characterized by the two main kinds of bifacial tools – handaxes and cleavers – and many varieties of these have been recognized. They are generally large tools and were often made, when technical circumstances dictated, from flakes which were shaped by retouch, sometimes by a hard and sometimes by a soft hammer. The large primary flakes from which the tools were made were obtained by several different percussion methods of which the proto-Levallois and Kombewa methods are the best known. These techniques make their first appearance with Acheulian assemblages that must be nearly 1.0 m years old but their occurrence is sporadic, e.g. in South Africa and the northwest Sahara, and implies a deliberate selection of these methods on the part of the tool-makers.

The percentage of bifaces in an Acheulian assemblage is variable, rarely exceeds 55% in unselected contexts and is usually much smaller. The remainder of the assemblage is composed of a variety of light duty equipment (scraping, boring and cutting tools on flakes and chunks) and heavy duty artifacts (core scrapers, choppers, spheroids, subspheroids and picks), together with the unmodified waste from their manufacture. The proportions in which these tool forms occur is very variable ranging from assemblages at one extreme with relatively high percentages of bifaces and few light duty tools to those at the other with many light duty pieces and no handaxes or cleavers. Between these two extremes a broad range of variability occurs including one facies in which the heavy duty component dominates (Kleindienst 1961). Several explanations for this variability have been proposed but it is this writer's belief that the primary reason is directly related to the kinds of activities in which the hominids were engaged. Of course, there are other factors involved (Isaac 1969) but always in hunter/gatherer societies form and function must be related and, if this is so, then the form of the tool can be expected to reflect the purpose for which it was used. Extreme instances could be a spheroid designed for pounding and a naturally backed knife for cutting. If, therefore, the full range of retouched and utilized tools in an Acheulian assemblage is examined for its behavioral, rather than its taxonomic, significance, it can be divided into seven or so "functional categories" – namely: (1) choppers, (2) spheroids, (3) light duty scraping equipment, (4) light duty boring equipment, (5) heavy duty scrapers and picks, (6) handaxes and cleavers and (7) flakes for cutting, their edges showing micro-chipping and use-polish. Only micro-wear studies of primary context pieces – still for the most part to be carried out – will determine how many of these are valid categories and to what extent some of them, or even individual tools, may have served more than one purpose, either with or without further modification. Partial confirmation for the validity of some of the categories comes from single context butchery sites such as those at Olduvai FLK NI Level 6 (*Elephas reckii*) and FLK NII (*Deinotherium*) (Leakey, M.D. 1971: 64-6, 85-6), Gadeb 8F (*Hippopotamus* sp.) (Clark and Kurashina 1979) and Mwanganda (*Elephas* sp.) (Clark and Haynes 1969). At these sites the stone artifacts are all flakes, light duty scrapers with choppers and polyhedrons and the inference is that these were the tools used in the butchery. Handaxes and cleavers are much less common at butchery sites, e.g. Torralba and Ambrona in Spain (Howell 1966), or are absent, e.g. Duinefontein in South Africa (Deacon 1976), so that bifaces may have been primarily designed for some other, as yet undetermined, set of activities that would not necessarily preclude their having sometimes been used for cutting meat as has been established for two of the handaxes from Hoxne (Keeley 1980: 143-7). Experiment also shows that handaxes and cleavers are very serviceable tools for dismembering large animal carcases (Jones 1981).

While composite functions must be envisaged for some artifacts, much of the Acheulian stone equipment is likely to have been used to produce other tools in materials which, other than in exceptional cases, have not survived. Examples are the stabbing spear from Clacton-on-Sea (Oakley *et al.* 1977a) and clubs and pointed digging-sticks of wood, or carrying trays of bark from the Kalambo Falls (Clark 1969: 162, 172). If this assumption is valid – that tool-form is related to the method of use – then this implies that some large sites and assemblages with a wide range of tool-forms may represent places where several sets of activities were undertaken, probably on a number of separate occasions, perhaps stretching over a number of years. Other sites where only a small number of tool-forms occur are believed to relate to short-term, single-purpose occupation.

Prehistorians have identified two parallel and evolving traditions in stone tool manufacture which are present from beginning to end of Acheulian times. These are the Acheulian biface tradition and the Developed Oldowan flake and chopper tradition previously referred to in some regions as the Hope Fountain tradition. While there is no problem with some assemblages in deciding to which tradition they belong, there are very many more that could be classified with either so that this dichotomy, like that between the Acheulian and Soan in India, is more a product of taxonomy than reality.

Moreover, only one kind of hominid – *H. erectus* and related fossils – is associated with both.

If the reasoning is valid that forms of stone tool reflect the different purposes for which they were intended, then it would seem to be more meaningful to group the various tool-kits of the late Lower and Middle Pleistocene within an all-inclusive Acheulian techno-complex. Within this several facies would be recognizable: biface-dominated (Acheulian *sensu strictu*), scraper-dominated (Developed Oldowan A, B and C) and heavy duty-dominated (unnamed as yet, but c.f. also "Sangoan"), etc. By the end of the Middle Pleistocene, therefore, the evolved Acheulian is manifest in assemblages with refined bifaces, often of diminutive size; in others where the characteristic tools comprise often finely retouched scrapers and other light duty equipment; and in others, again, where a mixture of bifaces, flake scrapers and other artifact forms occur. This is, in fact, the situation also in Europe (Tuffreau, this volume, 1979) where assemblages of similar composition are found and where it has been the custom to describe assemblages of Rissian and Eemian age as Acheulian, if bifaces are present, or as Mousterian if the assemblage is scraper-dominated.

It is within the general hypothesis relating to form and function summarized above, together with innovative technological change, not the least of which is the development of simple compound tools by means of different hafting methods, that the transition from Lower to Middle Palaeolithic is discussed here.

North Africa and the Sahara

The only region where a long and well-stratified sequence is found is the Atlantic coast of Morocco where Biberson (1961) has distinguished four stages of Pre-Acheulian and eight of Acheulian, all in relation to sediments and events associated with changes in sea levels and accompanying terrestrial formations. The evolved Acheulian (Stage VIII) is found within the Pre-Soltanian terrestrial, pink clay/breccia (Cap Chatelier Level D2, Sidi Abderrahman) that overlies the Harounian high sea level (=Tyrrhenian II of the Mediterranean) of early Last Interglacial times. The succeeding Ouljian transgression has been dated by Th/U to 75,000–95,000 years BP. There is every indication that this age is acceptable and, if this is so, then the final Acheulian on the Atlantic coast of Morocco is older than 95,000 years BP and must be dated within the Eem or immediately prior to this, as an age of 145,000–125,000 BP on Th/U is recorded for the Harounian beach (Biberson 1970).

The quantitive breakdown of this final Acheulian assemblage (*op. cit.*: 355-80) which is made predominantly from quartzite cobbles, suggests that it is probably a selected one that does not include most of the miscellaneous debitage. It is dominated (24.8%) by numbers of small – some very small – handaxes mostly cordiform, sub-triangular and ovate in plan form. Cleavers (0.6%) are, by this time, relatively unimportant and poorly made though the sub-rectangular form is represented. Fifty percent of the total consists of flakes of which one third probably come from specialized or unspecialized cores and the rest are biface trimming flakes. Only some 10% of the flakes exhibit retouch but a number of the others, perhaps as many as 30%, show evidence of utilization. Side-scrapers predominate, some being carefully retouched, and there are several sub-types of "point" or convergent scraper. Others of the small retouched tools resemble end- and end-and-side scrapers and there is one piece resembling a limace and another unique piece with scraper retouch and a tang – probably the earliest represented anywhere in north Africa, indeed in the world. The characteristic primary flake is either side- or end-struck but some 1% are blades – both regular, parallel-sided and long triangular examples. "Pebble tools" (9.7%) have been included among the shaped tools though the primary function of these pieces may have been as cores. Indeed, the commonest form of core at Cap Chatelier is a flaked cobble. Associated with these at Cap Chatelier, are Levallois and, in particular, disc or Mousterian-type cores. The cave sites at Sidi Abderrahman ("Grotte du Rhinocéros" and Grotto du Cheval") have produced assemblages similar to that from Cap Chatelier. The excavators have described these as "biface-bearing Lower Mousterian" (Mieg de Boofzheim and Plessis 1954; Biberson 1961: 393).

Assemblages of artifacts classified as belonging to Stage VIII of the Acheulian are also found (in secondary context) in the lower (2 m) terraces of the wadis (e.g. Oued el Khémis) and in association with the tufas of fossil springs (e.g. Aïn Hallouf). The same industry of small cordiform handaxes, flake scrapers and "Mousterian cores" as at Cap Chatelier occurs also at this inland site but a number of spheroids ("bolas") are also found here. The Quaternary sands and clays at the spring at Aïn Fritissa have also produced a number of evolved handaxes compatible with those from Stages VII and VIII of the Moroccan coastal sequence, presumably derived from the lower levels and overlain by a well-developed Aterian industry (Tixier 1959). The basal industry – in quartzite – at the Aterian mound spring of Tit Mellil (Antoine 1939; Biberson 1961: 381) may also belong to this evolved biface tradition.

The terminal Acheulian in Morocco is thus well-dated to the time (Pre-Soltanian) prior to the Last Interglacial (Ouljian) 6–8 m high sea level on the Atlantic coast, to the low terraces in the wadis and the tufas of some mound springs and is best known from the Cap Chatelier assemblage. It has a relatively high percentage of handaxes (25%) combined with retouched flake scraper forms based on various core types among which are pebble (angle) cores, Levallois and Mousterian disc forms. This terminal stage, therefore, combines a developed Acheulian and a Mousterian technology to the extent that it signals the end of the biface tradition and the elaboration of the Mousterian disc and the Levallois core methods (begun in Stage VII contemporary with Tensiftian dune deposits and correlated with the Riss). The strong technical associations of this evolved Acheulian and the single tanged scraper, limace and spheroids are significant indications that Middle Palaeolithic (Mousterian and Aterian) stone technology in the Maghreb was directly derived from the terminal Acheulian, the chief difference between the two being the almost total disappearance of the bifaces, the elaboration of the retouched flake scraper and point forms in the Aterian and a change in the relative importance of the raw material from quartzite to chert. In fact, the final Acheulian in the Maghreb is considered by one author as indistinguishable from the Mousterian of Acheulian Tradition in Cantabria (Freeman 1975: 731).

Turning to Algeria, the evolved Acheulian bifaces occur in association with old palaeosols on the plateau at Ouzidaïne near Tlemcen and at El Ma el Abiod near Tebessa (Balout 1955: 212-16, 224-31) but little more than this is known of their stratigraphic context. The assemblage from El Ma el Abiod in particular comprises finely-made lanceolate and cordiform handaxes together with retouched flake scrapers and points of Mousterian form but no cleavers. Balout considers it to be similar to the lower and upper occupation horizons at the fossil pan site at Sidi Zin, near Kef in Tunisia. Gobert

(1950) found here three horizons of evolved Acheulian made from limestone, flint and quartzite. These were sealed beneath a tufa containing a few artifacts described as "Mousterian". The fauna in the lower level is comparable to that of the Tensiftian and Pre-Soltanian while that from the tufa capping suggests a Soltanian (=Würmian) age. This is perhaps one of the very few sites in northern Africa where a record of the transition from the final Acheulian to the Mousterian is preserved. The Lower and Upper Acheulian levels are characterized by finely-made small lanceolate handaxes as well as small handaxes with a squared distal end, unifacially retouched like an adze. There are also pebble choppers, core scrapers, flakes and flake tools. In contrast, the middle level has no lanceolate (Micoquian) forms. Instead there are unifacial, ovate handaxes made on flakes; cleavers, often with retouched and edge-damaged bits; retouched flake scrapers; points of various kinds and discs. Whereas in the Acheulian horizons, flint and quartzite were used only for the small and light duty tools, all the so-called Mousterian artifacts from the tufa are made from these rocks. Side-scrapers predominate and thick-sectioned "points" and there is a small, crudely-made, almond-shaped handaxe. The above evidence from the Maghreb indicates that the terminal Acheulian was manifest in several different facies – one with lanceolate (Micoquian) handaxes and no cleavers (El Ma el Abiod and Lower and Upper Acheulian levels at Sidi Zin), one with cleavers and unifacial ovate handaxes (Middle level at Sidi Zin), one specializing in cordiform handaxes and few cleavers (Cap Chatelier) and others. All these contain well retouched scraper and point forms while some exhibit one or another specialty such as finely-made discs (Sidi Zin), spheroids (Aïn Hallouf), or "pebble tools" of different kinds (Cap Chatelier and Sidi Zin).

Three points that arise from study of these Maghrebian sites need to be stressed. Firstly, so far as the bifaces are concerned, the trend to smaller, often very small handaxes, the variability in biface form and frequency, and the generally reduced importance of the cleaver, as well as the equally varied finish to the handaxes which are sometimes very finely made, sometimes very crudely made. Secondly, the inseparable association of a light duty, flake tool component and sometimes of a heavy duty element – choppers and core scrapers – though the quantitative values of these components may differ considerably. Thirdly, the variability in the core types used to produce the flakes – unspecialized and specialized (Levallois and disc) forms – varying considerably in the proportions in which they are present in the assemblages. The flake tool component is always referred to as being of "Mousterian" type but it needs to be stressed that such a component is an equally intrinsic part of the late Acheulian in Africa; sometimes thick, unprepared, Quina-type flakes were preferred, at others the flakes come from Levallois and disc cores.

Whereas a change to drier climatic conditions and some tectonic movement in Tunisia (e.g. Gafsa (Balout 1955: 234-57)) may have had some localizing influence on the disappearance of the bifaces in the Maghreb, this can more certainly be seen as a world-wide trend that was taking place at a time of generally cold climate with warm oscillations in oxygen isotope stage 6 as defined in deep sea cores and a very warm climate in stage 5e (Eemian) (Shackleton and Opdyke 1973, 1976). The tradition of tool manufacture characterized in the Acheulian did not itself disappear, only the biface element – the handaxes and cleavers – ceased to be present. The remaining products of late Acheulian technology continued and took on a still more evolved and varied form in the different facies of the Middle Palaeolithic that are recognized in northern Africa.

The different stages of the Acheulian tradition and their evolution in the northwestern part of the Sahara have been traced by Alimen and Zuarte y Zuber (1978), in particular in and adjacent to the Saoura valley (see also Chavaillon 1964). The evolved Acheulian (Stages VI and VII) occurs in the sediments of the Final Ougartian episode which is equated with the Pre-Soltanian on the Moroccan coast and so with the final stage of the Acheulian there. Both stages VI and VII are characterized by well-made handaxes and cleavers, numerous side-scrapers, choppers, spheroids and Levallois flakes. The core forms, such as the proto-Levallois type found with the earlier Acheulian, now become rare and are replaced by typical Levallois and Mousterian disc cores; in Stage VI Levallois cores comprise 22% and disc 54% of all core forms. Assemblages of the final stage (VII) show considerable variation in biface and other forms as at Sidi Zin. Handaxes attain values of 60–80% of the shaped tools; at some sites lanceolates predominate, at others cordiform and limande forms are preferred. The characteristic cleaver has a U-shaped to sub-rectangular profile and is usually made on a Kombewa flake. Spheroids continue to be present. The disc core is now predominant but the Tachenghit form continues to be important at some sites (Alimen and Zuarte y Zuber 1978: 485-7). In the northwestern Sahara, as in the Maghreb, the variability in biface- and core-form, as well as other differences, make it possible to identify several local "facies" but whether these carry any general or chronological implications or are purely stochastic, is unknown.

On sedimentary and pollen evidence, the final Acheulian in the northwestern Sahara is contemporary with a time of fluctuating climate between sub- and semi-arid to arid, the occupation of this part of the desert having been brought to an end by an episode of severe aridity and the formation of the "Grands Ergs", the sands of which cover the habitation sites of the last Acheulian groups.

Many Acheulian assemblages are known from the Sahara (e.g. Arambourg and Balout 1955; Oussedik 1976; Clark *et al.* 1973; Alimen *et al.* 1963) but they are generally single context sites which, on typological and technical grounds, can be shown to belong to the Upper and later Acheulian. No more precise sub-division to identify a final or evolved stage is yet possible on stratigraphical or other evidence though certain sites in Nubia and the Western Desert may be presumed to lie close to the end of the Acheulian tradition there. These are all characterized by a complete, or almost complete, absence of cleavers.

The Acheulian assemblage from the occupation site of Arkin 8 in the Nile valley in Sudanese Nubia (Chmielewski 1968) appears to be late and has a very high percentage of small, ovate handaxes, bifacially flaked discs and choppers with a few amygdaloidal bifaces made from slabs of ferruginous quartzite and quartz nodules. A later assemblage, from Arkin 5 (*op. cit.*: 134-47), is characterized by bifacial lanceolate forms and a strong Levallois element. Also undated but possibly transitional stages are represented at Khor Abu Anga, near Omdurman (Arkell 1949) where, in fluvial and so secondary context, a biface industry occurred dominated by amygdaloidal, lanceolate and cordiform handaxes together with a few long lanceolates which recall those of the Lupemban Industrial Complex from Equatoria (see below). These Nile Acheulian assemblages are totally unlike those from the Western Desert where the artifacts are made from nodules of Eocene chert.

In the Western Desert at fossil spring sites in Dakhla Oasis (Schild and Wendorf 1977), at Bir Sahara (Wendorf and Schild 1980: 243-47) and at Kharga Oasis (Caton Thompson 1952: 54-73), Acheulian activity sites with many handaxes occur clustered round the old spring vents. At the two Dakhla sites (Schild and Wendorf 1977: 20) the handaxes are mostly amygdaloidal, sub-triangular or cordiform but there is also an important group of "backed bifaces" with asymmetric tip similar to those from central and eastern Europe and known there by the term "prondnik" among others. These have a flat, natural back and bifacial retouch on one or both side edges and are comparable to the bifacial "knife" forms distinguished with the Acheulian of East Africa by Kleindienst (1962: 89-90). The Middle Palaeolithic of western Germany contains many such backed knives and clearly the Quina-type bifacial side-scrapers of the Middle Palaeolithic of France belong in the same group of tools. These "prondnik" forms and the biased tip are found again with the Acheulian KO-10 site at Kharga Oasis but are quite absent at Bir Sahara. At the former, the amygdaloidal handaxes predominate while subtriangular and cordiform examples are in the majority at the latter site. Except for notched and denticulate pieces at the Dakhla sites, the light duty, flake tool element is of minor importance in all these assemblages. The Levallois technique is practically non-existent but discoid cores are present and are an important form at the Dakhla sites.

A characteristic of all these sites is the accumulation of many bifaces in the vents or "eyes" of the springs. Such accumulations suggest either that the artifacts were thrown there or that they slid down into an expanding vent and pool from the surrounding surface scatter. At the Dakhla sites there is much flaking waste from biface manufacture but at Kharga there is a comparatively small amount of debitage which suggests that the bifaces were made elsewhere. Schild and Wendorf (1977: 95-6) see a general time equivalence in the spring activity and associated high water table at all these sites (though the Bir Sahara spring is thought to be somewhat later) so that the more humid period in the Western Desert that made possible the later Acheulian occupation there was probably complex, consisting of more than one humid fluctuation, the first of which they tentatively equate with the Tensiftian period of dune accumulation on the Atlantic coast of Morocco. As in the western Sahara, occupation was brought to an end by an episode of severe aridity when the springs dried up and dunes and deflation basins were formed. With the climatic amelioration evidenced by reactivation of the springs and by water-laid sediments in the basins, these localities were re-occupied by groups making tools in the Mousterian tradition. Bifaces are no longer present and the artifact assemblages are mostly classified as Denticulate Mousterian, more rarely Typical Mousterian.

In Cyrenaica at the Haua Fteah Cave in the Jebel Akhdar (McBurney 1967), the lower part of the long stratified sequence below the Levallois-Mousterian belongs, with its warm fauna, within the time of the Eem or Last Interglacial and is, extrapolating from the radiocarbon sequence and carbon isotope evidence, probably 70,000–80,000 years old. The "Pre-Aurignacian" industry contained in these levels is, as yet, anomalous so far as north Africa is concerned though it clearly has affinities with the "Amudian" from Palestine and probably also with the Howieson's Poort facies of the Middle Stone Age in South Africa (see below). This "Pre-Aurignacian" blade industry appears to have little or nothing in common with the Acheulian or the overlying Levallois-Mousterian but, since the tips of three bifacially worked tools were found associated with the "Pre-Aurignacian", it may have been that handaxes were still being made at that time (McBurney 1967: 85-6). Although very different from the flint blades of the "Pre-Aurignacian", there are indications in Morocco (Biberson 1961: 372), as also in the Horn (Clark 1954: 165-68) and east Africa (Leakey, M. *et al.* 1969: 62), if not also in South Africa (Söhnge *et al.* 1937: 90), that blades in quartzite, lava and indurated shale become a feature, though not an important one, of some of the late Acheulian assemblages in these regions. It was, therefore, clearly within the capabilities of the makers of the late Acheulian to produce blades if circumstances made it advantageous to do so.

East Africa and the Horn

Turning to the Horn and East Africa, assemblages assigned to the late Acheulian have been dated by Th/U and Pa/U methods at Isimila, Tanzania, to c. 260,000 years BP (Howell *et al.* 1972). The same method has given an age of 180,000–200,000 BP for a handaxe and several flakes from an old marine shoreline at Dallol in the Afar Depression (Roubet 1969). From the Kalambo Falls, late Acheulian horizons have been dated to more than 60,000 years BP by radiocarbon and to more than 110,000 years BP by the amino-acid racemization method (Lee *et al.* 1976). Earlier Upper Acheulian from Olorgesailie and Kariandusi in the Kenya Rift are associated with sediments dated by the K/Ar method to 480,000 and 1,100,000 years BP respectively. All these assemblages are characterized by well-made handaxes and cleavers, a light duty flake scraper element and choppers and spheroids in varying percentages (Howell *et al.* 1962; Isaac 1977; Clark 1975). With the possible exception of the assemblages from the Kalambo Falls, none of these show any significant difference from what has come to be thought of as a standard East African Upper Acheulian. Certain other, as yet undated, assemblages suggest, on technological and stratigraphic evidence, that they belong in time to the terminal Middle or beginning of the Upper Pleistocene and so were late in the Acheulian tradition. Among these is that from the site of Kapthurin in the Lake Baringo basin of northern Kenya where a hominid mandible ascribed to *Homo erectus* was found. Acheulian living and factory sites with a well-developed Levallois component occurred there at two other horizons (Leakey, M. *et al.* 1969; Isaac 1982). Handaxes are ovate and sub-triangular in plan form and cleavers U-shaped with parallel or divergent side edges. These are associated with flake scrapers, choppers and retouched flakes of various forms made from lava. Half the cores are circular, radially prepared Levallois or proto-Levallois cores. Most of them have been struck and the flakes so produced retouched into bifaces. Significant is the presence of a blade and long flake element and 8.9% of the flakes can be so classified (Leakey, M. *et al.* 1969: 68). The industry is undated but an underlying lava has a K/Ar age of 230,000 BP while tuffs of the basal unit have been dated by the same method to 660,000 BP (Bishop 1972: 230-31). If the younger date is correct it suggests that the Kapthurin Acheulian assemblage lies near the end of the tradition.

Probably of comparable age is a late Acheulian in quartzite from the basal fluvial sediments of the streams at Hargeisa in northern Somalia where large and small Levallois cores are found in association with amygdaloidal and ovate handaxes and flake tools (Clark 1954: 160-69); the Somalia assemblages also contain a significant blade element similar to Kapthurin.

In Ethiopa the Acheulian dates back to 1.5 m years, but its time of disappearance is not yet fixed, except relatively, though it is thought to have taken place about 200,000 years ago. A Middle Stone Age in obsidian associated with old soil horizons and volcanic ash on the slopes of an extinct volcano at Gadamotta in the Ethiopian lakes section of the Rift (Wendorf *et al.* 1975) has been dated to 149,000 and 181,000 BP by the K/Ar method. This is a fully developed Middle Stone Age and, if these dates are correct, they suggest that the final Acheulian must be older. How much older is uncertain but the transition need not be very far removed from the older of these dates. The long and important stratified cultural sequence at Melka Kunturé in the Ethiopian highlands shows in the Final Acheulian (Garba III) a sudden decline in the number of handaxes and cleavers and an increase in the number of small bifacial artifacts, scrapers and borers, which now dominate the collections. These small bifacial pieces, made mostly from obsidian, show fully invasive retouch and are sometimes irregular, sometimes leaf-shaped. In addition, occurring first with the final Acheulian at Garba III are pointed, bifacial and unifacial pieces which have the appearance of "Mousterian" or "Stillbay" points. In fact, the upper level at the Garba III site is considered to mark the transition from the Final Acheulian to the Middle Stone Age (Hours 1976; Chavaillon *et al.* 1978).

Mention should also be made here of the collections of artifacts eroding from deeply weathered palaeosols on the slopes of Mount Kenya at Nanyuki (Leakey, L.S.B. 1931: 38-9) and another occurrence in similar geological context on the Dabat Plateau in northern Ethiopia (Clark 1945). Again, for typological/technical reasons, since they are undated, these are considered as being transitional to the Middle Stone Age and it is of interest that they occur at higher altitudes at a time of drying climate.

Before leaving East Africa it is also necessary to note the discovery of three human crania assigned to *Homo sapiens* (Leakey, R.E.F. *et al.* 1969) in sediments (Kibish Formation) in the lower Omo basin in southwestern Ethiopia, dated by Th/U to 130,000 years BP; further reference to these is made below.

Central and Southern Africa

There are no radiometric or other absolute dates for the late Acheulian/Middle Stone Age transition in Equatoria. However, since the early Middle Stone Age in the Zaïre basin is dated by radiocarbon to more than 40,000 years BP, the Acheulian there must be appreciably older. During the Middle Pleistocene, it would appear that the Zaïre basin and west Africa were covered by moist humid forest which probably extended over much of the Lake Victoria basin also so that hominid settlement – as distribution maps show – was confined to the peripheral, less densely vegetated parts. The Acheulian assemblages there have an important heavy duty element of pick-like forms, core scrapers, choppers and large modified flakes (Clark 1968; De Ploey and Van Moorsel 1963). At Nsongezi in the Kagera valley in southwest Uganda, Cole (1967) found a normal East African Acheulian stratified below a composite land surface rubble (M–N Horizon) containing an industry named Sangoan. Typologically, the Sangoan consists of heavy duty picks and core-axes together with the other tool forms referred to above and was first found on the hills overlooking Sango Bay on the west side of Lake Victoria.

At one time the Sangoan was recognized over a large part of the continent, mostly in the higher rainfall, more thickly vegetated regions, from the forests of west Africa (e.g. at Asokrochona, Ghana (Nygaard and Talbot 1976) and elsewhere (Wai-Ogosu 1973), the Zaïre and Victoria basins, to the savanna woodlands of south central Africa and down to the coastal evergreen thicket of Natal (Clark 1964). At the Kalambo Falls site, the Sangoan occurs above an unconformity and is stratified immediately above the Acheulian with radiocarbon dates suggesting an age of at least 40,000–50,000 years BP (Clark 1969: Appendix J). The heavy core-axes and picks are associated with a light duty, flake scraper element which makes up the most important shaped tool component. These are mostly various kinds of side-scraper on unprepared flakes resembling those associated with the Quina Variant of the western European Mousterian and are usually indistinguishable from those found with the underlying Acheulian horizons. Similar assemblages occur in Zimbabwe on the Charama Plateau (Cooke 1966) and in the lower part of Pomongwe Cave in the Matopo Hills in the western part of the country where the assemblage is dated to ⩾42,000 BP (Cooke 1963). In Angola, Zaïre and the Kagera valley, there occur rare, large bifacial, lanceolates which are characteristic of the early part of the Middle Stone Age industrial complex in Equatoria known as the Lupemban. It appears possible, therefore, that the Sangoan pick/core-axe element which is quantitatively very variable, is not a separate industrial entity but is related to some special activity within the final Acheulian/early Middle Stone Age similar to the heavy duty picks and scraper/adzes of the Australian aborigines where they are used only for heavy wood-working tasks. There is reason to accept the suggestion that the Sangoan picks/core-axes were also related to working wood and its by-products (Clark 1964). Whatever its function, there is manifest in the Equatorial higher rainfall regions, therefore, a technological innovation superseding the late Acheulian, that is absent from northern Africa and the more open, tropical savanna grasslands of the continent.

In South Africa it might be supposed that the occupation of caves took place towards the close of the Acheulian tradition, for example Montagu Cave in the south-western Cape (Keller 1973), the Cave of Hearths in the northern Transvaal (Mason 1962: 74-77, 159-169) or the Wonderwerk Cave at Kuruman (Malan and Cooke 1941; Malan and Wells 1943). The assemblages from these cave sites are generally considered to be late or final Acheulian but there is little other than typology to confirm this and the Acheulian and earliest Middle Stone Age occurrences at these sites, as elsewhere in southern Africa, are always separated by an unconformity and temporal hiatus of unknown length. In 1929 Goodwin and Van Riet Lowe recognized a "Fauresmith Industry" intermediate between the Acheulian and the Middle Stone Age and defined this on the basis of surface eroded finds of handaxes, flake scrapers, poorly-made cleavers and other tools in indurated shale on the high veld of the Orange Free State and the southern Transvaal (Goodwin and Van Riet Lowe 1929: 71-94). In the Vaal River sequence, Van Riet Lowe recognized three stages of the Fauresmith but the basis for this division was never published (Söhnge *et al.* 1937: 90-94), and no sealed primary context sites are known as yet. Three assemblages in secondary context may be representative of an evolved or final Acheulian: two are from the Vaal (Muirton (quartzite) and Inhoek 6 (indurated shale)) (Sampson 1972: 52-59) and the other, Rooidam (indurated shale), from a pan site near Kimberley (Fock 1968 and pers. comm.). These are all characterized by small, ovate, lanceolate and double pointed handaxes, rare and poorly-made cleavers and a large number of small scrapers and other small tools, together with disc cores. The artifacts in these assemblages are significantly smaller than those of the Upper Acheulian. The site at Rooidam has been dated by the Th/U and Pa/U methods to c. 174,000 years BP (Szabo and Butzer 1979) and if this date is compared with those for the Middle Stone Age from the Gadamotta sites in Ethiopia, it would appear (if both sets of results are reliable) that the interface between the end of the Acheulian tradition with its small bifaces and proliferation of small scraper forms, and the early part of the Middle Stone Age, should be placed at about 200,000 years ago in both East and South Africa. Confirmation for an age in this time range comes from cave sites on the south coast of South Africa. Here the earliest Middle Stone Age at sites like Klasies River Mouth and Herolds Bay Cave occurs immediately overlying and within the Last Interglacial beach and must be considered as being contemporary with at least the later part of this transgression and so to be dated at least 80,000-90,000 years ago (Deacon 1979). Even earlier dating evidence comes from the Border Cave on the boundary between Natal and Swaziland where correlation of litho-stratigraphic and sedimentary stratification allows identification of the Middle Stone Age levels, which are all older than 50,000 years BP, with oxygen isotope stages 4, 5 and 6, so placing the base of the Middle Stone Age deposits at c. 195,000 BP (Butzer *et al.* 1978).

All these early Middle Stone Age assemblages from South Africa are characterized by the production of long blades and radially prepared Levallois cores, together with the usual range of other debitage and core forms including discs. Equally characteristic is the rarity of retouched scraping edges though denticulate pieces occur and well-made points and scrapers in very small percentages are present at some sites (Deacon 1979: 91; Beaumont *et al.* 1978; Volman 1979). These may be compared with those found with the Acheulian/Middle Stone Age transition at Melka Kunturé in Ethiopia, noted above. Sub-Saharan sites differ, therefore, from those in north Africa where the earlier Middle Palaeolithic assemblages contain many retouched scrapers and points and are chiefly distinguished from the evolved and final Acheulian by the absence of bifaces and a greater use of the Levallois and disc core methods. However, in the absence of any reliable means of correlating successions in north Africa with those in the south, it is possible that the Mousterian and Aterian north of the Sahara equate, not so much with the early Middle Stone Age in southern Africa as with the later Middle Stone Age assemblages there in which retouched artifacts are appreciably more numerous. If this is so, then there is reason to consider the earlier production of percussion-struck blades in north Africa, seen in the "Pre-Aurignacian" at Haua Fteah and also perhaps, though later, in the Idfuan of non-Levallois facies on the Nile in Upper Egypt (Wendorf and Schild 1976: 244-51) to be complemented in the strong blade element associated with some of the early Middle Stone Age south of the Sahara, e.g. in South Africa (see above) and in Zambia at the Kalambo Falls (Clark 1974: 79). In particular, the succeeding Howieson's Poort facies of the Middle Stone Age in the interior and on the coast of South Africa, which is most probably 80,000–90,000 years old, emphasizes blade technology and the production of backed pieces (Deacon 1979; Butzer *et al.* 1978) and is more or less contemporary with the Cyrenaican "Pre-Aurignacian". The "Pre-Aurignacian" is between 70,000 and 80,000 years old and, associated as it is with a warm climatic episode and warm fauna, it most likely belongs within oxygen isotope stage 5a. The Howieson's Poort facies in South Africa, however, belongs to a time of cold climate and, if the ascribed age is correct (80,000–90,000 years BP) then it belongs in the preceding oxygen isotope stage 5b (Butzer *et al.* 1978). Table 1 sets out two alternative chronologies for correlating the late Middle and early Upper Pleistocene culture sequences in north, east and southern Africa, in relation to the oxygen isotope stages time-scale, the available dates and estimated dates for the stone industries.

It is hoped that the above review will have indicated that, although detailed descriptions of sealed, stratified and well-dated terminal Acheulian assemblages are generally lacking, the characteristics these have in common, whether in northern or southern Africa, show them to be, on the whole, very similar and to differ only in degree. Their common characteristics are: diminutive handaxes of variable form but with ovate, cordiform or sub-triangular and, less commonly, lanceolate shapes predominating; greatly reduced values for, or absence of cleavers; high values for retouched flake tools and use, or increased use, of the Levallois and disc core methods of flake production with, in some localities, the emergence of a blade element. The presence of bifaces is the main characteristic that distinguishes the terminal Acheulian from both the Middle Palaeolithic (Levallois-Mousterian and Aterian) of north Africa and the early Middle Stone Age south of the Sahara both of which, on the basis of the continuity in the lithic technologies, can best be seen as having evolved directly from the late Acheulian.

Still unclear as to its significance is the apparent anomaly in this evolutionary process presented by the "Pre-Aurignacian" and the other early blade assemblages from the Levant and the Howieson's Poort facies of the Middle Stone Age in South Africa. Are these assemblages activity-related and adapted to the processing of specific kinds of resources? might they have ethnic connotations? or might their presence be due to random selection by semi-isolated populations in which raw material may have played a not insignificant part? Any or all of these, or even other factors may be involved but, until new evidence becomes available, it is improbable that they will be better understood. What is becoming clearer, however, is that the "Pre-Aurignacian", Amudian, etc. of the Mediterranean are not an isolated phenomenon but are part of a geographically much more extensive trend prior to the onset of the Last Glaciation.

Hominid Fossils, Industrial Associations and Possible External Influences

It is equally important to examine the human fossils from the late Middle and early Upper Pleistocene for evidence of genetic changes that might show whether general evolutionary trends or more rapid variability were involved. This should enable a better assessment to be made of the cultural innovations of the Middle Palaeoltihic/Middle Stone Age and help us to determine whether these were predominantly local manifestations or whether some may have been modified by the effects of diffusion.

Trinkaus (this volume) has reviewed the fossil evidence from both northern and sub-Saharan Africa and, on the basis of all the available fossils, has concluded that "a gradualist model of hominid evolution is accurate for the period represented by archaic *Homo sapiens* in the fossil record and by the Lower and Middle Palaeolithic transition in the ecological record". Also apparent is the regional distinctness of the fossils from north, east and southern Africa which is surely to be expected when it is considered what vast distances separated the sparse and isolated populations in these three main regions of the continent. The crucial question, as Trinkaus points out, is a chronological one and for the majority of these fossils both their dates and associations are unsatisfactory. On faunal and cultural evidence, there is no doubt that a considerable time interval existed between the Salé, Broken Hill, Saldanha, Ndutu and Bodo fossils and those from Florisbad, Jebel Irhoud, Haua Fteah, Ngaloba, Omo-Kibish and Border Cave. As yet, however, this interval cannot be accurately measured. There is also no doubt about the Acheulian associations of the Saldanha, Ndutu and Bodo fossils; about the Mousterian associations of Jebel Irhoud and Haua Fteah; or about the early Middle Stone Age with the Border Cave and Klasies River Mouth Cave fossils. If the Mousterian of north Africa has an age generally comparable to that from the Haua Fteah associated with the north African Neanderthal remains, then this is likely to be in the general time range of $\geqslant$ 60,000–40,000 BP. The latest Acheulian on the Moroccan Atlantic coast is older than 95,000 BP, the date for the beginning of the Ouljian 6–8 m marine transgression, which is equated with the Tyrrhenian III (or late Last Interglacial) beach in the Mediterranean. If the preceding Harounian transgression in Morocco is the equivalent of the Tyrrhenian II which is considered to be of early Last Interglacial age and to date to c. 125,000 BP, then the Pre-Soltanian deposits with the final Acheulian would fall between the two (Butzer 1971: 24, 42). On the other hand, if the Harounian is the equivalent of the interstadial in the Riss, as has also been proposed (Jaeger 1975: 415), then the disappearance of the Acheulian would date to c. 200,000 years BP. The latter correlation seems the more probable. The Salé cranium and upper maxilla fragment could also be of this age or older, since they are certainly earlier than the Pre-Soltanian as the aeolianite in which they occur is overlain by a bone breccia with a Pre-Soltanian rodent fauna and the beach level to which the aeolianite is related is tentatively equated with the Anfatian transgression (Jaeger 1975: 400, 412).

It remains to be considered whether the "Pre-Aurignacian" and the Levallois-Mousterian of Cyrenaica and the Mousterian of the Maghreb could have been carried to north Africa as a result of population movement from outside the continent. There seems little substantive evidence to suggest movement from Mediterranean Europe at this time, either across the Straits of Gibraltar or via Sicily and the islands from southern Italy. However, Alimen (1975) has recently found evidence to suggest that earlier in the Middle Pleistocene, some movement may have occurred, though in the opposite direction. The Mousterian assemblages from the Maghreb (Balout 1965) with one of which the Jebel Irhoud skulls are associated (Ennouchi 1962) shows every indication of having been a natural development out of the terminal Acheulian and the characteristic tanged tools of the Aterian are unquestionably a northwest African invention (Ferring 1975). On the other hand, there is little morphological continuity between the Salé and Jebel Irhoud skulls (Trinkaus, this volume) though, given the appreciable time gap between them, this is not unexpected.

The early Upper Pleistocene industries in Cyrenaica present more of a problem, however. The "Pre-Aurignacian" has as yet no counterpart in any other region of northern Africa whereas its technological affinities with the contemporary "pre-Aurignacian" industries of the Levant from Amud, Adlun, Tabun and Yabrud (see Jelinek, this volume) are unquestionable. McBurney (1967: 130-31), in examining possible exotic origins for the Levallois-Mousterian of the Jebel Akhdar, has emphasized the similarity between the assemblages there and the contemporary industry in coastal Palestine. The succeeding Upper Palaeolithic industry, the Dabban, has also been shown to have affinities with contemporaneous industries in Palestine and the Negev (McBurney 1977: 15, 29). The only land bridge between southwest Asia and Africa is over the Isthmus of Suez and there is reason to suppose that there may have been movements of population between the two at various times from Middle and early Upper Pleistocene times up to the onset of the Last Glaciation. If it is valid to distinguish "culture areas" in prehistoric times on the basis of technology alone, then there is reason to postulate the existence of a circum-Mediterranean or, at least, a southeastern Mediterranean area at this time. That it did not spread up the Nile into Nubia is shown by the late persistence there of the Levallois method and presumably of the hominids responsible for it (the Singa skull may be a possible representative) to the exclusion of "foreigners". If, indeed, some hominid groups did move from southwest Asia into northeast Africa, their numbers must have been small since all writers stress the typical north African characteristics of the Neanderthalers from Jebel Irhoud and Haua Fteah. This distinctiveness presumably derives from hybridization with the autochthonous, evolved *H. erectus* lineage and from a long ecological and nutritional adaptation to the north African environments (Trinkaus, this volume p. 307). However, it is always dangerous to hypothesize too close or inseparable a relationship between genotype and culture and the Omo-Kibish skulls, in particular crania 2 and 3 which are essentially modern, carry the implication that by c. 130,000 BP, the East African gene pool had received input from some outside source.

Some credence is lent to the possibility of hominid migration into the Mediterranean coastal areas of north Africa by the fossil faunas there. Eurasiatic elements appear gradually in these faunas during the course of the Middle and Upper Pleistocene suggesting an "ecological filter" between their area of origin and north Africa. Also, the proportion of forms derived from Eurasiatic sources is highest during the Pre-Soltanian (Jaeger, 1975: 404-5) so that it is not impossible, though it remains unproven, that small groups of early *Homo sapiens* may also have found their way into North Africa at this time bringing cultural traits from southwest Asia.

Conclusions

In dealing with the African evidence, it must be stressed that the continent is an enormous, ancient land mass supporting many extremely varied eco-systems that can be expected to have given rise to a variety of adaptive patterns of behavior. Due to geographical, climatic and other difficulties, the hominid populations of the late Middle and early Upper Pleistocene would have had little or no possibility of inter-communication other than with contiguous groups. Over the millenia that are involved, therefore, communities that were both biologically and culturally distinctive can be expected to have developed in the very different eco-systems to which they had adapted. This distinctiveness is present both in the hominid fossils and in the stone industries in so far as the still very inadequate evidence for chronology and associations suggests. By the time of the disappearance of the final Acheulian tradition around 200,000 years ago, its assemblages were characterized by a tool-kit that, in northwest Africa, had little other than the bifaces to distinguish it from the succeeding Middle Palaeolithic. In the high rainfall/thicker vegetation regions of the tropics and subtropics also, only the heavy duty component (pick/core-axe and core scraper) and rare lanceolates which replaced the Acheulian bifaces, distinguished the earliest Middle Stone Age from the antecedent tradition. It was only in the Mediterranean coastal areas of Cyrenaica and, perhaps, also in the southern grasslands of the interior and the South African south coast, that the Middle Palaeolithic/Middle Stone Age assemblages showed more of a disparity. The "bladiness" of the early Middle Stone Age in the south and of the "Pre-Aurignacian" is already foreshadowed in some late Acheulian assemblages which include a weak antecedent blade element with the final Acheulian in Morocco, certain East African assemblages and the "Fauresmith" of the high veld. The technological transition between the Acheulian and the Middle Palaeolithic/Middle Stone Age is manifest in the disappearance of archaic forms and their replacement by new ones, rather than by any fundamental technological break. With the possible exception of Cyrenaica, this transition represents, therefore, a change within a continuing tradition due, no doubt, to improved methods of processing a wider range of resources. By evolved or final Acheulian times, handaxes were often quite small and, while many were well-made, numbers of others were irregular and had been formed by the removal of only a few hard hammer flakes giving the impression that they may have been used as cores. If this was indeed so, then it implies a fundamental change in the function of handaxes which may sometimes have ended up as being a source of raw material for the production of flakes until the perfection of the disc core method by the Middle Palaeolithic/Middle Stone Age hominids, invention and improved technology rendered biface production obsolete.

The evolutionary continuum suggested by the artifact assemblages from Africa is parallelled by the gradualistic model proposed by Trinkaus (this volume, p. 308) for hominid evolution in Africa – indeed throughout the Old World – at this time.

The extent to which there may be any connection between climatic fluctuation and the disappearance of the Acheulian biface in Africa remains uncertain. Indeed, at this time, no direct assertion seems possible beyond the general observation that the Acheulian had probably disappeared by the early part of the Last Interglacial and that the Middle Palaeolithic/Middle Stone Age was already well established by the time of the onset of the Last Glacial. In short, unlike at the end of this mode, it is impossible to see any clear-cut boundary at the beginning of the Middle Palaeolithic that separates it from the Acheulian, though a date around 200,000 years BP may not be too far from the time of the transition. It needs to be borne in mind, however, that the cultural evidence we have been considering – namely the stone implements – covers only one small, and possibly relatively insignificant part of the total material culture of these prehistoric populations. All their other products, in perishable materials, have not survived except for an occasional bone- or, even more rarely, wooden-tool. In so far as other behavioral traits and intellectual horizons may be postulated from surviving cultural phenomena, the expanding awareness of the Neanderthalers is observable in a number of practices. Belief in an after-life is attested by deliberate burial of the dead and the inclusion of grave goods. Care by the living of the handicapped and old is suggested by the finds at Shanidar and La Chapelle-aux-Saints. A bone flute fragment from Haua Fteah, the regular use of ochre for paint, or the inverted cranium in the stone circle at Monte Circeo, suggest involvement in intellectual and ritualistic pursuits. All these indications point to a more standardized and, therefore, more traditional pattern of behavior and social organization whose logical outcome was a more structured and, at the same time, a more rigid way of life which manifested itself in less variable tool-kits so that the changes observable by the end of the Acheulian document also, it would seem, the transition from a basically general pattern of life to the more regionally specialized socio-economic behavior of the Middle Palaeolithic/Middle Stone Age.

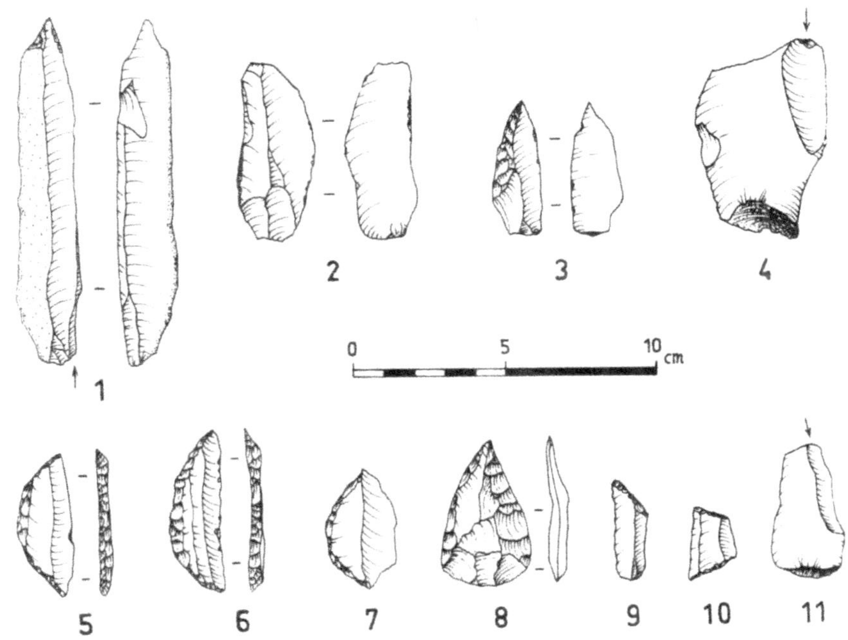

Figure 1 Pre-Aurignacian: 1, Awl-burin; 2,3 notched and backed blades; 4, proto-burin. Howeison's Poort facies: 5-7, backed blades; 8, unifacial point; 9, truncated blade; 10, trapeze; 11, burin. 1-4, Haua Fteah; 5,6, Nelson Bay Cave; 7-11, Howeison's Poort type site (after Clark 1978: 36, 40).

Age BP	Oxygen Isotope Stage	Climate	North Africa Alternative 1	North Africa Alternative 2	East Africa	South Africa Alternative 1 (Interior Plateau)	South Africa Alternative 2 (South Coast Caves)
0							
	1	Modern					
13,000			Various blade	Various blade	Later Stone Age Early Blade Industry (L. Zwai c. 27,000)	Later Stone Age	Later Stone Age Early Late Stone Age (Robberg: 18,000)
	2	Cold/Dry	Industries	Industries		Early Later Stone Age	
32,000			Early Upper Palaeolithic (Dabban c. 40,000)	Dabban ↑	End Middle Stone Age (?c. 40,000) ↑	Late Middle Stone Age 2	
	3	Cold/Wet	Levallois–Mousterian (c. 40,000-60,000) ↑				?
64,000		Temperate					
	4	Cold/Wet		Aterian Levallois-Mousterian ↓		Hiatus	
75,000			Pre-Aurignacian (Cyrenaica) (c. 60,000-80,000) ↓			Late Middle Stone Age 1	Middle Stone Age IV
	5a	Warm/Dry					
85,000							
	5b	Cold/Wet		↑		Howieson's Poort facies	Middle Stone Age III Howieson's Poort facies
95,000							
	5c	Warm/Wet		Pre-Aurignacian ↓		Hiatus	
105,000							Middle Stone Age II
	5d	Cold to Temperate Wet	?			↑	
115,000							
	5e	Warm/Wet					Middle Stone Age I
128,000					Omo-Kibish Crania (Homo s. s.) (c. 130,000)	Early Middle Stone Age ↓	
		Cold		(Acheulian, possibly continuing until 125,000)			
135,000							
	6	Temperate					?
145,000							
		Warm			Early Middle Stone Age Gadamotta (c. 180,000)		
195,000		Temperate					
	7	Warm/Wet	Terminal Acheulian	Terminal Acheulian	Terminal Acheulian (Dallol c. 200,000 ?Kapthurin c. 230,000 Garba III Isimila c. 260,000)	Terminal Acheulian (Rooidam c. 200,000)	? Terminal Acheulian

Table 1

Alternative Chronologies for the Lower Palaeolithic/Middle Palaeolithic Transition in North, East and South Africa

Based mainly on Shackleton and Opdyke 1973, 1976 and Butzer *et al.* 1978 for oxygen isotope data and correlations; on McBurney 1967 and Biberson 1970 for North Africa; on Street 1980, Wendorf *et al.* 1975, M. Leakey *et al.* 1969, R.E.F. Leakey *et al.* 1969, Chavaillon *et al.* 1979 and Roubet 1969 for East Africa; on Butzer *et al.* 1978, Beaumont *et al.* 1978, Deacon 1979 and Szabo and Butzer 1979 for South Africa.

Alternative 1 for North Africa shows McBurney's C^{14} and extrapolated ages for the Haua Fteah sequence in Cyrenaica. Alternative 2 shows ages based on the Moroccan Atlantic coast sequence and Th/U dates for Ouljian and Harounian beach levels and related industries which suggest possible alternative ages for the Cyrenaican sequence. Thus, the "Pre-Aurignacian" associated with a warm fauna would belong either in oxygen isotope stage 5a or 5c.

Alternative 1 for South Africa is based on the succession in Border Cave and Alternative 2 on the south coast caves in particular Klassies River Mouth Caves and Nelson Bay Cave.

If the age for the Harounian beach – 145,000-125,000 BP – is accepted, then the Acheulian in the Maghreb continued until c. 125,000 BP but, if this is not correct, then the Acheulian terminated at much the same time throughout the continent, i.e. at c. 200,000 BP.

References

Alimen, M.-H. 1975 Les "Isthmes" Hispano-Marocain et Siculo-Tunisien aux temps acheuléens *L'Anthroplogie* 79: 399-436.

Alimen, H, N. Chavaillon and R. Karpoff 1963 Nouveaux gisements paléolithiques dans l'Adrar des Iforas *Bull. Soc. Préhist. Fr.* 60: 352-63.

Alimen, H. and J. Zuarte y Zuber 1978 *L'Évolution de l'Acheuléen au Sahara nord-occidental (Saoura-Ougarta-Tabelbala)* CNRS. Meudon.

Antoine, M. 1939 Notes de Préhistoire marocaine XIV Un cône de résurgence du Paléolithique moyen à Tit-Mellil, près Casablanca *Bull. Soc. Préhist. du Maroc.* 12e année (1-4) – 1938.

Arambourg, C. 1963 Le gisement de Ternifine: l'Atlanthropus de Ternifine *Archives de l'Institut de Paléontologie humaine* Paris, 32: 37-190.

Arambourg, C. and L. Balout 1955 L'ancien lac de Tihodaïne et ses gisements préhistoriques *Actes du Congrès panafricain de Préhistoire* Algiers: 281-286.

Arkell, A.J. 1949 The Old Stone Age in the Anglo-Egyptian Sudan *Occ. Papers Sudan Antiquities Service* No. 6. Khartoum.

Balout, L. 1955 *Préhistoire de l'Afrique du Nord* Paris.

Balout, L. 1965 Le Moustérien du Maghreb *Quaternaria* 7: 59-78.

Beaumont, P.B., H. de Villiers and J.C. Vogel 1978 Modern man in sub-Saharan Africa prior to 49,000 years BP: A review and evaluation with particular reference to Border Cave *S. Afr. J. Sci.* 74: 409-19.

Biberson, P. 1961 *Le Paléolithique inférieur du Maroc atlantique* Pub. Service des Antiquités du Maroc, 17. Rabat.

Biberson, P. 1970 Index cards on the Marine and Continental Cycles of the Moroccan Quaternary *Quaternaria* 13: 1-76.

Bishop, W.W. 1972 Stratigraphic succession 'versus' calibration in East Africa. *In:* W. W. Bishop and J.A. Miller (eds.) *Calibration of hominoid evolution*: 219-46. Edinburgh and Toronto.

Butzer, K.W. 1971 *Environment and archaeology: An ecological approach to prehistory* 2nd Edition. Chicago.

Butzer, K.W., P.B. Beaumont and J.C. Vogel 1978 Litho-stratigraphy of Border Cave, KwaZulu, South Africa: A Middle Stone Age sequence beginning c. 195,000 BP *J. Archaeol. Sci.* 5: 317-41.

Caton-Thompson, G. 1952 *Kharga Oasis in prehistory* London.

Chavaillon, J. 1964 *Étude stratigraphique des formations quaternaires du Sahara nord occidental (Calomb-Bechar à Reggane)* CNRS Série Géol. No. 5. Paris.

Chavaillon, J. 1974 *L'Homo erectus* de Melka-Kunturé *La Recherche*, Paris, 49: 880-890.

Chavaillon, J., N. Chavaillon, F. Hours and M. Piperno 1979 From the Oldowan to the Middle Stone Age at Melka-Kunturé (Ethiopia): Understanding cultural changes *Quaternaria* 21: 87-114.

Chmielewski, W. 1968 Early and Middle Palaeolithic sites near Arkin, Sudan. *In:* F. Wendorf (ed.) *The Prehistory of Nubia* Dallas, Vol. I: 110-47.

Clark, J.D. 1945 A Kenya Fauresmith factory and home site at Gondar, northern Abyssinia *Trans. Roy. Soc. S. Afr.* 31: 29-37.

Clark, J.D. 1954 *The Prehistoric Cultures of the Horn of Africa* Cambridge.

Clark, J.D. 1964 The Sangoan culture of Equatoria: The implications of its stone equipment. *In:* E. Ripoli Perello (ed.) *Miscelanea en el Homenaje al Abate Henri Breuil* Barcelona, Vol. I: 309-25.

Clark, J.D. 1968 *Further Palaeo-anthropological studies in northern Lunda* Pub. Cult. No. 78. Museo do Dundo. Lisbon.

Clark, J.D. 1969 *Kalambo Falls Prehistoric Site I The Geology, Palaeoecology and detailed stratigraphy of the excavations* Cambridge.

Clark, J.D. 1974 *Kalambo Falls Prehistoric Site II.* Cambridge.

Clark, J.D. 1975 A comparison of the Late Acheulian Industries of Africa and the Middle East. *In:* K.W. Butzer and G. Ll. Isaac (eds.) *After the Australopithecines*: 605-60. Mouton, The Hague.

Clark, J.D. 1978 The legacy of prehistory: an essay on the background to the individuality of African cultures *The Cambridge History of Africa,* V. 2, Cambridge, pp. 11-86.

Clark, J.D. and C.V. Haynes 1969 An elephant butchery site at Mwanganda's village, Karonga, Malawi and its relevance for Palaeolithic archaeology *World Archaeology* 1: 390-411.

Clark, J.D. and H. Kurashina 1979 Hominid occupation of the East-Central Highlands of Ethiopia in the Plio-Pleistocene *Nature* 282: 33-39.

Clark, J.D., M.A.J. Williams and A.B. Smith 1973 The geomorphology and archaeology of Adrar Bous, Central Sahara: A preliminary report *Quaternaria* 17: 245-98.

Clarke, R.J. 1976 New cranium of *Homo erectus* from Lake Ndutu, Tanzania *Nature* 262: 485-87.

Cole, G.H. 1967 The Later Acheulian and Sangoan of southern Uganda. *In:* W.W. Bishop and J.D. Clark (eds.) *Background to Evolution in Africa* Chicago: 481-528.

Conroy, G.C., C.J. Jolly, D. Cramer and J.E. Kalb 1978 Newly discovered fossil hominid skull from the Afar depression, Ethiopia *Nature* 275: 67-70.

Cooke, C.K. 1963 Report on excavations at Pomongwe and Tshangula Caves, Matopo Hills, Southern Rhodesia *S. Afr. Archaeol. Bull.* 18: 73-151.

Cooke, C.K. 1966 Re-appraisal of the industry hitherto named the Proto-Stillbay *Arnoldia* (Rhodesia) 2(22): 1-14.

Deacon, J. 1976 Report on stone artifacts from Duinefontein 2, Melkbosstrand *S. Afr. Archaeol. Bull.* 31: 21-25.

Deacon, J. 1979 *Guide to archaeological sites in the southern Cape* Southern African Association of Archaeologists, Excursion June 30-July 4, 1979. University of Stellenbosch Occasional Paper, Stellenbosch.

De Ploey, J. and H. Van Moorsel 1963 Contributions à la connaissance chronologique et paléogéographique des gisements préhistoriques des environs de Léopoldville (Congo) *Studia Universitatis "Lovanium", Musée de Préhistoire* 19: 1-19.

Ennouchi, E. 1962 Un néanderthalien: L'Homme du Jebel Irhoud (Maroc) *L'Anthropologie* 66: 279-99.

Ferring, C.R. 1975 The Aterian in north African prehistory. *In:* F. Wendorf and A.E. Marks (eds.) *Problems in Prehistory: North Africa and the Levant* Dallas: 113-126.

Fock, G.J. 1968 Rooidam: A sealed site of the First Intermediate *S. Afr. J. Sci.* 64; 153-58.

Freeman, L.G. 1975 Acheulian sites and stratigraphy in Iberia and the Maghreb. *In:* K.W. Butzer and G. Ll. Isaac (eds.) *After the Australopithecines*: 661-744. Mouton, The Hague.

Gobert, E.G. 1950 Le gisement paléolithique de Sidi-Zin *Karthago* 1: 3-51.

Goodwin, A.J.H. and C. Van Riet Lowe 1929 *The Stone Age Cultures of South Africa* Ann. S. Afr. Mus. Vol. 27. Edinburgh.

Hours, F. 1976 Le Middle Stone Age de Melka-Kontoure. *In: Panafrican Congress of Prehistory and Quaternary Studies, 8th Session, Addis Ababa, 1971* Addis Ababa: 99-104.

Howell, F.C. 1966 Observations on the earlier phases of the European Lower Palaeolithic *Amer. Anthrop.* 68: 88-201.

Howell, F.C., G.H. Cole and M.R. Kleindienst 1962 Isimila: An Acheulian occupation site in the Iringa Highlands, Southern Highlands Province, Tanganyika. *In:* G. Mortelmans and J. Nenquin (eds.) *Actes du IVe Congrès Panafricain de Préhistoire et de l'Etude du Quaternaire* Section III Pré- et Protohistoire. Mus. Roy. de l'Afrique Centrale, Ann Sér. in 8° Sciences humaines, No. 40. Tervuren: 43-80.

Howell, F.C., G.H. Cole, M.R. Kleindienst, B.J. Szabo and K.P. Oakley 1972 Uranium-series dating of bone from the Isimila prehistoric site *Nature* 237: 51-52.

Howells, W.W. 1980 *Homo erectus*, who when and where: A survey *Yearbook of Phys. Anthrop.* 23: 1-23.

Isaac, G. Ll. 1969 Studies of early culture in East Africa *World Archaeology* 1: 1-28.

Isaac, G.Ll. 1977 *Olorgesailie: Archaeological studies of a Middle Pleistocene lake basin in Kenya* Chicago.

Isaac, G. Ll. 1982 The earliest archaeological traces. *In:* J.D. Clark (ed.) *The Cambridge History of Africa, Vol. I From the earliest times to c. 500 BC* Cambridge: 157-247.

Jaeger, J.-J. 1975 The mammalian faunas and hominid fossils of the Middle Pleistocene of the Maghreb. *In:* K.W. Butzer and G.Ll. Isaac (eds.) *After the Australopithecines* Mouton, The Hague: 399-418.

Jones, P.R. 1981 Experimental butchery with modern stone tools and its relevance for Palaeolithic archaeology *World Archaeology* 12: 153-65.

Keeley, L.H. 1980 *Experimental determination of stone tool uses* Chicago.

Keller, C.M. 1973 Montagu Cave in prehistory: A descriptive analysis *University of California Anthrop. Records* 28: 1-98.

Kleindienst, M.R. 1961 Variability within the Late Acheulian assemblage in eastern Africa *S. Afr. Archaeol. Bull.* 16: 35-52.

Kleindienst, M.R. 1962 Components of the East African Acheulian assemblage: An analytical approach. *In:* G. Mortelmans and J. Nenquin (eds.) *Actes du IVe Congrès panafricain de Préhistoire et de l'Étude du Quaternaire* Mus. Roy. de l'Afrique Centrale, Ann. Sér. in 8º Sciences humaines, No. 40, Tervuren: 81-112.

Leakey, L.S.B. 1931 The Stone Age cultures of Kenya Colony. Cambridge.

Leakey, M., P.V. Tobias, J.E. Martyn and R.E.F. Leakey 1969 An Acheulian industry and hominid mandible, Lake Baringo, Kenya *Proc. Prehist. Soc.* 35: 48-76.

Leakey, M.D. 1971 *Olduvai Gorge: Excavations in Beds I and II* 1960-1963. Cambridge.

Leakey, R.E.F., K.W. Butzer and M.H. Day 1969 Early *Homo sapiens* remains from the Omo River region of south-west Ethiopia *Nature* 222: 1132-1138.

Lee, C., J.L. Bada and E. Peterson 1976 Amino acids in modern and fossil woods *Nature* 259: 183-86.

Malan, B.D. and H.B.S. Cooke 1941 A preliminary account of the Wonderwerk Cave, Kuruman District *S. Afr. J. Sci.* 37: 300-12.

Malan, B.D. and L.H. Wells 1943 A further report on the Wonderwerk Cave, Kuruman *S. Afr. J. Sci.* 40: 258-70.

Mason, R.J. 1962 *Prehistory of the Transvaal* Johannesburg.

McBurney, C.B.M. 1967 *The Haua Fteah (Cyrenaica) and the Stone Age of the South-East Mediterranean* Cambridge.

McBurney, C.B.M. 1977 *Archaeology and the Homo sapiens sapiens problem in northern Africa* Derde Kroon-Voordracht, Amsterdam.

Mieg de Boofzheim, P. and C. Plessis 1954 Sidi Abd-er-Rahman, Grotte du Rhinocéros *Bull. Soc. Préhist. du Maroc* n.s. 7-8: 29-53.

Nygaard, S. and M.R. Talbot 1976 Interim report on excavations at Asokrochona, Ghana *West Afr. J. Archaeol.* 6: 13-19.

Oakley, K.P., P. Andrews, L.H. Keeley and J.D. Clark 1977a A reappraisal of the Clacton spearpoint *Proc. Prehist. Soc.* 43: 13-30.

Oakley, K.P., B.G. Campbell and T.I. Molleson (eds.) 1977b *Catalogue of fossil hominids, Part I, Africa* (2nd Edition) London.

Oussedik, O. 1972 Les bifaces acheuléens de l'erg Tihodaïne (Sahara central algérien): Analyse typométrique *Libyca* 20: 153-161.

Pilbeam, D.R. 1975 Middle Pleistocene hominids. *In:* K.W. Butzer and G.Ll. Isaac (eds.) *After the Australopithecines* Mouton, The Hague: 809-856.

Roubet, C. 1969 Essai de datation absolue d'un biface-hachereau paléolithique de l'Afar (Ethiopie) *L'Anthropologie* 73: 503-24.

Sampson, C.G. 1972 *The Stone Age industries of the Orange River Scheme and South Africa* Nat. Mus. Bloemfontein, Mem. No. 6.

Schild, R. and F. Wendorf 1977 *The prehistory of Dakhla Oasis and adjacent desert* Polish Academy of Sciences, Warsaw.

Shackleton, N.J. and N.D. Opdyke 1973 Oxygen isotope and palaeomagnetic stratigraphy of equatorial Pacific core V28-238: Oxygen isotope temperatures and ice volumes on a 10^5 and 10^6 year cycle *Quat. Research* 3: 39-55.

Shackleton, N.J. and N.D. Opdyke 1976 *Oxygen-isotope and Palaeomagnetic stratigraphy of Pacific core V28-239: Late Pliocene to Latest Pleistocene* Geol. Soc.

Amer. Mem. 145.

Söhnge, P.G., D.J.L. Visser and C. Van Riet Lowe 1937 *The geology and archaeology of the Vaal River basin* Geol. Surv. Mem. 35, Pretoria.

Street, F.A. 1980 Chronology of late Pleistocene and Holocene lake-level fluctuations, Ziway-Shala Basin, Ethiopia. *In:* R.E. Leakey and B.A. Ogot (eds.) *Proc. 8th Pan-African Congress of Prehistory and Quaternary Studies, Nairobi, 1977* Nairobi: 143-46.

Szabo, B.J. and K.W. Butzer 1979 Uranium series dating of lacustrine limestones from pan deposits with Final Acheulian assemblages at Rooidam, Kimberley District, South Africa *Quat. Research* 11: 257-60.

Tixier, J. 1959 Les industries lithiques d'Aïn Fritissa (Maroc oriental) *Bull. Archaeol. Maroc.* 3: 107-244.

Tuffreau, A. 1979 Les debuts du Paleolithique moyen dans la France septentrionale *Bull. Soc. Préhist. Fr.* 76: 140-42.

Volman, T.P. 1979 The Middle Stone Age sequence at Nelson's Bay Cave. *In:* J. Deacon *Guide to archaeological sites in the southern Cape,* Southern African Association of Archaeologists, Excursion June 30-July 4, 1979. University of Stellenbosch Occasional Paper, Stellenbosch: 65-67.

Wai-Ogosu, B. 1973 Was there a Sangoan industry in West Africa? *West Afr. J. Archaeol.* 3: 191-96.

Wendorf, F. and R. Schild 1976 *Prehistory of the Nile valley* New York.

Wendorf, F. and R. Schild 1980 *Prehistory of the Eastern Sahara* New York.

Wendorf, F., R.L. Laury, C.C. Albritton, R. Schild, C.V. Haynes, P. Damon, M. Shaffiquillah and R. Scarborough 1975 Dates for the Middle Stone Age of East Africa *Science* N.Y., 187: 740-2.

THE LOWER/MIDDLE PALAEOLITHIC TRANSITION IN THE HUNSGI VALLEY, PENINSULAR INDIA

Katragadda Paddayya

The Hunsgi Valley forms part of the drainage system of the river Krishna and lies in the southwestern part of the Shorapur Taluk of Gulbarga District, Karnataka State. The writer's investigations during the last one decade or so have brought forth evidence proving that the area was continuously occupied by prehistoric man right from the Lower Palaeolithic stage to the Iron Age of the 1st millenium B.C. Following preliminary observations about the present environments of the valley and a brief study of the Lower and Middle Palaeolithic cultures, the present paper will draw attention to the limited but suggestive evidence arguing in favor of the local origin of the Middle Palaeolithic culture from the preceding Acheulian culture.

The Area

The Hunsgi Valley is an irregularly shaped basin of erosional origin. It measures 15 km long (east and west); the breadth (north-south) varies from 4 to 10 km. It is enclosed by tablelands on the north, west, south, and to some extent, on the eastern side also; these rise to a maximum height of about 75 m above the valley floor. The plateau on the west forms part of an extensive land surface (> 500 m above sea level) serving as a water-divide between the Hunsgi and Don streams, both being tributaries of river Krishna. The plateaus on the remaining three sides are merely offshoots of this land surface. The valley floor is a gently sloping peneplained surface. In the west it has an elevation of 490 m above sea level and slopes down to a level of 430 m at Hunsgi in the east, where the entire drainage is integrated into a single stream called the Hunsgi nullah.

Granite, shales and limestones, and the Deccan Trap are the main rock formations of the area. The tablelands flanking the valley are made up of shales and limestones, and the Deccan Trap. The valley is partly occupied by shales and limestones and partly by granite. The importance of the occurrence of these varied geological formations in such a small area is that their junctions give rise to artesian springs, which form the only perrenial source of drinking water in the entire valley.

The climate of the area may be characterized as semi-arid, with an annual rainfall of about 700 mm. Almost the whole of this precipitation is received during the monsoonal period (July to October). Owing to agricultural operations, overgrazing by animals and other activities, the valley has almost completely been denuded of its natural vegetation cover. However, the presence even till today of small patches of scrub jungle at favorable points along the stream courses and elsewhere in the valley is a sure indication that in the past, when human interference was but minimal, the valley supported a much thicker cover of vegetation. An interesting aspect of these relict patches is that they still yield a variety of wild plant foods.

As mentioned previously, the valley constitutes the headwater zone of a small stream called the Hunsgi nullah, which joins the Krishna after an easterly course of about 30 km. It is this lack of direct relationship with a major river and the insulating effect created by the tablelands encircling the valley which prompt us to regard the area as a micro-geographical unit, with the attendant advantages for studying the past human cultures on a regional basis. The Hunsgi nullah has four main affluents, which in turn are replenished by numerous feeders descending from the plateaus. All these are ephemeral and carry water only during the wet season. The ankle-deep flow in the bed of the Hunsgi nullah is the only source of drinking water available on the surface during the dry season. This flow takes its origin from a group of artesian springs (lying between the villages of Isampur and Devapur) issuing along the course of the northern affluent of the Hunsgi nullah. The flow lasts throughout the year and does not dry up even during periods of severe drought in the area. More important, the presence of thick accumulations of travertine along the Hunsgi nullah shows that these springs are of geological antiquity; the occurrence at some sites of the Acheulian levels on top of the travertine deposits proves that the springs were active during the time period covered by the Stone Age cultures.

The Lower and Middle Palaeolithic Cultures

The writer's intensive studies since 1974 and previous initial surveys have made known a considerable number of Lower and Middle Palaeolithic sites (Figure 1). The former tradition is represented by the Acheulian culture. The nature and distribution of sites and the character of lithic industries of the two cultures are briefly commented upon below.

Of the over 40 Acheulian occurrences discovered till the field season of 1978–79, half a dozen are associated with fluviatile deposits and hence may be called secondary sites. The remaining occurrences are connected with occupation activities and range from 3,000 to 7,000 sq. meters in extent. Detailed surface surveys and excavation at two of the localities have made it possible to identify occupation levels (20 to 35 cm thick) extending up to one meter below surface. One striking feature about the distribution of these sites is that they are confined to the valley. The nearest group of Acheulian sites outside the valley lies about 60 km to the southwest (Pappu 1974). The valley thus constitutes a good instance in Indian prehistory of the correspondence of natural areas with culture areas. Secondly, the summits/shoulders of the tablelands flanking the valley have so far failed to produce any evidence of Acheulian occupation; all the known sites occur in the valley proper. From this it would follow that, although they must have been exploited in respect of food resources, the tablelands were avoided by the Acheu-

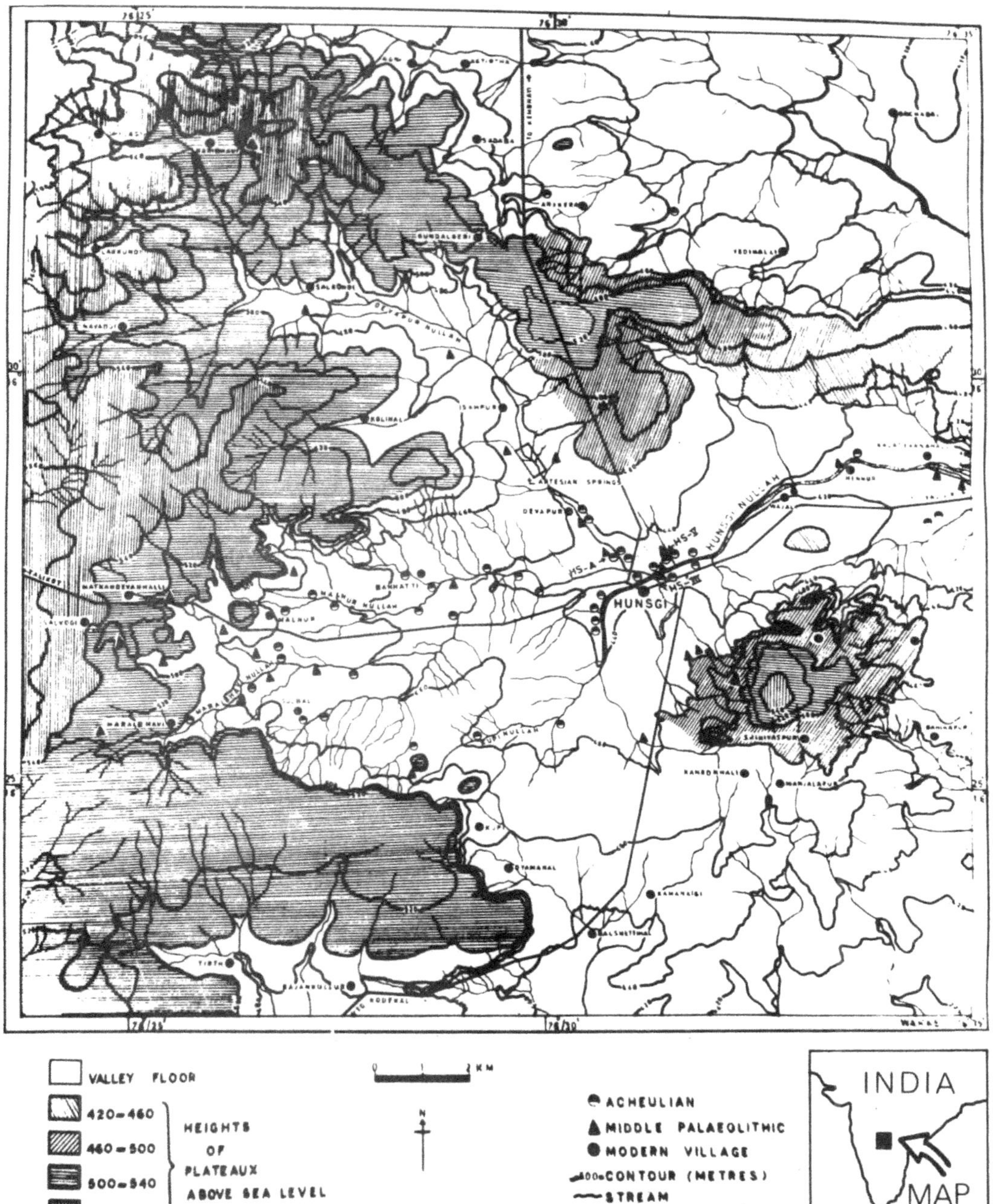

Figure 1 Acheulian and Middle Palaeolithic sites of Hunsgi valley in relation to topography and drainage.

lian man for occupation purposes. Absence of perennial sources of water could be cited as one of the inhibiting factors. Moreover, limestone required for tool-making is already available on the valley floor either as nodules and slabs in the foothill area or as pebbles in stream beds; as such, there was no need for the Acheulian groups to exploit the limestone formations of the tablelands. As we will note shortly, this land-use pattern undergoes changes from the following Middle Palaeolithic phase onwards.

The third important aspect of these sites concerns their distribution within the valley floor itself. As many as 20 localities occur as a cluster within a radius of 2 to 3 km from the village of Hunsgi. The remaining sites are randomly distributed across the entire valley floor. Based on the evidence of artesian springs referred to earlier, the writer puts forward the view that the site-cluster around Hunsgi resulted from the coming together of most or all of the Acheulian groups of the valley during the dry season, as the shallow flow of water fed by the springs would form the only secure source of drinking water during this part of the year. As against this, the randomly distributed sites of the valley floor would mark the dispersal of the groups during the ensuing wet season when water as well as plant and animal foods would be available in plenty all over the valley.

Coming to the lithic industry, 12 surface assemblages and three from excavation have been studied in detail so far and permit the following observations (for details, see Paddayya 1977a, 1977b, 1979).

1. The use of limestone as the principal raw material is one of the most interesting aspects of the land-use pattern of the Acheulian groups. The employment of this otherwise soft rock betrays their adaptability to the locally available resources; its selection appears to have been forced by the non-availability or rarity of quartzitic rocks most commonly utilized in other parts of the country. Although other rocks like granite, sandstone, dolerite and chert are evidenced in some of the assemblages, artifacts made of these rocks are limited in number. For example, out of a total of 291 specimens found in an area of 22.75 sq. meters in Tr. 3 at Locality V near Hunsgi as many as 261 specimens are made of limestone and the rest of other rocks.

2. Cleavers, handaxes, knives, choppers, picks and polyhedrons are the chief artifact types of the industry. Flake-tools like scrapers, especially those made on flake blanks, are either absent or limited in number. The assemblage from the 1975 excavation in Trench 3 at Hunsgi V is cited below as an example of the composition of the industry.

I. Shaped Tools	**No.**
1. Cleavers	28
2. Handaxes	18
3. Knives	14
4. Choppers	9
5. Picks	8
6. Polyhedrons	10
7. Spheroids	4
8. Scrapers	15
9. Flakes with prepared butts	2
10. Backed tool	1
II. Utilized Pieces	
1. Anvil stones	3
2. Hammerstones	5
3. Utilized flakes	15
4. Utilized tabular pieces	2

III. Dèbitage	
1. Cores	5
2. Worked nuclei	20
3. Flakes	5
4. Waste products	127
Total	**291**

3. The frequent employment of stone hammer technique for preparing heavy-duty tools (as seen in the irregular outlines and thick cross-sections), the preponderance of pointed forms among the handaxes, and the common occurrence of types like choppers, picks and polyhedrons all show that we are dealing with an early variant within the Acheulian tradition.

As for the Middle Palaeolithic culture, the valley has made known 31 sites so far. Unlike in the case of the Acheulian sites, we now see the occurrence of sites on the summits and shoulders of tablelands, thereby implying changes in the land-use patterns. Raw material seems to have played a key role in this regard, for chert, which replaces limestone as the medium of working, is totally lacking on the valley floor, whereas the limestone and Trap formations of the escarpments contain extensive veins of this material. The Middle Palaeolithic groups were therefore forced to exploit these sources. Of the 7 sites discovered thus far on the tablelands, some may be grouped as workshops on account of the presence of large quantities of dèbitage. The rest produce only a limited number of artifacts; it is probable that they served as short-duration camp sites connected with the procurement and treatment of animal and plant foods.

The remaining 24 sites are spread all over the valley floor. Eleven of these are occupation sites; in many cases there is clear evidence that chert nodules were brought from the plateau veins and worked upon at these localities. Unlike in the case of the Acheulian occurrences, it has not been possible to identify regular occupation levels, as the small size of artifacts and agricultural activities have resulted in widespread dispersal of the material. At the remaining 13 localities the artifacts are found in fluviatile deposits. At Locality-A near Hunsgi (HS-A) these deposits overlie an Acheulian tool-bearing gravel. From the local origin of these deposits and the fresh condition of artifacts, it is clear that they were derived from nearby occupation sites.

Another noteworthy feature of the distribution of these sites is that they occur far and wide outside the valley, more particularly in the area called the Shorapur Doab, lying to the east (Paddayya 1971, 1974). This wide dispersal seems to have been facilitated by the availability of chert in the form of surficial gravel spreads and veins in geological strata.

As with the raw material, the lithic industry of the culture differs markedly from that of the Acheulian culture. It belongs to the flake-tool tradition inasmuch as 80% to 90% of the finished tools are fashioned on flake blanks obtained by means of the plain flaking and prepared core (Levallois) techniques. These blanks are transformed into finished types by treating them with edge-retouch and such other forms of secondary working. Scrapers (with several sub-types), points and borers or perforators are the chief types of the industry. Among a total of 328 specimens from nine sites studied in detail thus far, scrapers have been found to account for 76%, points for 11%, borers for 10% and combination types for 3% (Paddayya 1974: 9-10).

Evidence for a Transitional Phase

From the foregoing account it is sufficiently clear that the Acheulian and Middle Palaeolithic cultures of the valley are separate entities, each with its own characteristic features. The differences between them in respect of raw material as well as typo-technology are such that they would render the postulation of any evolutionary relationship between the two seemingly unthinkable. But a closer scrutiny of the evidence shows that such a relationship does exist between the two cultures. There are a few sites belonging to either of these cultures which persuade us to posit a transitional stage.

Among the Acheulian sites the assemblages obtained from Trench 1 at Locality III (HS-III) and Trench 1 at Locality V (HS-V) near Hunsgi, and the alluvial deposit at Chennur seem to represent an evolved stage of the culture. The former has been studied in detail (Paddayya 1977b: 386 and Table 1). It was obtained from a 1.25 m thick colluvial gravel consisting of granite and limestone blocks set in a matrix of whitish calcareous earth. It consists of 507 specimens falling under the following types:

	No.
I. Shaped Tools	
1. Cleavers	39
2. Handaxes	15
3. Choppers	11
4. Simple points	4
5. Bifacial points	2
6. Knives	7
7. Scrapers	45
8. Borers	7
9. Burinate tool	1
II. Simple Artifacts	
1. Cores	41
2. Worked nuclei	21
3. Flakes	106
4. Flake-blade	1
5. Blades	10
6. Miscellaneous pieces	21
7. Waste products	176

Two features are striking about this assemblage. First, the use of chert as the principal raw material; about 70% of the artifacts are made of this material, including some of the heavy-duty tools like handaxes and cleavers. The other feature concerns the occurrence of true scrapers, borers and points which are characteristic of the following Middle Palaeolithic culture. Although no statistics can be provided here, the other two assemblages share these features. We thus appear to have in these assemblages a late stage of the Acheulian culture, already anticipating the Middle Palaeolithic culture.

Likewise, there are a few Middle Palaeolithic sites (e.g., the occupation site at Salgondi and the river gravel site near Kangonhalli and Devapur) which betray elements of the earlier tradition by way of producing a small number of handaxes and cleavers. Their presence gains in meaning when it is realized that they are made of limestone.

Thus, the evidence provided by raw material and typo-technology suggests a basic continuity of tradition from the Acheulian to the Middle Palaeolithic stage. There are two other considerations which support this view. Certain basic similarities exist between the land-use patterns of the two cultures, *viz.*, preference shown to the valley floor for occupation purposes and the location of sites along stream courses, and the occurrence of the remains of both cultures at a few sites (e.g., Hunsgi Localities V and A). Hence we seem to be justified in proposing the view that the two cultures form part of an evolving tradition. It must be admitted that there is no stratigraphical evidence supporting this view, but such evidence can hardly be expected to be preserved on open-air sites of the kind examined in the present study.

In this connection it is important to note that the Hunsgi valley is one among several regions in Peninsular India witnessing evolution of the biface tradition into the flake-tool tradition. The Renigunta region in Andhra Pradesh may be cited as an example of these complexes of open-air sites (Murty 1966: 111). One of the Middle Palaeolithic sites of this area (Locality V) contains well-shaped handaxes and cleavers, respectively, making up for 23% and 6% of the total assemblage; it probably represents an intermediate phase between the Acheulian and Middle Palaeolithic cultures. Still more recently, excavations in one of the rockshelters at Bhimbetka in the Narmada valley have made known clear-cut stratigraphical evidence attesting the gradual transformation of a very late Acheulian tradition into a typical Middle Palaeolithic culture (Misra 1978); the transformation consists in the disappearance of bifaces and corresponding increase in the number of flake-tools. Similar but less clear evidence is available at the nearby rockshelter of Adamgarh (Joshi 1978). In the light of these data it is very probable that the flake-tool traditions of the Middle Palaeolithic phase developed from the local biface traditions simultaneously in several regions of the country. We therefore need to have a closer look at hypotheses which have sought to derive the Indian Middle Palaeolithic from as far afield as Western Europe via West and Central Asia (Sankalia 1974: 205).

In conclusion, it is interesting to speculate about the larger issue of why this transition or change from core-biface tradition to flake-tools should have occurred at all. The use of fine-grained materials like chert, which yield perfect conchoidal fracture, must have conferred great advantages at both primary flaking and secondary working levels. Moreover, on account of their widespread occurrence in geological contexts and portability, the employment of these materials must have facilitated the occupation of new areas. Likewise, the replacement of large tools by flake-tools must have offered its own advantages: a) the possibility of turning out a large number of tools from a given block of raw material; b) improved efficiency and manual control in regard to both household and outdoor activities; c) their amenability to the processes of 'recycling'; and d) portability of finished tools over considerable distances. These advantages are such that their importance can be recognized independently by any human group by trial and error method. It will therefore be no wonder if our studies lead to the conclusion that this transformation took place in many parts of the Old World separated in space and, to some extent, even in respect of time.

References

Joshi, R.V. 1978 *Stone Age Cultures of Central India* Poona, Deccan College.

Misra, V.N. 1978 The Acheulian industry of rock shelter III F-23 at Bhimbetka, Central India – a preliminary study *Australian Archaeology* 8: 63-106.

Murty, M.L.K. 1966 *Stone Age Cultures of Chittoor District* Unpublished Ph.D. dissertation of Poona University.

Paddayya, K. 1971 High level gravels and related Palaeolithic sites of Shorapur Doab *Quartär* 22: 95-110.

Paddayya, K. 1974 The Middle Palaeolithic Culture of the Shorapur Doab *Journal of Indian History* 52: 1-19.

Paddayya, K. 1977a An Acheulian occupation site at Hunsgi, Peninsular India: a summary of the results of two seasons of excavation (1975-6) *World Archaeology* 8: 344-55.

Paddayya, K. 1977b The Acheulian culture of the Hunsgi valley (Shorapur Doab), Peninsular India *Proceedings of the American Philosophical Society* 121: 383-406.

Paddayya, K. 1979 Excavation of a new Acheulian occupation site at Hunsgi, Peninsular India *Quartär* 29-30: 139-155.

Pappu, R.S. 1974 *Pleistocene Studies in the Upper Krishna Basin* Poona, Deccan College.

Sankalia, H.D. 1974 *Prehistory and Protohistory of India and Pakistan* Poona, Deccan College.

PEBBLE-CORE AND FLAKE ELEMENTS: PROCESS OF TRANSMUTATION AND THE FACTORS THEREOF

A CASE STUDY OF THE TRANSITION FROM LOWER TO MIDDLE PALAEOLITHIC IN INDIA

Asok K. Ghosh

The initiation of research work in Indian prehistory, more especially of stone age and of course of palaeolithic industry, was made by Foote (1866) in 1863. The same year, 1863, made a hallmark with the publication of *The geological evidences of the antiquity of man* by Charles Lyell. Just four years before, in 1859 Boucher de Perthes' handaxes from the terrace gravels of the Somme valley established the authenticity of the tool-making Pleistocene man in Europe, and made a real breakthrough in terms of recognition of prehistory as a distinct discipline.

The forwarding comments are pregnant in many ways. First, the relative antiquity of work on "the occurrence of stone implements" in India. Secondly, the credibility of the work during the time period when the emergence of the discipline had just taken place. Thirdly, the caution which was made in fixing up at least the cultural chronology of the discovered stone implements. Foote, being a geologist, of the Geological Survey of India, along with his great quantum of experience and knowledge in the field-geology of the peninsular India, put forward a surfeit of data on "geological features of the implement-bearing formations." He even presented a number of sections with the sequence of beds. Despite all the details, he never labelled the term palaeolithic as a criterion for identification of the stone implements. In the said work, the conventional nomens of tool typology, viz. handaxes, cleavers, etc., were not used, perhaps intentionally. Instead of the same, the implements were described mostly on the basis of working edge. Later, he (Foote 1901, 1916) used a relatively better term, palaeolithic, but not further into lower, middle and upper palaeolithic. In naming the tool type specimens – the terms: axe, spearhead, narrow pointed type, broad pointed type, etc. were taken into consideration. Mention was also made about core and flake tools. The illustrations of blades and biserrated blades in the catalogue of Foote were of course the indication that they belong to post-lower palaeolithic period.

The term middle palaeolithic came into vogue much later, but even at the beginning the said term was not well-accepted. In the chapter entitled "middle palaeolithic", Burkitt (1921) also used a sub-title in parenthesis, noting that this period was "often classed with the lower palaeolithic by some writers who do separate a middle palaeolithic era." Even much later, in the European situation, the term middle palaeolithic received the recognition with the distinct presence of the association of Mousterian industries.

During the last fifty years, a considerable progress has been made with accelerated speed on the varied problems related especially to middle palaeolithic in Europe. Within the total problem, the major coverages include distribution, chronology, broad palaeoenvironment, origin, variation, technology, tool typology, and above all the holistic culture. An analytic synthesis of the above works, centering round European middle palaeoltihic has been very efficiently brought out by Bordes (1961a, 1961b).

Juxtaposition of Terminologies in India

Almost all the initial works on Indian stone age cultures were carried out by European scholars. This was mostly due to contemporary political conditions and administrative and academic situations. Naturally, the European scholars imposed their own terminologies in various sectors, including the expressions necessary in stone age nomenclatures in India. As a result, like Europe, the palaeolithic period in India was divided into lower, middle and upper palaeolithic eras. Probably, no attention was paid with a view to finding out the justification of the transplantation of European terminology in so-called Indian counterparts. The event of exactly the same nature also took place with African stone age cultures. After a lapse of quite a long time, the realization was understood and accordingly a new concept of stone age terminology came into existence with the use of the terms, viz. Early, Middle and Late Stone Ages. In course of working with the same terms, further modifications were made with the inclusion of transitional stages, as First Intermediate and Second Intermediate. Again, for conspicuous variabilities among the industries of the same time period or cultural phases, the local or regional nomens were given due importance. In a complex of this kind European middle palaeolithic and African Middle Stone Age are not coterminus. Rather, two independent chrono-cultural nomens were put forward to indicate the separateness on the one hand, and to explain the two situations with proper care on the other.

In Indian condition, the first deviation from the monotony was made by Burkitt and Cammiade (1930), and the possibility of the same was made due to differential context and application of appropriate methodological approach. A rich collection of stone tools from Kurnool by Cammiade was analyzed by Burkitt. In default of stratigraphic information, the classification was made, primarily on the basis of shape and size – and each class was considered as a series. No matter whatsoever, all the large-sized tools, as handaxes, for example, belong to one specific series. On the contrary, a single type, even the handaxe, may belong to any era of the palaeolithic period, more especially in Indian palaeolithic culture. At the same time, considering the demerit, it is also felt that during the period under consideration faced by a number of limitations, there was hardly any way out.

The difficulty in adoptation of European terminology went on persisting and to get rid of the constraints of the problem, the once-used 'series' concept was again promulgated by Sankalia (1956) after a quarter of a century for the purpose of distinguishing the sequential ordering of lithic industry from western India. In spite of its very little difference from the former one, the major drawbacks could not be ruled out.

The air of uneasiness, and perhaps a little bit of unawareness, with terminology prevailed in Indian situation for quite a long time. In the meantime the flow of suggestions was continued. One such suggestion was made by Subbarao (1958) in which his three-fold division comprised Early Stone Age, Middle Stone Age and Late Stone Age. Of course, he paid special attention to Middle Stone Age. The suggestion, however, did not make any solution. Rather, the original age-old problem with only difference of change of source. Instead of Europe, these terminologies were borrowed from Africa. As stated earlier, in Africa, like India, the problem of terminology was also identified in course of the use of European terminologies. The new concepts of terminology and even the nomens are in accordance with the regional needs. It does not mean in any way that in place of European, the African forms of expression will be appropriate or better in India. Again, in the case of Africa, the concept and the details of terminologies were modified in the course of the works in the same continent. In reality, no improvement was made in India in the sector of stone age division terminological units. As a result, the communication of device of the discipline and also of the works are hampered to a large extent.

It is worth mentioning to point out the terminological nomenclatures adopted by De Terra and Paterson (1939). On the basis of their survey, collection and examination of stone tools recovered from the terraces in the Soan valley, they employed the terms as Pre-, Early-, Late, and Evolved-Soan. Of course all these terms were not used by the pioneering authors, but in the course of further works a framework emerged on terms fitted with chronological ordering. A revision of the same has been proposed with logical propositions (Ray and Ghosh 1981). The use of nomens of specificity in the Soan culture complex is indicative of the fact that it was not possible to fit in the tool sequence within the framework of conventional European terms, more so the generalized formulation is provided with the basis of broad chronology associated with techno-typology of tools. Other means were also searched for to avoid the ambiguity in terminology which have resulted in the coinage of local terms to designate the specific industry as Madrasian (Menghin 1930), Soan (De Terra and Paterson 1939) and Nevasian (Sankalia and Banerjee 1958). These three cultural terminologies are provided with varied bases of characteristics. For Madrasian the connotation is "the Abbevillio-Acheulian industry of south India around Madras" (Krishnaswami 1947). The Soan includes the total industrial complex of the Pleistocene stone age culture in the Soan valley, in the northern part of the Indian sub-continent. No such clear indication or explanation is found to be present with the term Nevasian. The attempt to justify the coinage of the term became conspicuously futile one after another. This is very much marked by the continuous statements of the proposer, Sankalia. Instead of dropping the term Nevasian, he attempted to reinforce the same for all kinds of varied contexts, and as a result the term itself has mostly lost its identity. In his own words (Sankalia 1962: 43) the name Nevasian has been described as:

> "Since the tools occur in gravel, overlying the earlier deposit and underlie later deposits containing microlithic industries, it was thought advisable to designate the culture as "Middle Palaeolithic" (Sankalia *et al.* 1960). However, the term presupposes the existence of an Upper Palaeolithic culture which is not yet well attested from many regions. It was therefore suggested that a term of much wider connotation, like the "Middle Stone Age" (Subbarao 1958) be applied to it. This is also found to be unsatisfactory. Hence, as is the common practice with archaeologists and anthropologists, it is proposed to call the culture by the type site – Nevasa – "Nevasian" (Banerjee 1957). This name prefixed with the regional name as "Karnatak or Orissa Nevasian" will give a correct idea about the nature of culture, without involving questions of its place in time or any relation to the Middle Palaeolithic Industry of Western Europe or North Africa."

In his statement the attempt of correlation with Western Europe is not very satisfactory, even when the manuscript was written by Sankalia (1962). Since then in many of his writings Sankalia has attempted to include Nevasian, but nowhere has he achieved success. On the other hand, the complexity has been intensified. To indicate my own comment, the relatively recent work of Sankalia (1974: 144) may be referred which includes:

> "... it has been proposed to call the assemblage and the cultural stage it signifies "Middle Palaeolithic," or "Nevasian" after the type site of Nevasa."

He further continues his argument in the same context as:

> "It has been urged that the term "Middle Palaeolithic"* should not be used for this industry, for it immediately makes us compare it with the namesake in Europe, and the term "Middle Stone Age" be used instead. In the first place, the term "Middle Palaeolithic" may by usage remind one of the European, but these are most simple, common terms, ..."

Finally, Sankalia (1974) made a propitiatory conclusion, as

> "Until better, more positive, evidence is available to place the culture later (or earlier?), this *new* Stone Age culture should be called "Indian Middle Palaeolithic" or "Nevasian" (italics mine).

Further elaboration has not been made with intention, because with more examples the focal point of interest might be lost. From a brief summary of the ideas on terminology on Indian stone age cultural division, it is clear that there is a total chaos on this issue. This does not mean that better attempts are absolutely absent. Meaningful suggestions also came in the scene, but the acceptance of the same was not totally possible, and this may be due to certain personal factors, rather favor for self doctrine.

It is proposed to put forward a brief outline of the chrono-cultural framework of the palaeolithic culture of India, along with logistic terminologies. Otherwise, the aim and content of the present paper can hardly be explained.

In Search of a Valid Terminology

From the very outset, the present author was inquisitive for a proper understanding of the broad outline of Indian prehistory. With a view to satisfy his concept and the related ideas, a great amount of data from different parts of Indian sub-continent were collected, and all of them were analyzed in the perspective of historical treatment. This has naturally helped for a meaningful clarification of the whole problem. In the total

* There is in fact no logical contradiction between the term "Middle Stone Age" and "Middle Palaeolithic", as pointed out by Sankalia (1974: 144) can seldom be accepted.

process, the lacunae of the terminology were identified. The net result has rendered immense help both for assimilating and bringing out a valid and rational model.

In the broad framework, there are basically two compartments. One, on the critical assessment on borrowed terminology, and the other on the examination of suitable titles. In fact, some details on both the facets have already been discussed in some other papers of the present author. It is, perhaps, essential on part of the author to present a brief summary of author's comments in these two relevant issues.

With regard to transplantation of terminology from other countries the main demerits are:

a. Each region is provided with its own identities, and as such what is true to one region may or cannot be exactly true to another region.

b. The terms, viz. lower, middle and upper palaeolithic, of Europe may be suitable for Europe in very many ways, despite this the flows are not completely absent. In such a three-fold terminology – the general assumption is of equal division. In reality, these three divisions are not of equal time dimension, and not even similar in any way.

c. The above-mentioned three terms are being used in Europe for a very long period of time, and with the use of these terms the significations are made on the total contents. But if the same terms are used in other context with dissimilar condition – the basic connotation cannot be changed. This is the reason for which European terminology was not apt in Africa, and as a result new nomens were searched for in view of the existing need of the region. In Africa the concept was totally different. Early, Middle and Late Stone Ages have no counterparts with European terms. In this consideration, the footnote of Sankalia (1974) mentioned earlier also in this paper is sheer meaningless.

d. Borrowing of the terminology must be avoided, as the criteria of the terms in two different regions can seldom be similar. The characters on which the terminology are based are of great importance as they could signify the causative factors of terminology. The European glacial episodes are not of the same nature with the African pluvial phases.

The salient features of the new terminological index have been pointed out here to show the relevance of the same in meaningful propositions. In other words, the following points are the reasons to account for the coinage of new nomens. They are:

a. The basic differentiation of the two major cultural stages, viz. stone and metal stages, is based on the materials – stone and metal. The broad divisions of the stone age, as old (palaeo), middle (meso) and new (neo), have not been altered. But the further sub-divisons within the palaeolithic period have been re-named as: Pebble-Core, Flake and Flake-Blade elements, with the consideration of form or nature of raw materials mostly used.

b. In case of type-diagram it is noticed that there is a number of marked characters of change. The major forms are the emergence, survival or continuity, disappearance and at times reappearance. In this case the types have been identified in terms of the form of raw material on which they are made. A synthetic approach has been found to be useful for the purpose of identification of one stage from the other(s).

c. The model of cultural chronology, amalgam of typology, technology and the utilization of form of raw material, has been both compared and substantiated with the available stratigraphic evidences. In such an exercise, the broad trend has been given importance, and deviating variations have been mentioned. The deviations may be explained in terms of local variant.

Both pros and cons of the terminologies of older transplanted and new emergent forms have been shown here to evaluate the demerits of the former and the merits of the latter. These critical estimations also show that the search of an appropriate terminological scale was indeed essential, without which even the communication on palaeolithic materials could not have been possible. If necessary, this model may easily be compared with the models of either Africa or Europe or both. Finally, of course indicating the necessary and relevant reasons, the terminologies as Pebble-Core, Flake and Flake-Blade elements will be used here, as already been used earlier. Earlier tests have confirmed the hypothesis not only with one geo-cultural region, specifically Singhbhum (Ghosh 1970a) – the area of initiation, but also with other regions, especially in the north (Ray and Ghosh 1981), central (Ghosh 1961) and south (Ghosh 1973a).

The author's own idea has been dealt with the corroboration of samples and data earlier (Ghosh 1969, 1971a, 1971b, 1973). In this connection mention will be made of some other scholars who found the similar situation from their works, but they did not make any deflection from the conventional idea. For a representative illustration, Sankalia's own comment (Sankalia 1964: 372) may be quoted:

> " . . Thus, though the flakes are in a majority in any collection, and the Middle Stone Age industry or culture might be designated a "flake culture", still it should be prerequisite, as it was the only flake culture, like the Clactonian and the Levalloisian".

Both Clactonian and Levalloisian are more related to technology than culture, and thereby Sankalia cautiously avoided the terminology which he felt better and followed the existing conventional one.

Lower and Middle Palaeolithic vis-a-vis Pebble-Core and Flake Elements

Lower and middle palaeolithic industries, and even at times cultures, are not really the exact and strict counterparts of Pebble-Core and Flake elements. In spite of the fact, the two sets of assemblages in a set chronological framework may be compared. Comparisons are only possible when the materials to be compared are not exactly identical, and the separation is based on the facts of similarity and dissimilarity. In the present case, with two sets, there are both affinities and differences.

In this connection some of the hindrances or basic limitations of Indian stone age, more especially of the palaeolithic period, are to be mentioned. The major ones are related to chronology. The finds from the terraces do not always represent the materials as contempoary with the said terrace formations. The stratigraphic evidences are not much sufficient. Moreso, the works from many of the sites and even regions do not indicate the necessary data to examine the process of transformation from one element to another, from Pebble-Core to Flake element, as from lower to middle palaeolithic.

Neither the Pebble-Core element nor the Flake element will be described and discussed in great detail in the present undertaking. Such attempt is somewhat unnecessary, rather only those points will be taken greater care which are provided with the bearing on the process of transformation from one element to the other.

For clear understanding of the concerned problem, the basic facts of Pebble-Core and Flake elements are noted here and the extension of the same will be made with two other dimensions – the distribution and the stratigraphic feature. Further, another backdrop will be included in the treatment and that too is on environment, especially which facts are available and nothing more greater speculation.

The main focus of interest will be laid on the area of transformation or overlapping in all possible ways, as of technology, typology, assemblage and differential complexes. At the same time, endeavors will be made to search for the factors, the process and total results.

Pebble-Core Element

This may be considered as the earliest unit of the stone age culture in India which initiates the palaeolithic period. Neither the absolute nor even the relative date of the same is available which may be considered as reliable. Despite the fact, the vertical extension of the same indicates that this element was in existence for a very long time, and the length of this period is far greater than any period of other elements. As the duration was quite high the dynamicity was high, but the process was slow. This is the dawn of culture and thereby the type and the technology were primeval at the onset. It is presumed that the basic tool type in India which may also be considered as the earliest is the chopper with free flaking technology. Later, this type, probably, gave rise to a distinct different type, known as handaxe. From the typo-technological study – the relationship between handaxe and the chopper may be ascertained and on the basis of the same it is logical to comment that both the types are provided with a techno-typological genetic relationship between them without sudden change in the form of aberration.

In the same way, some amount of relationship may be drawn between handaxe and cleaver, especially at the initial stage of the emergence of cleaver. If chopper, handaxe and cleaver are set one after another – the successive relatedness may be brought out. On the other hand no such connection may be established between chopper and cleaver, if the intermediate type, handaxe, is taken out. The typological lineage of this kind is also apt between Pebble-Core and Flake element, and for such reason the transformation between the two elements is bridged by the area of overlapping. The other predominant type within the Pebble-Core element is the scraper, and the genesis of the same is laid in chopper, of course the modification has found to be marked in case of scrapers which came late, and are in association with handaxes. In this case the total techno-typology of handaxe has influenced the morphology of scrapers. This may be one of the reasons for the diverse types of scrapers, and this type is found to be present throughout in the stone age for its greater and useful functional implications.

In the stratigraphic context the Pebble-Core element covers three stages, viz. lower gravel bed, lower silt bed and upper gravel bed. Of course within such an extension, the same element made a good deal of transformation. In spite of the alteration, the basic pattern did not change much and thereby the same element did not shift the identity. In intensive examinations, the microunits may be worked out, and they have been found out with the name as Abbevellian, Acheulian I, Acheulian II and Acheulian III.

During the Pebble-Core element, a majority of the tools were made on pebble-core or a-pebble core. In the course of knapping, large flake debitages came out, a greater part of the same became waste while smaller parts were perhaps utilized for functional use, either without or with prior workings on the flakes. In this case the flakes are cohesive with the Pebble-Core element, but they are not really the integrated part of the total complex, rather more a by-product. The flakes were not knapped with any intention of making tool types, but at times used as *ad hoc* tools. In such case the inclusion of the flake as a constituent of culture complex is an open question. These flakes are more the debitages than produced flakes for functional operation as blanks or tools.

The flakes thus detached were primarily waste products, later some of them were utilized, further later a few of them were worked and utilized. The last stage of the process gradually made a complex movement and the prototype of other tools, such as points, borers, knives, etc., came into being. If the total assemblage materials of the Pebble-Core element at its fag end is examined, the earlier mentioned types may be identified. Of course, they will not pose themselves as accurately representative types of similar tools found in the later stages.

Flake Element

In all possible ways, this element may be considered as a distinctly different stage, came after the Pebble-Core element. A greater amount of preparation of the same was actually made in the background stage, and the full-fledged form came into being as a specific element. The existence of this element in time depth is comparatively much lesser than the Pebble-Core element. The distinctiveness of the Flake element is found in the form of typological assemblage, more especially the incidence of specific types in terms of percentage and even in proportion. There are marked falls of certain types as chopper, handaxe and cleaver, but they did not extinct totally. On the other hand, the rise or increase of other types, as scraper, point, borer, knife, etc., is found in conspicuous form. On the basis of the evidence it may be suggested that the prototypes of the previous element became very specific types during this element.

Proper heed was not paid to this phase, either as a chrono-cultural unit or a period of changed techno-typology of the tools. In terms of stratigraphic position, this element is found in the junction level of upper gravel bed and loose upper clay. Though the extension of time coverage of this period of dictinction level between the end of one sediment and the beginning of other sediment cannot be worked out. But as the position of the layer is situated on a higher level, and it marks the end of a wet period – the state of such environmental episode could have been continued for quite a long time. It is not only the change of the later climatic condition but the magnitude and intensity of the same which were really of great value, because those extra events were responsible for the deposition of other types of clastic sediments. Again, after the fag end of the wet period of gravel-bearing activity, there was probably a stage during which the same gravel bed at its top was exposed with some amount of weathering activity.

The above critical observations in relation to typology, involved technology and even broad stratigraphy (with the exception of terrace sequence in the north) are well agreeable with most of the works on palaeolithic in India. Especial mention may be made with respect to the areas as north (de Terra and Paterson 1939; Mohaptra 1974), northeast (Krishnaswami and Soundara Rajan 1951), west (Misra 1967), east (Ghosh and Basu 1969; Ghosh 1970b), central (Khatri 1961, 1962; Sen and Ghosh 1963; Joshi 1964, 1978) and south (Ghosh 1973a; Pappu 1974), etc. Even in course of their works from all the regions, most of the works found, and expressed, the difference – but the apt expression was seldom made.

The above-mentioned synoptic outline has been dealt with a view to making the situation prepared for exploring the area or region of transition and transformation. In the theoretical sense the area of overlapping or transition is expected to have some of the common characters from the preceding and succeeding stages, and in addition to the same there must be some other character with the help of which the area will be identified from other two extreme areas. There is a difference between Pebble-Core and Flake elements, and at the same time some commonness are also present between the two. In the same way, the area of transformation is different from either the Pebble-Core element or the Flake element or even from both of them. The amount of similarity or difference depends on the position of area of transformation in relation to two specific elements, Pebble-Core and Flake. After all the mobility in the space of transformation is variable, it is highest in the central portion and gradually decreased at two ends. In the expression wave length, the early phase of transformation is more akin to Pebble-Core element and the late phase is tending towards equalization with the Flake element. For intensive understanding the realization of modulation within the area of transformation can never be escaped. The transition is, however, a specific in its own way.

A Glimpse of Essential Data

The present paper is not supplying all the data related to the concerned problem – the task is neither necessary nor possible. Despite this, a very quick survey of the relevant materials is included herewith. It may be noted that in the present case that part of Pebble-Core element and Flake element has been put forward which has something to do with the process of transformation. About the said two elements brief information has already been discussed earlier in this paper.

About distribution, there are five major areas in this sub-continent where there are concentration of industries of Pebble-Core element. These five areas are located in the northwestern, northeastern, eastern, central, western, and southern parts (Figure 1). The areas have been identified on the basis of clustering of large number of sites within the individual area. The areas in turn are indicative of the habitation of hominids with Pebble-Core element for an immense period of time. Incidentally, it may also be mentioned that in all these regions the final phase of the said element is found to occur.

In case of the next elemental (industrial) phase the distribution is found to be extended, though of course the earlier areas are found to be continued with habitation. In this distribution of sites, and their amalgam into areas there are two differences from

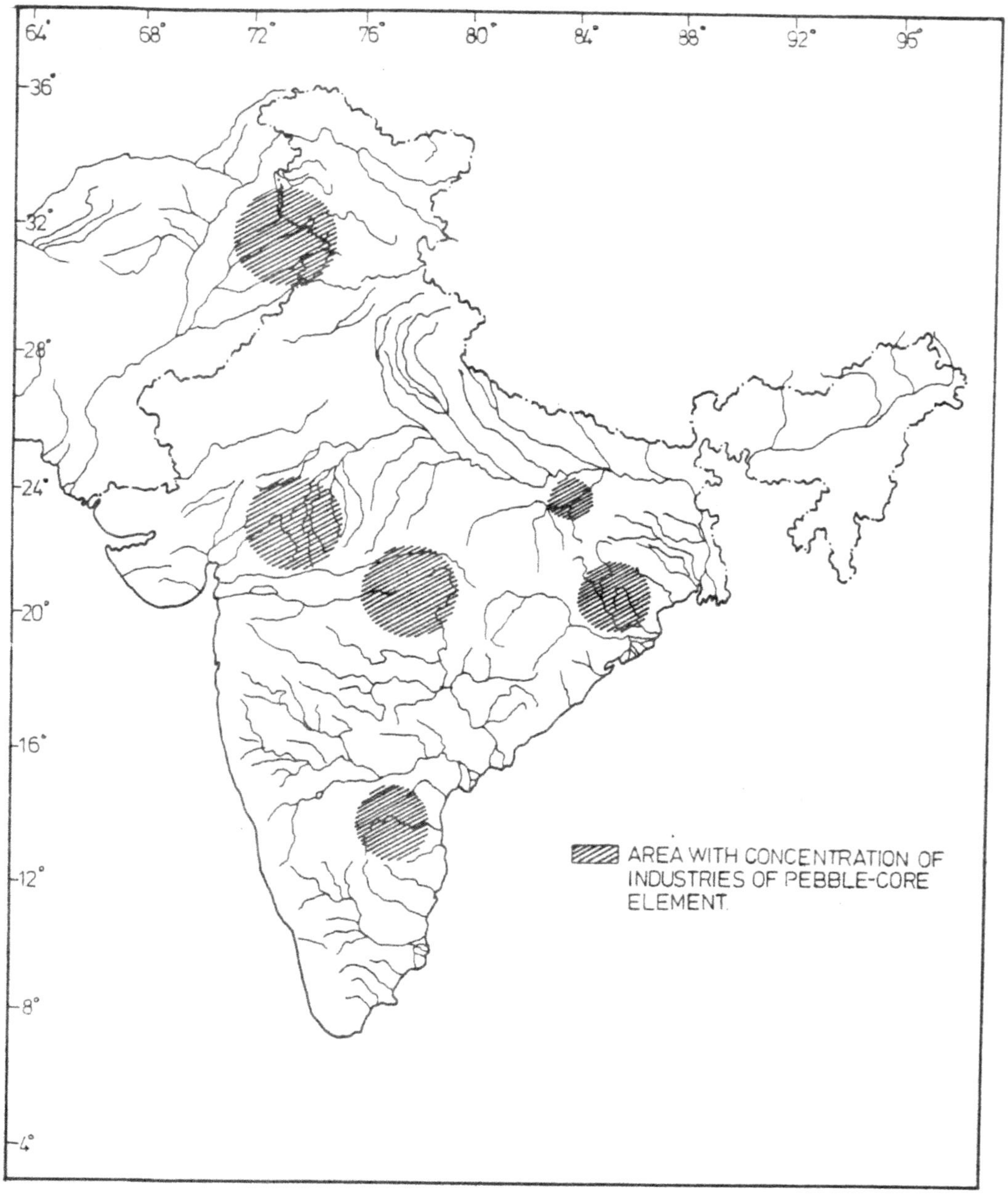

Figure 1 The distribution of Pebble-Core element.

the earlier one. First, the coverage of spatial dimension is greater in comparison to Pebble-Core areas. Secondly, there are two additional area units – one slightly north of central area and the other is situated between western and southern areas. All the seven areas, out of a total of eight, are situated in the Peninsular part. The seven adjoining areas in the Pensinsular part have been considered together to indicate the overall distribution pattern in the said region (Figure 2).

The enlargement of distribution pattern must have been taken place through a process of mobility of tradition and migration of the hominids. The diffusion of such mechanism must have occurred as a result of certain factors which are acutely responsible for the slow but continuous change. This indeed the process of transformation from one element to another.

It is also noticed that with the extension of the industries of Flake element, addi-

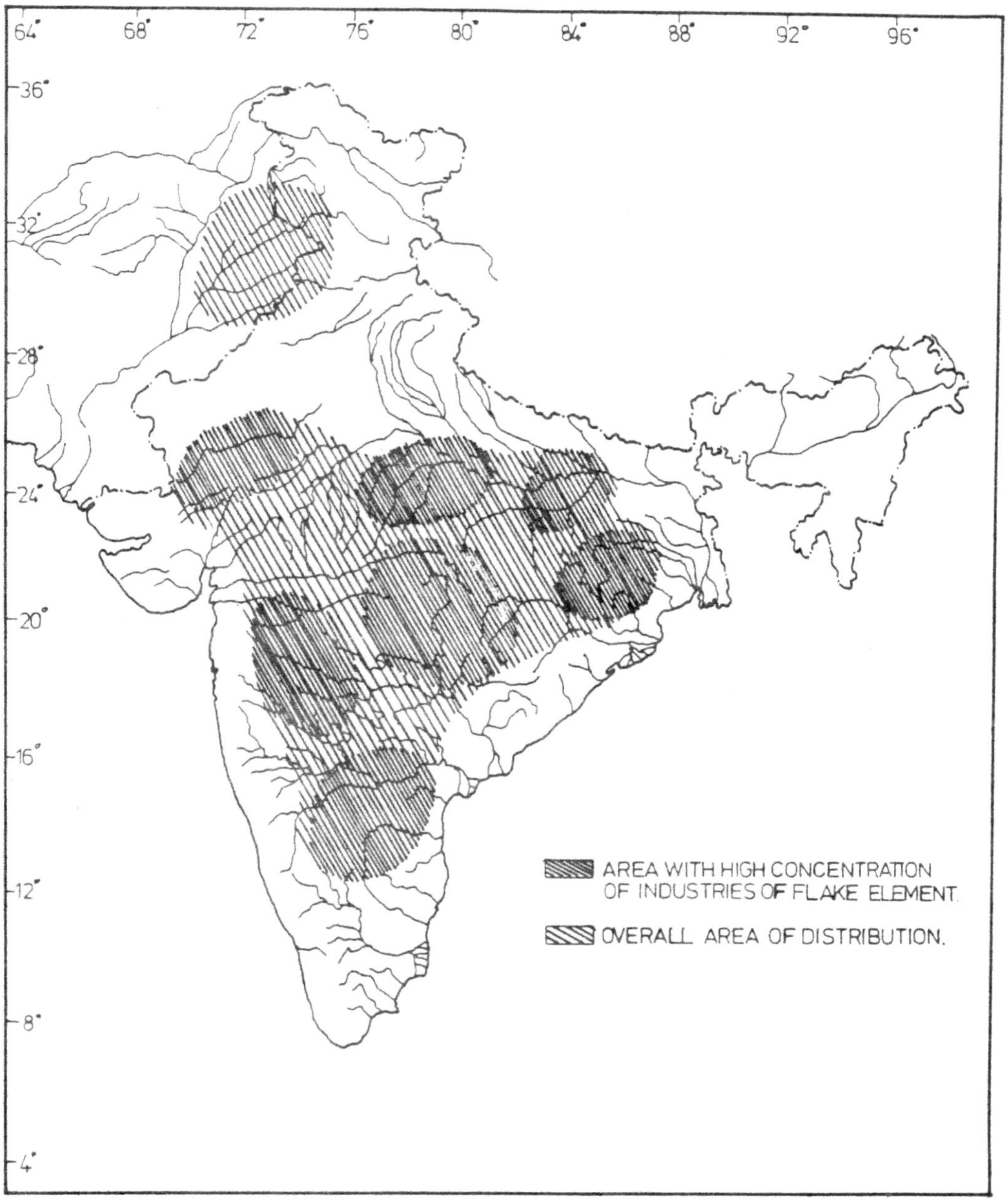

Figure 2 The distribution of Flake element.

tional areas came under the coverage of habitable ground. From the facts, it may be assumed that the later element in its total content had the power of ecological balance. The encounterance of this kind was made through changed technology and typology.

In understanding the geo-stratigraphy, which will indicate an approximate idea about the relative chronology of the transition between Pebble-Core and Flake elements, some of the basic data require to be included here, and these too are on the basis of the areas under consideration. In northwestern India, the initial phase of T_2 through the process of aggradation marks the period of transition. The situation is more or less similar in the vicinity areas, mainly in the Himachal Pradesh (Lal 1956; Mohaptra 1974).

In the Wagan valley of Rajasthan (Misra 1967) the transitions may be located at the contact zone of pebbly gravel and overlying mottled clay.

At so-called type site Nevasa (Banerjee 1957) tools of the Flake element, of course with a different nomenclature, have been described to come from the layer of comparatively loose and small gravel, designated as gravel II.

In most of the southern regions, the transition is either found on the higher level of the secondary laterite deposit or in the level of intersection of the secondary laterite and gravel beds. This difference of stratigraphic units and the occurrence of the transition between the two elements may be explained in terms of local geomorphology, i.e. elevated area and low-lying valley. But in terms of time scale both the occurrences are contemporary, of course within a broad range of variation (Ghosh 1973b).

In the eastern part, which includes the states of West Bengal, Bihar and Orissa, the overall stratigraphic chronology is more or less similar. In this case, the further eastern part, viz. Assam, has been excluded for the paucity of reliable and reported data. Out of two gravel and two clay beds, the transition is slightly older than the contact zone of upper gravel and upper clay. In cases, the same transition is found to occur in the upper portion of the detrital laterite deposit (Ghosh 1970a).

In the central region, the presence of the area of transition between two elements is more or less the same as found in the eastern part. Here too, the end phase of the upper gravel marks the broad temporal area of transition (Sen and Ghosh 1963).

A little deviation is found to occur in the northeastern part, centering around Mirzapur (Krishnaswami and Soundara Rajan 1951) as the transformation appears to start within the bed of upper gravel, not further up.

From a holistic viewpoint, it may be said that the end of Pebble-Core element does not distract much with regard to time period and so also the beginning of Flake element. In such a circumstance, the transition between the two elements can hardly be of completely different time period. The contemporaneity of this new nature may be explained in two ways. One, during the transition the climate was in a state of the setting in of desiccation. This pan-Indian climatic change initiated the hominids to opt for a change-over. Of course, the change of such nature went through a slow process. This also fits in with the hypothesis of transformation. Secondly, the intra- or inter-industrial relationship was of better form and thereby the diffusion of traditional innovation quickly set in.

About typology of tools, and technology to some extent, an intensive examination was carried out with a large set of samples which covers a continuous spectra starting with Pebble-Core element and ending with Flake element. In such an exercise, the morphological attributes and metric variables have also been considered. The result of the analysis reveals that the most predominant types in the Pebble-Core element are: chopper, handaxe, cleaver, scraper, and a few less conspicuous types. On the other hand, the prevalent types in Flake element are: point, borer and knife, again other few more types.

In the area of transition, three different specific situations are met with. First, a trend of disappearance – as in the case of chopper. Secondly, a position of continuous existence – as in the case of handaxe, cleaver and even scraper. Thirdly, the context of new emergence – as of points, borers and knives. In no case of disappearance, continuance and emergence, the position with all the types is the same. As a matter of fact, the total amalgam of the three forms, presence and absence, is the generalized resultant of transition from one element to another (Figure 3).

When technology is superimposed on this changing pattern of typology, it is found that this aspect follows the same principle. Free flaking technique is dwindling out, controlled flaking is continuing and secondary trimming is emerging.

As scraper, as a blanket type, is the most common – the differentiation within the

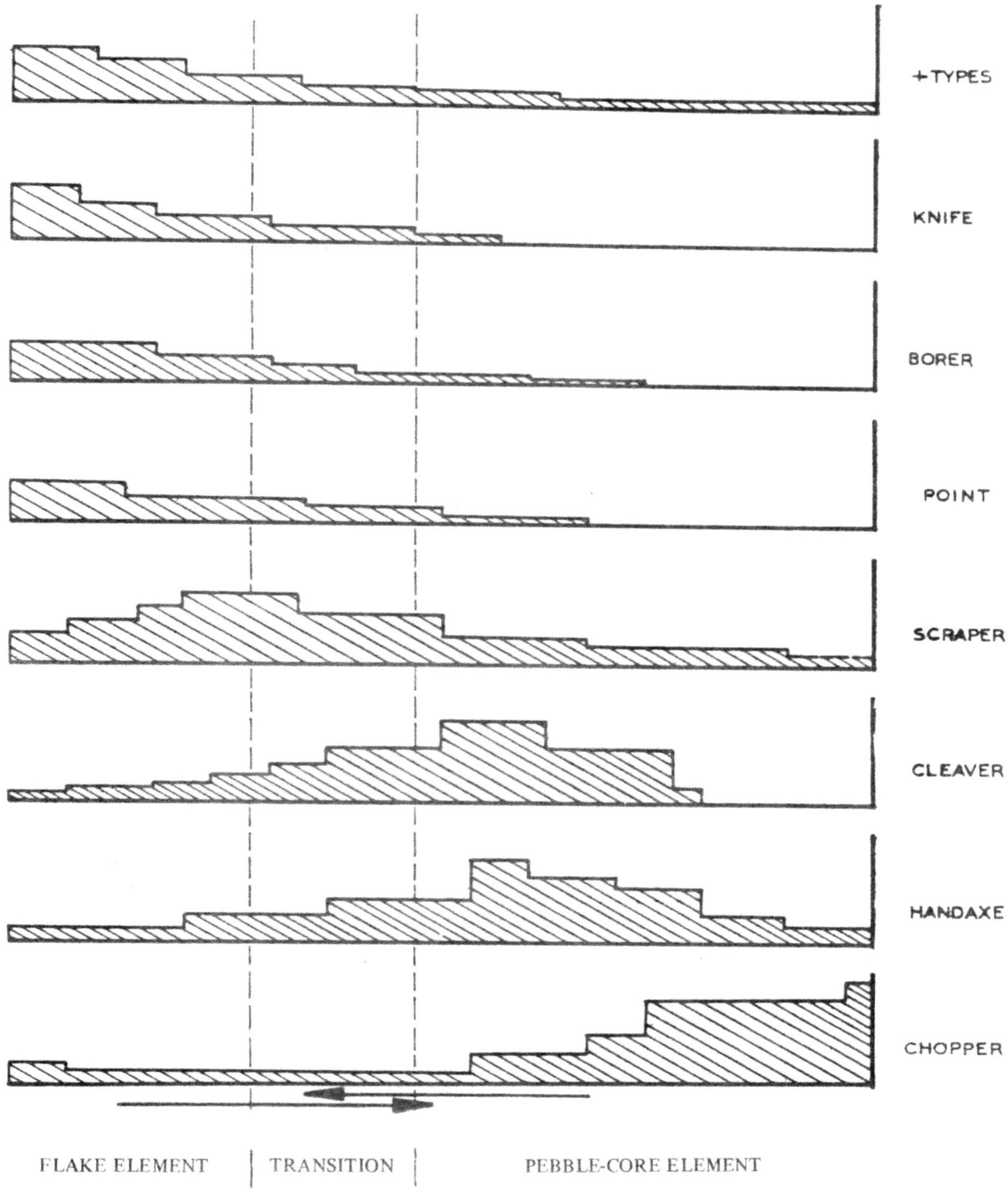

Figure 3 Type diagram covering Pebble-Core and Flake elements and the area of transition.

same broad type may briefly be said in the present context. Broadly there are three varieties of scrapers; they are side scrapers, end scrapers and round scrapers. In the case of the side variety – the side may be straight (single/double), convex, concave, concavo-convex. The area of scraping area in the case of the end scraper may be considered as a character of variation. The combination of side-cum-end scrapers are also found. All the scrapers are found in the whole spectra, but during the Pebble-Core element, they are mostly made on core, and they are in the majority of cases on flake during the Flake element. In the area of transition, the proportion of flake scrapers is greater. Again, in this case even the core scrapers are provided with better working area with the utilization of better technical skill. Here the type scraper may be considered as a marker for the purpose of evaluation of status in two major elements. The above points, as distribution, stratigraphy, typology and technology, are the indicators of spatial distribution, temporal coverage, and the technological capability and capacity both for making the tools and using them with the expectation of appropriate results. In the course of displaying the information and the analysis, it has been attempted to show that the area of transition of industrial complexes cannot be well demarcated, and it is mainly due to the overlapping situations in all respects. At the same time it may be suggested that transition is the initial stage towards a major changeover.

Factors, Process and Results of Transmutation

If the end phase of Pebble-Core element and the onset of Flake element are considered as separate entities, the situation of the two elements in their specific phases may be considered as somewhat allelomorphic to each other. This is primarily due to the fact that they occupy the same relative position on homologous cultural phenomena, and in the developmental process the produced effects are different. In organic mutation pertaining to living cells, the change is sudden, but normally the event is very infrequent – though it can be speeded up through induction. In the present case, the context is super-organic and the changes are scarcely deleterious.

The transmutation process of culture is provided with a number of causative factors, and in cases they are interlinked with one another. Population explosion of cultural actors may perhaps be the prime factor. To cope with this gradual changed situation, the survival fitness must have been enhanced. Both the factors, related to hominid population, of increase of offspring and survival fitness may be explained with the help of natural selection. For enhanced survival fitness two characters are of immense importance. They are: furtherance of security, and greater exploitation of food resources. These two characters again gave rise to betterment of livelihood. With population explosion, there is the necessity of wider spatial coverage which will accommodate the population, and the area of food resources will be extended. The cumulative resultant will give rise to spread of mobility. The results of above chain actions are revealed from the two maps (Figures 1 and 2) which display the extension of area of industrial complexes from one element to another.

The above factors, process and results are speculative in nature, albeit with the thought process of rational and logistic approach. The furtherance of the same idea may be made with other sets of data, which are concrete. The transmutation which took place in course of changeover from one element to another is also revealed in the stage of transition of the assemblages of industrial complexes. As a matter of fact, it is essential to examine the reflected resultants of behavioral forms of the actors of transitional stage.

Among many distinctive characters which differentiate the industries between Pebble-Core and Flake elements, the sum total of techno-typological features of the tools is an important marker. The marker may again be examined through different traits. Most of the traits are integrated to one another. Perhaps the most spectacular trait is the size diminution of tool types, at least for the majority of them. This has directly affected in the area of economy of raw materials. Simultaneously, further economic benefit related to raw materials has been made with the utilization of more raw materials, as from quartzite and quartz to different varieties of colloidal silica. Thereby the demand on quartzite and quartz has been decreased, and the replacement has been made with the help of other varieties of raw materials.

The processual development of tool making technology and the cumulative results of technological operations have accentuated the total technological change. Further change in this sector has been made possible with the utilization of new raw materials. To work on new raw materials new technology came into being. All the above-mentioned basic characters described above, and some additional points which will be discussed later, have been arranged with a purposeful view to generate the idea in the form of a model of interacting process (Figure 4).

A large-sized tool, as handaxe, might have been used for a number of functions, as probably digging, cutting, scraping, etc. The service of multiple functions cannot be retained with the diminution of size. Naturally, to serve all the functions, a series of tools are to be prepared, and this idea brought out the multiplicity of types. The concept, individual type of specific function, led to specialized function of each type. If the characters such as size diminution, economy and utilization of raw materials, multiplicity of types and specialized functions are properly scrutinized – they point to another character which is increase in working area. All these characters in their turn, independently or conjointly, indicate another pointer, related to proper distribution of energy and labor. A number of characters, related to typology, technology, shape and size, rather than broad morphology, multiplicity of type, economy and utilization of raw materials, specialized function and increase in working area, have demonstrated through an illustration (Figure 5) and that too is at least indicative of the fact that the total raw material necessary for the production of one handaxe will produce at least four tools of different types.

Finally, spread of mobility was further made favorable with the load decrease of tool kit. With the diminution in size on the one hand, and multiplicity of types with specialized function on the other – the kit of tools was provided with a number of types for serving various purposes without being heavy (Ghosh 1973c). The same kit with the types was at par with the existing needs of the actors, both in the area of defense and offense (Ghosh 1973d)

The transition which has been pointed out here, placed between Pebble-Core and Flake elements, may not be much different from the transition between lower and middle palaeolithic cultures in the sense of generalization. After all, each greater area is having its own identity. To mention one of such identities in India, the example may be shown from the Mousterian of Acheulian Tradition (Ghosh 1976).

The transition which has been emphasized here is related to culture, the constituents being typology and technology, in the relative and broad framework of chronology. The environmental transference has also been touched within the scope and possibility. About the actor, the hominid population, either of the transition or of the industrial elements, the data are completely absent. Despite this it may be suggested that it is not unlikely to propose the transition of the actors which is the linkage between *Homo erectus* and *Homo sapiens neanderthalensis*. After all, the total cultural manifestations depend on the total biology of the actors in response to the prevailing needs and the adaptation and adjustment of the same. The evolution of hominid through the transition of industrial stages may perhaps be postulated (Ghosh 1971c) with the help of total assemblages and the micro-variation within the fractional stages. It is simply hoped that more data in the near future will be of immense help to clear out the problem.

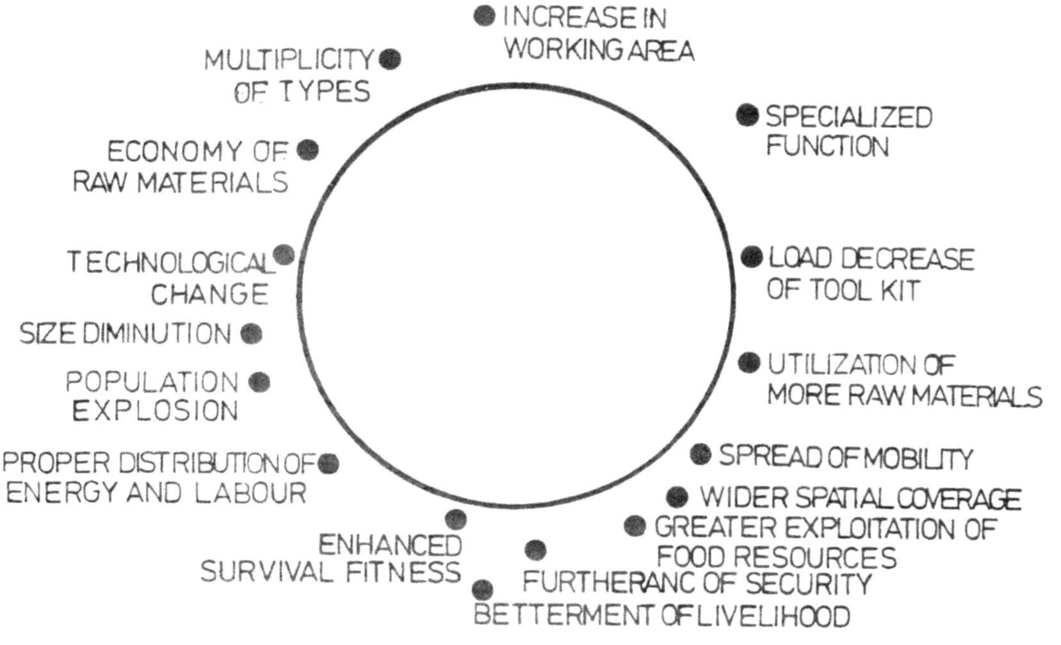

Figure 4 Model of causative factors and emergent results during transition from Pebble-Core to Flake element.

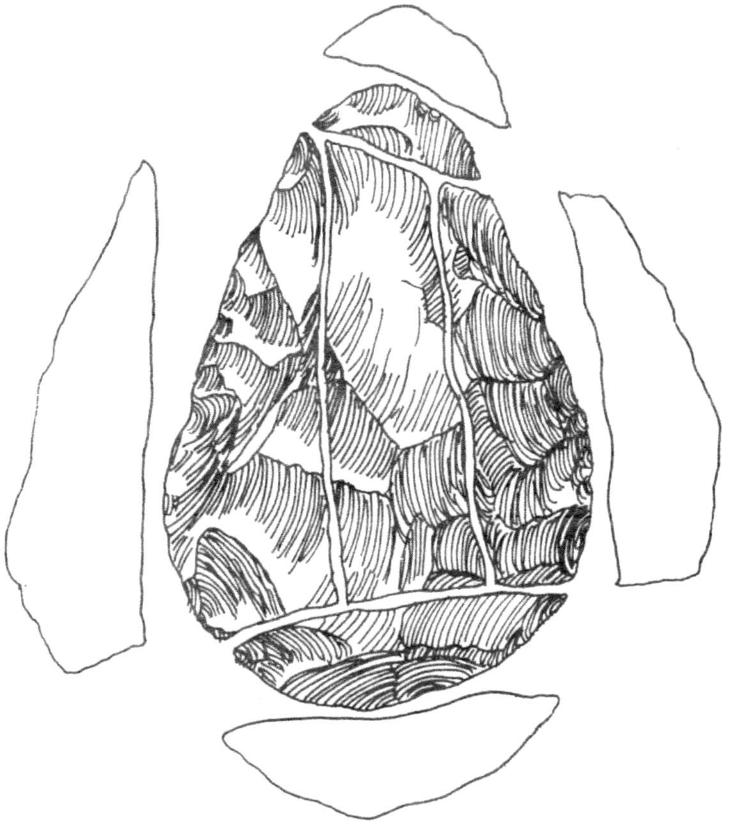

Figure 5 Diagrammatic representation of the economy of raw material, multiplicity of types from a single type, and increase in working area with very specialized functions.

References

Banerjee, K.D. 1957 *Middle palaeolithic industries of the Deccan* Doctoral dissertation, Poona University and Deccan College.

Bordes, F. 1961a *Typologie du paleolithique ancien et moyen* Bordeaux: L'Institut de Prehistoire de l'Universite de Bordeaux.

Bordes, F. 1961b Mousterian culture in France *Science* 134: 803-810.

Burkitt, M.C. 1921 *Prehistory – a study of early culture in Europe and the Mediterranean basin* Cambridge: Cambridge University Press.

Burkitt, M.C. and L.A. Cammiade 1930 Fresh light on the stone age in south-east Asia *Antiquity* 4: 327-329.

De Terra, H. and T.T Paterson 1939 *Studies on the ice age and associated human cultures* Washington, D.C.: Carnegie Institution.

Foote, R. 1866 On the occurrence of stone implements in laterite formations in various parts of Madras and north Arcot districts *Madras Journal of Literature and Science* 3(2): 1-35.

Foote, R. 1901 *Government Museum, Madras – catalogue of the prehistoric antiquities* Madras: Government Press.

Foote, R. 1916 *Madras Government Museum – the Foote collection of the Indian prehistoric and protohistoric antiquities* Madras: Government Press.

Ghosh, A.K. 1961 A new lithic site in Mandla district *The Eastern Anthropologist* 14(3): 209-215.

Ghosh, A.K. 1969 Developmental pattern of palaeolithic culture in India. *In:* Mireille Ters (ed.) *Etudes sur le Quaternaire dans le Monde,* VIIIe Congres INQUA. Paris, Union International pour l'Etude du Quaternaire. pp. 1033-1038.

Ghosh, A.K. 1970a The palaeolithic culture of Singhbhum *Transactions of the American Philosophical Society* N.S. 60: 1-68.

Ghosh, A.K. 1970b An analysis of Mayurbhanj industry *Journal of the Indian Anthropological Society* 5(1-2): 233-248.

Ghosh, A.K. 1971a Palaeolithic industries of eastern India in the light of recent discoveries. *In: Actes du VIIe Congres International des Sciences Prehistorique et Protohistoriques* Tome I, Praha. pp. 160-164.

Ghosh, A K. 1971b Ordering of lower palaeolithic tradition in south and southeast Asia *Archaeology and Physical Anthropology in Oceania* 6(2): 87-101.

Ghosh, A.K. 1971c Flake and Flake-Blade industries in India in the context of human evolution. *In:* F. Bordes (ed.) *The origin of Homo sapiens, Paris: Unesco.* pp. 95-100.

Ghosh, A.K. 1973a Emendation of Madrasian industry. *In: Actes du VIIIe Congres International des Sciences Prehistoriques et Protohistoriques,* Tome II, Beograd. pp. 39-48.

Ghosh, A.K. 1973b Chrono-cultural perspective on the palaeolithic industries of India. *In:* Ghosh, A. and D.P. Agrawal (eds.) *Radiocarbon and Indian Archaeology* Bombay: Tata Institute of Fundamental Research. pp. 23-31.

Ghosh, A.K. 1973c Environment as revealed from tool kit – a study on Indian prehistoric context *Paper Commissioned for UNESCO/INQUA Symposium on prehistoric man, his material culture and natural environment* Christchurch. pp. 1-14.

Ghosh, A.K. 1973d Prehistoric and natural environment – a study on their relationship in Indian context *Paper Commissioned for UNESCO/INQUA Symposium on Prehistoric man, his material culture and natural environment* Christchurch. pp. 1-12.

Ghosh, A.K. 1976 *Mousteroid facies in the palaeolithic culture of India* Academy of Sciences of USSR, Novosibirsk. pp. 84-89 (in Russian).

Ghosh, A.K. and A.N. Basu 1969 An appraisal on the palaeolithic industries of Mayurbhanj in the light of recent discoveries *Science and Culture* 35(9): 476-479.

Joshi, R.V. 1964 Acheulian succession in central India *Asian Perspectives* 8(10): 152-163.

Joshi, R.V. 1978 *Stone age cultures of central India* Poona: Deccan College.

Khatri, A.P. 1961 Stone age and Pleistocene chronology of Narmada valley, central India *Anthropos* 56: 519-530.

Khatri, A.P. 1962 Origin and development of series II culture of India *Proceedings of the Prehistoric Society* 28: 191-208.

Krishnaswami, V.D. 1947 Stone age India *Ancient India* 3: 11-57.

Krishnaswami, V. and K.V. Soundara Rajan 1951 The lithic tool industries of the Singrauli basin, district Mirzapur *Ancient India* 7: 40-65.

Lal, B.B. 1956 Palaeoliths from the Beas and Banganga valleys, Punjab *Ancient India* 12: 58-92.

Menghin, O. 1930 *Weltgeschichte der steinzeti* Vienna.

Misra, V.N. 1967 *Pre- and protohistory of the Berach basin, south Rajasthan* Poona: Deccan College.

Mohaptra, G.C. 1974 Lithic industries of Himachal Pradesh. *In:* Ghosh, A.K. (ed.) *Perspectives in palaeoanthropology* Calcutta: Firma K.L. Mukhopadhyay. pp. 199-212.

Pappu, R.S. 1974 *Pleistocene studies in the upper Krishna basin* Poona: Deccan College.

Ray, R. and A.K. Ghosh 1981 Position of palaeolithic culture of north India *Proceedings of the International Conference on Neogene/Quaternary boundary* Geological Survey of India, Calcutta, pp. 143-150.

Sankalia, H.D. 1956 Animal fossils and palaeolithic industries from the Pravara basin at Nevasa, district Ahmednagar *Ancient India* 12: 3-52.

Sankalia, H.D. and K.D. Banerjee 1958 The middle palaeolithic culture of the Deccan Karnatak and central India *J. of the Palaeontological Society of India* 3: 158-169.

Sankalia, H.D., S.B. Deo, Z.D. Ansari and S. Ehrhardt 1960 *From history to prehistory at Nevasa (1954-56)* Poona: Deccan College.

Sankalia, H.D. 1962 *Indian archaeology today* Bombay: Asia Publishing House.

Sankalia, H.D. 1964 Middle stone age culture in India and Pakistan *Science* 146: 365-375.

Sankalia, H.D. 1974 *Prehistory and protohistory of India and Pakistan* Poona: Deccan College.

Sen, D. and A.K. Ghosh 1963 Lithic culture-complex in the pleistocene sequence of Narmada valley, central India *Rivista di Scienze Preistoriche* 18(1-4): 3-23.

Subbarao, B. 1958 *The personality of India* Baroda: M.S. University.

IDENTIFICATION OF TRANSFORMATION FROM LOWER/MIDDLE PALAEOLITHIC WITH TYPO-TECHNOLOGICAL MARKER

Ranjana Ray

Studies on stone age cultures of India started over a century and throughout the whole period, the historical course which has given rise to the cumulative form is indeed significant. The initiation of studies were made by Europeans and as a matter of fact their ideas and concepts were imposed upon Indian situation. Indian materials were primarily thought of as European counterparts, in a separate geographical regimen. The total effect brought out to express Indian tool typology, involved technology, chrono-cultural stage and other related issues in terms of European model. The European influence is quite conspicuous in the earlier terminological concepts of Lower, Middle and Upper Palaeolithic cultures. Difficulties were met within the context of correlation between materials from Europe and other parts of the Old World. The situation on stone age culture became more enigmatic with the time and only terminologies went on accumulating. Sankalia (1956) brought in the series concept after Burkitt and Cammiade (1930). Sankalia proposed a three-fold series instead of the four-fold classification of Burkitt and Cammiade. In the context of terminology the influence of African classification also came into existence (Subbarao 1958). The lithic tradition of India is divided into Early, Middle and Late Stone Ages. Both the European and African terminologies have their own meaning and usages in terms of stratigraphy, chronology and typo-technology in the same area. The series concept is no longer in practice but both the African and the European terminologies are used simultaneously in Indian archaeology, Lower Palaeolithic or Early Stone Age and Middle Palaeolithic or Middle Stone Age. The workers (Misra 1962: 122) felt the difficulty but no suitable concept was devised until Ghosh (1970) put forward a meaningful proposition. The concept agrees well with the traditional scheme. The whole range of the Pleistocene period is divided into three chrono-cultural elements, *viz.* Pebble-Core, Flake and Flake-Blade. The earliest element, Pebble-Core, has evidenced the development of chopper and biface within its span of time. Core and Pebble are the main forms of raw material. The flakes detached in the process of working the cores form an integral part of the element. The elements which follow the Pebble-Core are Flake and Flake-Blade dominated by flakes and blade-like flakes and finally blades, as form of raw material. All the three elements can be identified on the basis of techno-typology and they follow geological stratigraphy and chronological ordering. The Pebble-Core element covers the major part of Pleistocene. This element emerged simultaneously all over India at a transitional phase by the end of the lower Pleistocene and the beginning of the middle Pleistocene. The element passed through different environmental periods, two wet phases and a dry phase in a cyclic nature. Despite this fluctuation of the environment no major change in tradition took place. A gradual development in typo-technology is marked.

Conspicuous change of tradition from Pebble-Core to Flake and Flake-Blade element is marked. The latter two elements have shorter time span, compared to the earlier cultural element, spreading through the upper part of the upper Pleistocene. The Flake element precedes the Flake-Blade element and succeeds the Pebble-Core element. The element has an assemblage of distinct techno-typological traits. The Flake element is stratigraphically identified by Ghosh (1970) in eastern India coming from the junction of upper loose gravel and upper clay, middle part of the upper Pleistocene very tentatively.

In this paper the terms Pebble-Core and Flake element are used instead of Lower Palaeolithic or Early Stone Age and Middle Palaeolithic or Middle Stone Age respectively. The Early and Middle Stone Ages in Africa have two intermediate stages, which are not found in Indian sequence. The Middle Palaeolithic stage is identified in Europe by distinct flake type and technique which is not comparable to India. There is no correlation whatsoever of the European terminology on the one hand and Indian terminology on the other. The three-fold divisions in both the situation do not approve of any stagewise correlation from one region to another. In this paper an attempt is made to identify the transition of tradition from one phase to another, the earlier Pebble-Core element, and the later Flake element. Both tool typology and involved technology are the important criteria of these elements. The analysis of the typo-technological assemblages of both the elements in an intensive way, reveal the transition of one element to the other.

Sites and Stratigraphy

The Indo-Pakistan subcontinent of South Asia has got a wide distribution of both Pebble-Core and Flake element sites. First systematic study was carried out by De Terra and Paterson (1939) in Panjab along the Soan valley. Similar geo-environmental region was explored by Sen (1955), Lal (1954) and Mohaptara (1976). Western part of Rajasthan has yielded cultural remains along the valley of Berach and its tributaries, Gambhiri, Wagon and Kadmali (Misra 1967). The Berach and Gambhiri are rich in Pebble-Core element while the Wagon and Kadmali have yielded larger number of tools of Flake and Flake-Blade elements. The state of Gujrat has yielded sites along the valley of Sabarmati, Mahi, Orsa, etc. Todd (1939) discovered both the elements at Salsette island near Bombay. Upper Krishna Valley in the Maharastra and Karnataka state (Pappu 1974) has yielded important sites. Narmada valley in central India is very important, a number of sites are discovered in the area (De Terra and Paterson 1939; Sen and Ghosh 1963; Joshi 1978). Nevasa and Navdatoli (Sankalia *et al.* 1960) are important stratified sites in the area. The south Malaprabha and Ghataprabha basins in Karnataka have yielded cultural remains (Joshi 1955). States of Tamilnadu and Andhra are provided with a number of sites. Foote (1866) collected palaeoliths for the first time in India, in this region. Afterwards scholars worked and large number of sites like Vadamadurai, Attirampakkam (De Terra and Paterson 1939; Krishaswami 1938; Ghosh 1973), Giddalur (Soundara Rajan 1952), Renigunta (Murty 1968) have been discovered. On the east Singrauli basin has yielded sites in the districts of Banda, Allahabad, Rewa, Hamirpur, etc. (Krishnaswami and Soundara Rajan 1951). South Bihar had been extensively worked out (Ghosh 1970). A number of sites are found in Orissa together with sufficient data on stratigraphy (Bose and Sen 1948; Mohapatra 1962). The western part of West Bengal, the plateau region is rich in sites, mainly the districts of Midnapur (Ghosh 1962) and Bankura (Sen *et al.* 1963). The above account shows that the Pebble-Core and the Flake and Flake-Blade elements have a wide distribution in India. It is worth mentioning in this context that the Flake element is provided with a wider distribution than the Pebble-Core element.

It is an enormous task to compare the stratigraphic sequences from all parts of India. On the basis of geology, geomorphology, and the palaeoclimatic condition, reconstructed regionally, it may be said that there are two major divisions. The first one is the extra-peninsular region in north-western India, and the other the peninsular region. In the northwestern part of India, the geo-stratigraphy is based on terrace formation. The area was under the influence of Himalayan glaciers and was within the periglacial region. The terraces are aggradational and degradational, formed under the influence of glacial and interglacial sequences respectively. There are altogether five terraces (De Terra and Paterson 1939; Mohapatra 1976). The Pre-Soan industry is coming from TD, the Boulder Conglomerate of upper Siwaliks. This bed belongs to second glacial stage. The terrace T_1 of second interglacial period yielded industry called Early Soan with Sub-stages A, B and C. The terrace T_2 belonging to third glacial stage yielded industries of Late Soan A and B with a continuity into third interglacial of the terrace T_3. The Pre-Soan and Early Soan belong to the Pebble-Core element and the Late Soan belongs to Flake element.

The whole of peninsular India which was within the a-glaciated zone follow a common norm in stratigraphic sequence with cyclic alternation of gravel and silt, two in each case. The pattern of erosion and deposition, volume of the particles of the sediment vary according to the local topographic condition. The stratigraphic sequence had been most carefully reconstructed in eastern India (Ghosh 1966: 154). The Pleistocene sequence consists of compact secondary pebbly laterite underlain by mottled clay. Over the mottled clay lies a lower gravel conglomerate. The lower gravel conglomerate is underlain by lower silty clay, over which lies a bed of upper loose gravel. The gravel and silt phases represent wet and dry conditions respectively in a cyclic pattern. The Pebble-Core element is found throughout the deposits of lower gravel conglomerate, lower silty clay and upper loose gravel. The Flake element is coming from the junction bed of upper loose gravel and upper clay or from the upper part of secondary laterite. The overlying deposit of upper clay yielded Flake-Blade element. The Flake element shows the transition of the preceding culture into succeeding.

Typo-Technology

Typology of the industries belonging to Flake element includes a variety of types. All the types cannot be considered as completely new. Rather, it should be expressed that typologically this industry is a combination of old, i.e. the preceding Pebble-Core element, and new forms. The mixed assemblage may be due to different reasons, as survival, extinction and inclusion. Without going into detail it may be suggested that some of the earlier typological traits became useful in the succeeding period and those types were assimilated and continued in the process. In this case this involves functional necessity. Some traits resulted into a change for acceptance into another tradition. The Flake element emerged from the Pebble-Core element and in the process of development Flake tradition played an important role. Besides the role of tradition, the skill and mastery of technology are also important criteria.

An intensive study of the published materials on Flake element in India has revealed the existing typology of the element. A number of sites have been selected for the study of transition of the Pebble-Core into Flake element. The selection depends on several factors, such as those which have got sufficient stratigraphic evidence, those which have yielded both the Pebble-Core and Flake element in succession and which have the frequency distribution of types recorded. In other words, the important areas are considered here. On the western part of India, south Rajasthan sites (Misra 1967), are taken into consideration. The sites of Nevasa (Sankalia *et al.* 1960), upper Krishna basin (Pappu 1974) in the states of Maharasthra and Karnataka are within the south-western part of India. In central India sites along the Damoh area and Narmada valley (Joshi 1961, 1978) are taken into account. On the east, Singhbhum (Ghosh 1970) and Orissa (Mohapatra 1962) have revealed important data. On the north the Late Soan (De Terra and Paterson 1939; Paterson and Drummond 1962; Mohapatra 1976) is considered.

The main types are different varieties of scrapers, point, borer and burin. Besides these there are a number of flakes, flake-blades and blades, both utilized and unutilized. A number of cores, handaxes, cleavers and choppers make up the total collection. An examination of the individual types of the area considered will reveal the exact nature of survival, emergence and change of the types in the process of transition from one element to the other.

Flake

The flake which was constantly found in every level of Pebble-Core element became the dominating trait in the Flake element, with considerable change in typo-technology. Most of the artifact types are made on flake. Besides flake as a waste product, large number of flakes occur as unretouched pieces, but with definite sharp edges. A number of these show mark of utilization (Sankalia *et al.* 1960; Mohapatra 1962; Misra 1967; Ghosh 1970). Levallois flakes are profuse in number, Clactonian elements are not absent and Mousteroid element in a restricted sense (Ghosh 1974) is also found.

Flake-Blade

These are regarded as blade-like flake by Mohapatra (1962). Ghosh (1970) and Joshi (1978) considered these as flake-blades. These are flakes having morphological similarity with blades, but the technology of blade is not usually used for their manufacture. Most of the flake-blades are not types but blanks on which types are made. The flake-blades are the types which belong predominantly to the Flake, Flake-Blade element. The blade coming from the Berach basin (Misra 1967: 30) has more affinity to blade-like flake than true blade.

Blade

True blades, both morphological and technological as defined by Movius *et al.* (1971) appeared in this stage but their number of occurrence is less compared to other types (Joshi 1961; Mohapatra 1962; Misra 1967). These are at least twice as long as wide, with more or less parallel edges. Some of the blades are with the back blunted and some have both the surfaces left unretouched sharp. The number of occurrence of this type increases in the Flake-Blade element (Ghosh 1970).

Core

The group of cores considered here are not tools by themselves. Large number of cores are found from India (Joshi 1961, 1978; Mohapatra 1962; Misra 1967; Ghosh 1970; Pappu 1974). The nature of flake scars on the cores show prepared technique. The tortoise and discoidal cores, diagnostic of Levallois technique are significantly present. Unprepared Clacton cores are also quite frequent. Cores showing the removal of blades are also found in less frequency. Blade cores in eastern India are mostly found a little later than the Flake element. Joshi (1961: 32) has referred to some tools as nodules. These are elongated cores with chisel and pick-like edges.

Scrapers

Movius (1949) defined scrapers as small choppers, made on core, flake and even blade. The working edge is usually rounded, semi-oval or almost straight in form and is with secondary working. Scrapers are the predominant tool type of the Flake, Flake-Blade element throughout India. Scraper is found in the preceding Pebble-Core element but its frequency of occurrence is less significant, compared to the chopper and bifaces. Another basic difference between the scrapers of the two elements is that flake became the main form of raw material in Flake element as opposed to core in the Pebble-Core element. In the Flake element scrapers show greater variety of forms and may be divided into a number of sub-groups. Various criteria may be taken into consideration for the sub-grouping. On the basis of the form of raw material there are scrapers made (i) on core/pebble, (ii) flake and (iii) blade; (ii) is predominating. Several criteria are taken for further sub-grouping of scrapers; they are, location of working area in respect to bulb of percussion wherever possible, and form of working area.

(a) Side Scraper

This is the most dominating type in the Flake element all over India. Location of working edge is either on a single edge or on both. Form of working edge are either straight or convex. Mohapatra (1962) identified a further sub-type of scraper, the point-shaped scraper. From the description and illustration these appear to be nothing but side scrapers with convex edge on flake.

(b) Concave Scraper

This type is a variety of side scraper where the form of working edge is concave. This is also known as hollow scraper (Sankalia *et al.* 1960; Mohapatra 1962). Concave scraper is rare in the Pebble-Core element. In the present context Mohapatra (1962: 216) found a few hollow scrapers from Early Stone Age assemblage in Orissa.

(c) Round Scraper

This type has working edge all around the periphery. These are made both on flake and core. This type was found in the Pebble-Core element but frequency of occurrence was less in comparison to that in the Flake element, as is revealed in the present study.

(d) End Scraper

These are the types which have their working edge along one of the shorter axes. In fact when made on a flake or flake-blade or blade the working edge is located either proximally or distally. This type is mostly found in Flake element. Only a few are found in the Pebble-Core element in south Rajasthan (Misra 1967). Within the Flake element, end scrapers made on flake outnumber those made on core.

(e) Steep Scraper

These probably emerged in the Flake element. These are made on core/nodule or split pebble. The steep scraping edge is prepared by controlled flaking. Joshi (1978) further subdivided these into unifacial and bifacial sub-types.

(f) Side-cum-End Scraper

These scrapers are found mainly in the Flake element. Types made on flake predominate over types made on core.

Point

These are the types whose lateral edges converge at one end. This type is made both on core and flake. Those made on flake predominate. Next to side scraper, point-on-flake has major frequency of occurrence. Excepting a few points identified by Mohapatra (1962: 219) from Orissa, no account of it is found from the Pebble-Core element in the present context. The point-on-core has a morphological similarity with small handaxes but in type and technique they vary from the latter. The point-on-flake is further sub-grouped according to their form as leaf-shaped, sub-triangular. These are made both on prepared and unprepared core method. Fine retouches are often found along one or both the lateral margins. Tanged and shouldered points made on flake occur in small number in the Flake element. This sub-type may be considered as a emergent type.

Borer

The borer has a projected working edge, degree of projection is always more than point. The projected end is usually worked all round. Borers are made on core, nodule or on flake. Borers are more commonly made on core. This type is found in all the major areas taken into consideration here. Probably this type too is an emergent one in the Flake element.

Scraper-Borer

This type is found by Sankalia *et al.* (1960) at Nevasa, Mohapatra (1962) in Orissa, Pappu (1974) in the upper Krishna basin. These are further subdivided into two groups, one made on flake and the other made on nodule. This type was not found in the Pebble-Core element. One or more scraping edge together with a projected boring edge are found in a single tool. Both the scraping and the boring edges are with fine retouch.

Burin

This type is found in eastern India not in the Flake element but in the Flake-Blade element (Ghosh 1970). Mohapatra (1962: 113) found a few burins made on flake and on nodule. The Singhbhum burins are made on blade. The burins have the characteristic screwdriver-like edge formed by the intersection of burin facets. This type is not identified in any other region considered in this context. Probably this type emerged from the Flake element.

There are types which are always taken as belonging to Pebble-Core element only, such as handaxes, cleavers and choppers. There was a trend among the earlier workers who classified the collection mainly on the basis of typology, to include the above-mentioned types into the Pebble-Core element without even considering the stratigraphy of their occurrence. Misra (1967) and Ghosh (1970) identified the types which belong stratigraphically to the later cultural element, the Flake element. There is an overall size diminution of the types, refinement of technique and a marked fall in frequency of occurrence. It is possible that future systematic work may reveal further evidence of the continuity of traits from the Pebble-Core element in the form of handaxe, cleaver, in other areas besides South Rajasthan and Singhbhum.

Handaxe

This type is less in number. Handaxes show better workmanship and are made both on core and flake but the latter sub-type is more in number. The sub-types that are found in the Flake element are elongated, oval, ovate, cordate and almond.

Cleaver

Cleavers are small in number and are made on flake. Nowhere else except Singhbhum, cleaver is found in Flake element (Ghosh 1970). They are very small in number, smaller in size, made by a refined technique and only U-butt type is present.

Choppers/Scrapers

These are the types identified in the Flake, Flake-Blade element by Joshi (1961) and Ghosh (1970) from central and eastern India respectively. These are in fact morphologically similar to the choppers of Pebble-Core element but are made on smaller pebbles, with mostly fine-grained raw material like chert. Both unifacial and bifacial sub-types are identified. Although Joshi (1961, 1978) found them in the Middle Stone Age in general, Ghosh (1970) found them specifically in the Flake-Blade element, not in the Flake element.

Study of typology of the Flake element reveal the lack of standardization of typological nomenclature in India. Ghosh (1973) made a cluster analysis on the types of this element. Further work on this line wlll be more useful in future. The above account shows the nature of types continuing, changing and emerging in the process of transition of Flake element from the Pebble-Core element. The transition involves technological development as well. The change of technology involved a change in the selection of raw material. Fine-grained materials like chert, chalcedony, agate, opal were selected instead of quartzite and quartz. The selection depended upon availability and in areas like south Rajasthan (Misra 1967) and Adamgarh (Joshi 1978) quartzite and sandstone formed the main raw material.

Main techniques involved in the manufacture of tools are Acheulian, Clactonian, Levalloisian, Mousterian and punch. All these techniques excepting Mousterian and punch are found in the Pebble-Core element. As the earlier techniques seem to continue, the technological traits are not completely new.

Acheulian

The Flake element in India has a genetic relationship with the Acheulian of Pebble-Core element. The controlled flaking and cylinder hammer technique continued into the Flake element from the Pebble-Core element. The presence of handaxes, cleavers and various core tools also shows the persistence of Acheulian element.

Levalloisian

Levallois is the most dominating technique of the Flake element. A study of the flakes, flake-blades, flake tools and distinctive cores reveals this. The Flake element is dominated by flake tools, scrapers of different varieties, of which side scrapers preponderate. Levallois technique played a very vital role in the development of these tools. This technique was present in the Pebble-Core element. Towards the later part of the Acheulian phase of Pebble-Core the Levallois technique emerged. The Late Soan phase of north India was considered by De Terra (De Terra and Paterson 1939) as Soan industry proper because this phase saw the emergence of Levallois technique. In this context it may be said that in the extra-peninsular region the development of Levallois is somewhat later than that in the peninsular India. Paterson (1940) considered the technique in northern India as Levallois-Mousterian. Fine retouch along the working edge is a development of technique in the Flake element.

Mousterian

Ghosh (1974) made an intensive study of the Mousterian in India. His study revealed that there is a technique in the Flake element which he would consider as Mousteroid of Acheulian Tradition. Both Levallois and Acheulian played an important role in its development. In south Rajasthan (Misra 1967), Maharasthra (Sankalia *et al.* 1960), Karnataka (Pappu 1974), central India (Joshi 1961, 1978), Orissa (Mohapatra 1962) and in Panjab (De Terra and Paterson 1939; Mohapatra 1976) no true Mousterian element is found, which may be compared to the Mousterian of Europe.

Punch-Blade

This technique is found in a small frequency in South Rajasthan, Maharasthra, Karnataka, central India and Orissa. In eastern India (Ghosh 1970) this is a later technological innovation into the Flake-Blade element. In Soan Valley, the Dhok Pathan and Pindi Gheb industry, which De Terra and Paterson (1939) very doubtfully correlated to T_4 and which Movius (1949) termed as 'Evolved Soan' yielded the punch-blade technique in a small frequency. Until more stratigraphic evidence is available this technique is considered as an innovation of the Flake-Blade element.

The typo-technological analysis of the Pebble-Core/Flake element transition shows that types like handaxes, cleavers, scrapers continued into the Flake element. The handaxes, cleavers, scrapers, although a continuity, is somewhat different in the typo-technological sense (Fig. 1). These are mostly on flake, smaller in size and neatly made. The sub-types of scraper have increased in the Flake element. Side scraper, round scraper and the end scraper are found in the Pebble-Core element in a smaller frequency. Steep scraper, side-cum-end scraper and scraper-borer are the new sub-types in this element. Borer, burin and blade are also the emergent types. The presence of burin in this element is not yet well established. Considering the nature of occurrence in the Pebble-Core element point may be regarded as an emergent type than a survival. Tange point emerged during this phase. Acheulian and Levallois, the major techniques of the Flake element, are the continuity and development from the Pebble-Core element. The Mousteroid although a new trait in this phase emerged out of the joint influence of Acheulian and Levallois in a new setup. In conclusion it may be said that the Flake element in India is a cumulation and development of adaptive mechanism. This phase is set in a different climatic regimen and the cultural element which continued from the Pebble-Core into the Flake underwent change and development due to necessity in the new environment. The development of skill and bio-cultural capability played a vital role in the transfor-

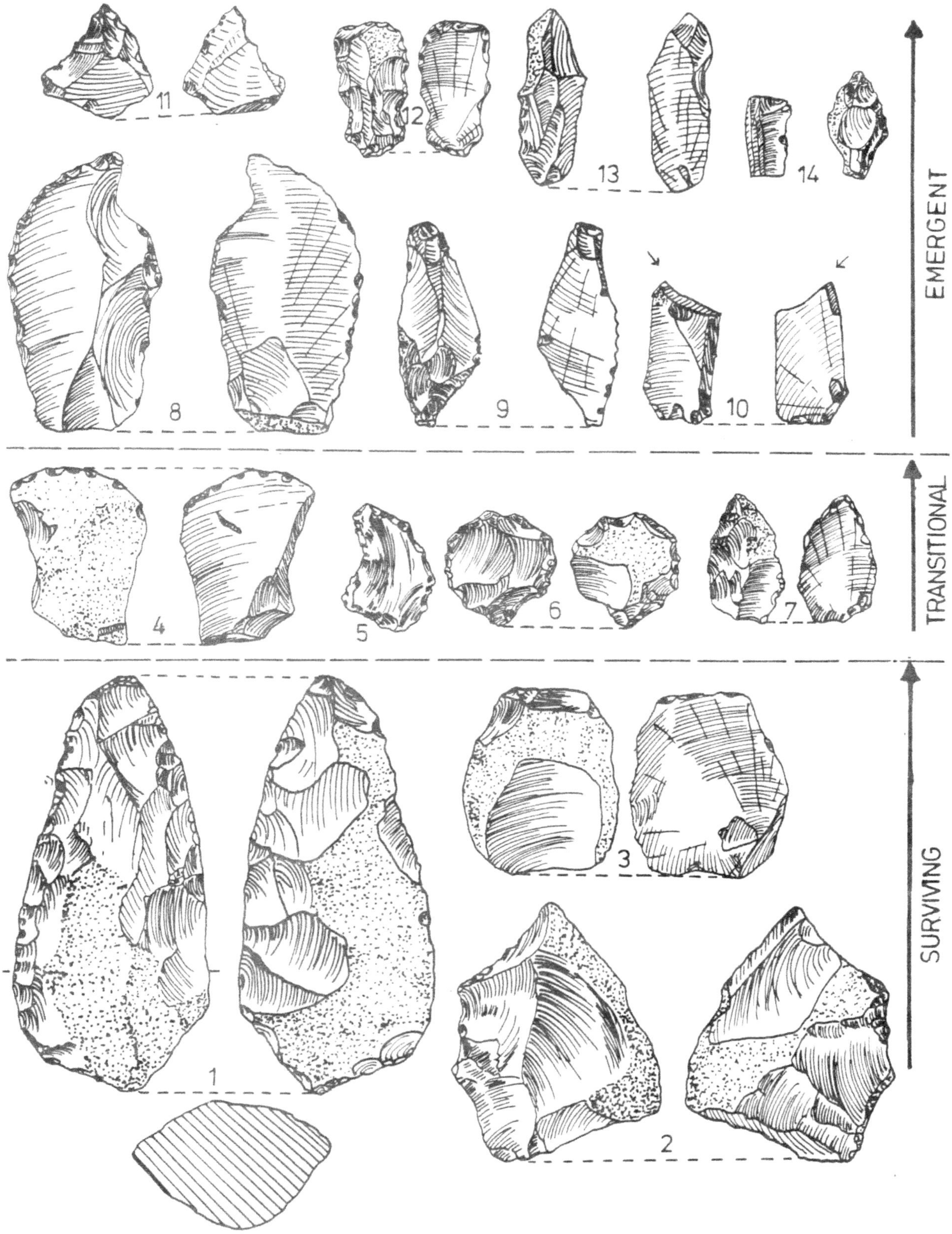

Figure 1 Types surviving from Pebble-core to Flake element. 1, Handaxe; 2, Side-scraper; 3, Flake; Types transitional from Pebble-core to Flake element: 4, End scraper; 5, Concave scraper; 6, Round scraper. Types emergent to Flake element: 8, Scraper-borer; 9, Steep scraper; 10, Burin; 11, Borer; 12, Side-cum-end scraper; 13, Flake-blade; 14, Blade; 15, Tanged point.

1-3, Misra 1967; 4, 8-10, 13, 14, Mohapatra 1962; 5,15, Sankalia *et al.* 1960; 6, 12, Pappu 1974; 11, Joshi 1961.

mation. Study on the Flake element in India is incomplete. Further data on chronology, typology, technology and palaeoecology is needed. A standard inventory of the typology of this phase based on attribute analysis will be useful to the workers.

References

Bose, N.K. and D. Sen 1948 *Excavations in Mayurbhanj* Calcutta.

Burkitt, M.C. and L.A. Cammiade 1930 Fresh light on the stone age in south-east Asia *Antiquity* 4: 327-329.

De Terra, H. and T.T. Paterson 1939 *Studies on the ice age in India and associated Human cultures* Carnegie Institute, Washington, D.C.

Foote, Robert Bruce 1866 On the occurrence of stone implements in laterite formations in various parts of Madras and north Arcot districts *Madras Journal of Literature and Science* 3(2): 1-35.

Ghosh, A.K. 1962 Discovery of prehistoric stone implements from north-west Midnapur, West Bengal *Science and Culture* 28: 338-339.

Ghosh, A.K. 1966 Implementiferous laterite in eastern India. *In:* Sen, D. and A.K. Ghosh (eds.) *Studies in Prehistory* Calcutta, Firma K.L. Mukhopadhyay, pp. 151-162.

Ghosh, A.K. 1970 The palaeolithic cultures of Singhbhum *Transactions Amer. Phil. Soc.* (New Series) 60(1): 1-68.

Ghosh, A.K. 1973 Emendation of Madrasian Industry. *In: Actes du VIIIe Congres International des Sciences Prehistoriques et Protohistoriques* Tome II. Edited by Grga Novak, Beograd, pp. 39-48.

Ghosh, A.K. 1974 Denticulates in India: examination of a type concept *J. Hong Kong Archaeological Soc.* 5: 47-56.

Ghosh, M. 1973 *Studies on Flake and Flake-Blade elements in India* Ph.D. Thesis, Calcutta University.

Joshi, R.V. 1955 *Pleistocene studies in the Malaphabha basin* Poona, Deccan College.

Joshi, R.V. 1961 Stone age industries of the Damoh area, Madhya Pradesh *Ancient India* 17: 5-36.

Joshi, R.V. 1978 *Stone age cultures of Central India* Poona, Deccan College.

Krishnaswami, V.D. 1938 Environmental and cultural change of Prehistoric man near Madras *J. Madras Geographical Association* 13: 58-90.

Krishnaswami, V.D. and K.V. Soundara Rajan 1951 The lithic tool industries of Singrauli basin, district Mirzapur *Ancient India* 7: 40-65.

Lal, B.B. 1954 Palaeoliths from Beas and Banganga Valley, Punjab *Ancient India* 12: 58-92.

Misra, V.N. 1962 Problems of terminology in Indian Prehistory *The Eastern Anthropologist* 15(2): 113-124.

Misra, V.N. 1967 *Pre and Protohistory of the Berach basin, south Rajasthan* Poona, Deccan College.

Mohapatra, G.C. 1962 *The stone age cultures of Orissa* Poona, Deccan College.

Mohapatra, G.C. 1976 Geo-tectonic developments, sub-Himalayan lithic complex and post-Siwalik sediments *IX Inter. Congrès UISPP, Colloque VII*, pp. 31-59, Nice.

Movius, H.L.Jr. 1949 The lower Palaeolithic cultures of southern and eastern Asia *Transactions Amer. Phil. Soc.* (New Series) 38(4): 376-383.

Movius, H.L.Jr. and A.S. Brooks 1971 The analysis of certain major classes of upper palaeolithic tools: Aurignacian scrapers *Proc. Prehistoric Society* 37: 253-273.

Murty, M.L.K. 1968 Blade and burin industries near Renigunta on the south-east coast of India *Proc. Prehistoric Society* 34: 83-101.

Pappu, R.S. 1974 *Pleistocene studies in the upper Krishna basin* Poona, Deccan College.

Paterson, T.T. 1940 Geology and early man, 1 and 2 *Nature* CXLVI: 12-15 and 49-52.

Sankalia, H.D. 1956 Animal fossils and palaeolithic industries from the Pravara basin at Nevasa, district Ahmednagar *Ancient India* 12: 3-52.

Sankalia, H.D., S.B. Deo, Z.D. Ansari and S. Ehrhardt 1960 *From history to prehistory at Nevasa (1954-1956)* Poona, Deccan College.

Sen, D. 1955 Nalagarh palaeolithic culture *Man in India* 35(3): 177-184.

Sen, D. and A.K. Ghosh 1963 Lithic culture-complex in the Pleistocene sequence of Narmada Valley, Central India *Rivistadi Science Preistoriche* 18(1-4): 3-23.

Sen, D., A.K. Ghosh and M. Chatterjee 1963 Palaeolithic industry of Bankura, West Bengal *Man in India* 43(2): 100-113.

Soundara Rajan, K.V. 1952 Stone age industries near Giddatur, district Kurnool *Ancient India* 8: 84-92.

Subbarao, B. 1958 *The Personality of India* Baroda, M.S. University.

Todd, K.R.U. 1939 Palaeolithic industries of Bombay *J. Royal Anthrop. Inst.* LXIX: 257-272.

Discussion

G. Freund: Last year I had the opportunity to visit the most important museums of India and to view a lot of material in the Deccan College at Poona. I am familiar with Indian literature too, so I am very glad to have heard such a summary today. Thank you very much. But I think that when we see only the technological and the typological point of view, we can always make a tradition. We have the traditions from pebble culture to handaxe culture to Middle Palaeolithic in the form of Mousterian, but what we do not have is a good stratigraphy, as you mentioned. What we need is the fauna, and we have practically none. The terraces in the Deccan are not well developed, nor well dated. We have some rock shelters; and there are new excavations in the several valleys you showed on the map. The rivers always flow from west to east, as you showed. Most of the sites are in the upper valleys. We hope that the recent excavations in the Amada Valley, Pravara Valley, Krishna Valley and others will give us new data. Much work has been done by my German colleague Kowinus, who has worked in Poona for several years.

My impression is that our Indian colleagues do not agree on the question of whether there is a tradition from the pebble culture to the handaxe culture, or are they two separate branches – the pebble culture on the one side, with the transition you told us about to flake tools, and the handaxe culture on the other side. What I saw in Poona was an enormous quantity of handaxes and cleavers, in an assemblage which also contained side scrapers and other flake tools, that is what the Poona colleagues call Acheulian and what is normally represented in the valley sites and in some of the rock shelters. When stratigraphy exists, it seems that this Acheulian develops into a late Acheulian in the European sense. On top of this there is sometimes a Middle Palaeolithic of Mousterian character, but with handaxes, and then Mesolithic. What we heard today was a theory on technology and typology, but I think that what we need is a good chronology based on geology, palaeontology and C14 dates, which we do not have today.

R. Ray: I agree with Prof. Freund that we need more geological and chronological data in the sense of absolute dating. One of the things that is always confusing to people is the Himalaya, which was under peri-glacial conditions. There is the Soan chopper-chopping tool tradition of Movius and, as you know, he has included this element into the chopper-chopping tool world of South-East Asia. But in a recent work, bifacial technique was found in the Soan and there is still confusion whether it can truly be included with the chopper-chopping tool element.

A.J. Jelinek: Concerning the points that you have mentioned: are these bifacial points, or are they Levallois points with the flake industry?

R. Ray: They are mostly Levallois points, but there are bifacial points also.

G. Bosinski: Levallois technique is widespread, and of course, it will be relevant to our Congress to ask whether Levallois technique defines Middle Palaeolithic or is an independent occurrence. It is a rather complicated technique and you tried to make us believe that all this came out independently – in the tradition from the Early Soan pebble-core complex to the Late Soan flake element. But if I understood in the right way, the flake element is marked by Levallois technique. Was it something to mark a new period, such as Middle Palaeolithic? What do you think?

R. Ray: All over India the flake element is marked by Levallois technique. What is found so far is that Levallois technique was also present in the Upper Acheulian but has become dominant in the flake element where most of the tools are made on flakes of Levallois technique. So Levallois has a generic relationship in the pebble-core element.

5

THE ORIGIN OF MODERN MAN

THE FIRST *HOMO SAPIENS SAPIENS* IN THE NEAR EAST

Bernard Vandermeersch

Two different human types lived in the Near East during the Mousterian period. On one side, the Tabun, Amud and Shanidar fossils, generally called Neanderthals, are sometimes designated as Near Eastern Neanderthals. On the other side, the Skhul and Qafzeh hominids are generally thought to be *Homo sapiens sapiens* and are what I would call "Proto-cromagnons", as the term is applied by Professor Clark Howell.

Some important questions have arisen from the simultaneous existence of these two populations which are morphologically different but occur in the same area, as well as in the same cultural environment. I would like to discuss some of these.

- Is there any phylogenetic relationship between the two populations? If not, we are left with three other questions.
- What is the chronological relationship between these two populations? Are they contemporaneous or not?
- What is the origin of the Near Eastern Neanderthal?
- What is the origin of the Proto-cromagnons?

The first question I will consider is whether any filiation does exist between the Near Eastern Neanderthals and the Skhul and Qafzeh hominids. Did the Proto-cromagnons evolve from the Neanderthals? This looks difficult to accept for two reasons. The first one is provided by their comparative morphologies. When I studied the adult fossil remains from Qafzeh, I was surprised not to find any of the specialized Neanderthal features among them. The common features of these two populations did already exist with *Homo erectus*. So we are dealing with ancestral features. In fact, the first *Homo sapiens sapiens* are morphologically closer to evolved *Homo erectus* than to the Neanderthals. Fewer morphological transformations are required to proceed from the evolved *Homo erectus* to *Homo sapiens* than to proceed from Neanderthal to *Homo sapiens sapiens*. If we do accept an evolution from Neanderthals to *Homo sapiens sapiens*, then we have to admit that some of the features which characterize *Homo erectus* were transformed into Neanderthal characteristics and then returned to their primitive state, which is difficult to accept palaeontologically.

The Neanderthals represent a divergent and very specialized branch in our history. Their role in recent human evolution was minor or non-existent. A further argument against the Neanderthal-sapiens lineage is the recent discovery at Saint-Cesaire, a rock-shelter in South-West France, of a Neanderthal skeleton in a Chatelperronian layer. Therefore, in Europe, at the beginning of the Upper Palaeolithic, Neanderthals and modern men were contemporaneous. For these reasons I do not believe in the evolution from Neanderthal to modern man.

The second question we have to discuss concerns the contemporaneity of Neanderthals and sapiens in the Near East. I will not discuss the geological and palaeontological evidence concerning the Mousterian age of Qafzeh and its relationship to the Tabun Mousterian. I would like only to point out two things: For some anthropologists the Skhul hominids represent an interbreeding between Neanderthals and *Homo sapiens*. This is difficult to prove but such an assessment would involve simultaneous existence of the two populations.

My second point concerns the Tabun hominids. Garrod wrote that the Tabun I skeleton was found in layer C and that the Tabun II jaw was discovered 85 cm below, in the same layer. Professor Jelinek, on the other hand, thinks that the burial was in layer D and that the jaw, even being found below it, was in layer C. Unfortunately, we are today missing the evidence to work out the true stratigraphic position of Tabun II. Nevertheless, the Tabun II jaw does show modern morphological characteristics, among which are the typical chin, the height of the body sloping down backwards, and the short retromolar space. It would be very important, therefore, to determine the precise stratigraphic position and the age of Tabun II. For the moment, I have to consider Tabun I and II as more or less contemporaneous.

The third question concerns the origin of the two Near Eastern populations. The Neanderthals are only known in Europe and in the Near and Middle East. They do not exist in East and South Africa, nor in Asia. In addition, we have discovered Preneanderthals only in Europe. Recent works prove that there is only one human lineage in Europe between the Mindel and Würm glaciations: the Neanderthal one. Thus, the idea of Presapiens in Europe must be discarded.

If all the Preneanderthals are European, we must conclude that the Near Eastern Neanderthals did migrate here. This migration can be approximately dated. Indeed, if we compare the Near Eastern Neanderthals with the European ones, the former are more like those from the Riss-Würm-Saccopastore, for instance – than those from the Würm. Therefore, we can accept the idea that some Neanderthals left Europe for the Near East at the end of the Riss or during the Riss-Würm interglacial.

A further question concerns the origin of the Proto-cromagnons. The skulls of Qafzeh show some archaic features, for example, the morphology of the occipital bone of Qafzeh 3. These features, which vary among the specimens, prove that the Proto-cromagnons derived from evolved *Homo erectus*. The Zuttiyeh skull is the only fossil which remains chronologically between Qafzeh and *Homo erectus*. Recently, Gisis and Bar-Yosef have shown that this fossil does not come from a Mousterian but from an Acheulian of Yabrudian facies level. Accordingly, this fossil would belong to a period between 90,000 and 140,000 years.

This fossil exhibits some archaic features, like the thickness of the vault, the general robustness, and the large eyebrow; and some evolved features, including vertical forehead and the large frontal capacity. But it lacks any of the specialized Neanderthal features, such as an extension of the maxillary region, the disposition of the nasal bone, and the large frontal sinuses.

The differences between the Zuttiyeh skull and the Qafzeh ones seem to be relatively small. Furthermore, the morphological differences between Zuttiyeh, Skhul and Qafzeh are less important than those between Zuttiyeh and the Near Eastern Neanderthals. Therefore, I think that the Zuttiyeh skull is probably the most ancient *Homo sapiens sapiens* in the Near East.

To summarize, I think that the Near Eastern Neanderthals are not autochthonous to the region. It is possible, furthermore, that we have an evolution to modern man since the end of the middle Pleistocene.

Discussion

E. Trinkaus: I am going to talk about a slightly earlier time period in my presentation. At the moment I have what are partly questions and partly comments to make on this talk, since we have a difference of opinion on some of the material.

I would like to make a couple of points, which are more general comments than specific morphological ones. First, there is an assumption that there was contemporaneity in the Near East of the Skhul and Qafzeh material with the Near Eastern Neanderthals, and that there was contemporaneity in Europe between the Châtelperronian Saint-Césaire skeleton and the Aurignacian-associated anatomically modern humans. There has been considerable discussion this week to suggest that the contemporaneity of the two groups of Near Eastern Mousterian hominids is open to question. Although it is clear that the Neanderthal specimen from Saint-Césaire was very close in time to some of the Early Aurignacian specimens, for example the anatomically modern frontal bone from Velika Pećina in Yugoslavia that is dated to about 34,000 years ago, it is uncertain whether they were in fact contemporaneous. I think that this is something that one of the prehistorians who knows the French archaeological record of that time period in detail might care to comment on.

With regard to my second point, I personally find it difficult to accept the existence of two genetically distinct human populations in the same region of the world adapting to the same cultural pattern, which in the Near East was represented by the Mousterian tradition. It is certainly possible to have one form of human in Europe and another form in the Near East and a third form in Africa for a long period of time. But to suggest that for thirty to fifty thousand years we had two genetically distinct populations within the area of the Levant, I find biologically very difficult to accept.

P.M. Masters: Is the Tautavel skull, in Southern France, a pre-Sapiens?

B. Vandermeersch: No. For me it is a pre-Neanderthal.

B. Arensburg: The main problem is whether we are speaking of different types, different groups, or different species. I think this is the main point.

EVOLUTIONARY CONTINUITY AMONG ARCHAIC *HOMO SAPIENS*

Erik Trinkaus

During the late Middle and early Upper Pleistocene there was a shift in human technology, the transition from the Lower to the Middle Palaeolithic. As the other papers in this symposium demonstrate, there is a lack of consensus as to the exact timing and precise nature of this transition, both within and between geographical regions. Yet, there is little doubt that there was a shift at this time in human prehistory.

The question that I would like to address concerns not the lithic industries, but the hominids who were responsible for leaving behind such a profusion of tools. Were there significant biological changes between the manufacturers of the lower Palaeolithic and those who made the Middle Palaeolithic? That is, do we find a gradual modification of form during this transitional time, or is there evidence for a period of rapid evolution? Since there were geographical differences in Pleistocene human morphology, this problem will be examined separately for six areas of the Old World.

The Fossil Sample

The portion of the hominid fossil record with which I am concerned includes those specimens which are generally included within the species *Homo sapiens* but are more archaic than, and morphologically distinguishable from, anatomically modern humans. These fossils will be referred to collectively as "archaic *Homo sapiens*." They include the Neanderthals and Early Neanderthals of Europe and the Near East and specimens of similar evolutionary grades in eastern Asia and Africa.

Reference will also be made to certain fossils (e.g. Arago, Broken Hill, Ngandong, Petralona) which have been referred to both *H. erectus* and *H. sapiens*. The assignment of these specimens to precise taxonomic categories is probably a futile exercise, since they represent populations close to the boundary which human paleontologists have placed between these two chronospecies. Given the mosaic nature of hominid evolution and the incompleteness of most of the fossils, any taxonomic designation given to these fossils would be subject to debate (Howells 1980; Wolpoff 1980b). For the purposes of this discussion, these problematical specimens will be included within *H. sapiens*, although it is fully recognized that some of them may be better accommodated within *H. erectus*.

In the discussion of archaic *H. sapiens* fossils and the evolutionary patterns they delineate, a modified version of the *Homo sapiens* grade system devised by Stringer *et al.* (1979) will be used (Table 1). In this scheme, grade 1 represents the earliest archaic *H. sapiens* and includes specimens which have been placed in *H. erectus*. Grade 2 includes most of the Early Neanderthals or their equivalents. Grade 3a, as defined by Stringer *et al.* (1979), comprised only the Neanderthals, or "Classic" Neanderthals, of Europe and the Near East. Grade 3a will be used here to include both the Neanderthals of Europe and the Near East and hominids of a similar evolutionary grade from Africa and eastern Asia; this use is not intended to obscure the important morphological distinctions between these early Upper Pleistocene geographical samples but is designed to recognize their general similarity in evolutionary grade. Grade 3b includes anatomically modern humans. As with any system which breaks a possible continuum there are individuals which fall between the categories. This grade system is therefore merely a heuristic device to order the fossil record.

Chronological Considerations

The chronological ordering of late Middle and early Upper Pleistocene hominids continues to present difficulties. Each year sees the reordering of specimens in time as the geology and the fauna and flora of sites are reexamined and as radiometric dates become available. In addition, it has become apparent in recent years that the traditional glacial-interglacial sequences of the European Pleistocene cannot be maintained (Butzer and Isaac 1975; Kukla 1977). Regional stratigraphic sequences (Butzer and Isaac 1975) and oxygen isotope records (Shackleton and Opdyke 1976) make it clear that many more than four to six glacial periods existed during the Pleistocene. The recent data still support the traditional subdivisions of the European Upper Pleistocene, the last interglacial (Riss-Würm or Eem; oxygen isotope stage 5), the last glacial (Würm or Weichsel; oxygen isotope stages 2 to 4) and the postglacial (Holocene; oxygen isotope stage 1). However, the relative ages of various Middle and Lower Pleistocene glacial and interglacial deposits are very much open to debate (see Kukla 1977, for a review).

It is possible to place many of the European fossils from the last glacial and interglacial reasonably confidently in time (e.g. Vandermeersch 1965; Wolpoff 1980a). However, many of these chronological assessments are tenuous, especially for the specimens which were discovered during the last century and the early decades of this century (Stringer and Burleigh 1981). The relative positioning of older European fossils and specimens from other regions of the Old World is considerably more tenuous. Reasonable assessments of their geological ages can nontheless be made for most of these fossil specimens, if one bears in mind the uncertainties involved. An approximate chronological or ordering of Middle and early Upper Pleistocene hominid fossils will therefore be made for each region of the Old World.

Phyletic Gradualism versus Punctuated Equilibria

It has been argued (Gould and Eldredge 1977; Kurtén 1981; Pilbeam 1975; Stanley 1979) that hominid evolution has been characterized by punctuated equilibria, as opposed to phyletic gradualism. A model of punctuated equilibria may well be representative of hominid evolution as a whole, but acceptance of this interpretation does not dictate which of these two macroevolutionary models typified the segment of hominid evolution represented by archaic *H. sapiens*. Significant gradual evolution can take place within periods of relative stasis (Stanley 1979). Therefore, either model may be appropriate for this relatively short period of hominid evolution.

European Archaic *Homo sapiens*

For the Middle and Upper Pleistocene the most complete hominid fossil record comes from Europe. A number of partial skeletons exist for the early last glacial Neanderthals, since they frequently buried their dead. However, all of the Middle Pleistocene and last interglacial fossils are incomplete. As a result, the morphology of the Neanderthals is well known, but that of their predecessors can be built up only as a composite of numerous specimens.

The oldest known inhabitants of Europe, some of whom may be as old as 500,000 years, provide a base line for consideration of the more recent archaic *H. sapiens*. These hominids are represented by the fossils from Arago, Bilzingsleben, Mauer, Montmaurin, Petralona and Vértesszöllös. They are similar to *H. erectus* in their large, pneumaticized cranial bases, angulated occipital bones with transverse tori, robust supraorbital tori, heavy alveolar regions and robust mandibular corpori (Billy and Vallois 1977; Lumley 1975; Lumley and Lumley 1979; Schoetensack 1908; Stringer *et al.* 1979; Thoma 1966; Vlček 1978). These features have led several workers to place these specimens within *H. erectus* (Lumley and Lumley 1979; Vlček 1978; Wolpoff 1977). Yet these hominids exhibit features which point towards the more recent European specimens and away from *H. erectus* (Howells 1980; Stringer 1980). There is an expansion of the cranial vault which is reflected in higher cranial capacities, decreased postorbital constriction and less occipital angulation. There is a decrease in the mean dimensions of the premolars and molars and an increase in the anterior dental dimenions (Wolpoff 1975). This is accompanied by a slight anterior positioning of the dentition in the facial skeleton, which sets these hominids apart from east Asian Middle Pleistocene hominids (Aguirre and Lumley 1977; Stewart 1970).

The only diagnostic postcranial remains associated with these early *H. sapiens* is a left ilium and ischium from Arago, Arago 44 (Lumley and Lumley 1979). Its surface morphology is virtually modern, like those of later Neanderthals, but in cross-section the ilium presents a massiveness which is exceeded only by the *H. erectus* ilium from Olduvai, OH 28 (Day 1971).

The succeeding group of European hominids falls between grades 1 and 2, morphologically and chronologically (Lumley 1978). It consists primarily of the partial crania from Steinheim and Swanscombe. Once considered modern in aspect (Vallois

1958), these specimens are quite archaic (Hublin 1978; Stewart 1961a; Stringer 1978; Wolpoff 1980b). The relative roundness of their cranial vaults is due in part to the extensive crushing of the Steinheim piece (Weinert 1936; Wolpoff 1980b) and in part to their being female. Cranio-facial sexual dimorphism was pronounced among Middle Pleistocene hominids (Aguirre and Lumley 1977; Lumley 1975; Wolpoff 1980a), although perhaps not as pronounced as has been suggested (compare Wolpoff 1980b, and Stringer 1981). Despite their archaic morphology, these fossils exhibit features which anticipate the later Early Neanderthals and Neanderthals. Steinheim and the probably contemporaneous mandible from Atapuerca show some mid-facial prognathism (Aguirre and Lumley 1977; Weinert 1936). And Steinheim and Swanscombe exhibit occipito-mastoid crests and supra-iniac fossae (Hublin 1978). The latter feature is diagnostic of the Early Neanderthals and Neanderthals (Hublin 1978; Santa Luca 1978).

The Early Neanderthals of the penultimate glacial and last interglacial show a steady decrease in characteristics reminiscent of *H. erectus* and an increasing prevalence of Neanderthal features (Hublin 1978; Lumley 1973; Piveteau 1970, 1976; Stringer 1978; Vandermeersch 1978). There is an increase in cranial capacity, associated with parietal expansion and decreased occipital curvature. The several known occipital bones all exhibit supra-iniac fossae, and the few known temporal bones are Neanderthaloid. In addition, the facial skeleton exhibits increasing amounts of the mid-facial prognathism which dominates later Neanderthal skulls. This is apparent especially in the elongation of the mandibles, the anterior placement of the dentitions and the associated retromolar spaces.

Postcranial remains are rare for these hominids, but an ilium from the Grotte du Prince (Lumley 1972) and a femoral shaft from Ehringsdorf (Weidenreich 1941) agree with a Neanderthal affiliation for these fossils. The ilium is long, suggesting an enlarged pelvic aperture similar to those of the Neanderthals (Trinkaus 1976a). The femur shows the antero-posterior flattening and medial reinforcement characteristic of Neanderthal femora (Trinkaus 1976b), but this morphology is also found in *H. erectus* (Day 1971).

A transition between these Early Neanderthals and the Neanderthals of the early last glacial is provided by part of the Krapina sample (Smith 1976). These hominids, which date to the end of the last interglacial, are largely indistinguishable from the later Neanderthals in their morphological pattern. This is evident in their cranial vaults, mandibles, dentitions, scapulae, hand phalanges, pelves and femora (Musgrave 1977; Smith 1976; Trinkaus 1978; Wolpoff 1979). Yet, their supraorbital tori are heavier and more projecting (Smith and Ranyard 1980), and their dentitions, although exhibiting the typical Neanderthal anterior dental enlargement, both absolutely and relative to the cheek teeth, are the largest known for any sample of *H. sapiens* (Wolpoff 1979). Nontheless, the Krapina last interglacial sample blends with both the older Early Neanderthals and the later Neanderthals.

This sequence of last interglacial and Middle Pleistocene European hominids leads directly into the Neanderthals, or "Classic" Neanderthals, of the early last glacial. The morphological features of these fossil hominids have been extensively described (e.g. Heim 1978; Howells 1978; Hublin 1978; Musgrave 1971; Patte 1955; Stringer 1978; Trinkaus 1976a, 1976b; Wolpoff 1980a), so that only a few features, for which the earlier fossils provide comparative data, will be mentioned.

The Neanderthal cranio-facial region is dominated by mid-facial prognathism. This is reflected in the anterior placement of their nasal regions and dentitions and the associated large maxillary and mandibular retromolar spaces. The effect of the mid-facial prognathism is accentuated by their large anterior teeth, which increased the breadth of the lower face. Posteriorly, Neanderthal crania are characterized by a nuchal region which lacks a true occipital torus and an external occipital protuberance but which possesses a supra-iniac fossa. These are but a few of the morphological features which differentiate the Neanderthals from more recent, anatomically modern hominids, but they are particularly useful in comparisons of the Neanderthals with earlier European and non-European archaic *H. sapiens* fossils. Unfortunately, the host of postcranial traits, which appear unique to the Neanderthals, are of little use in this discussion, since postcrania are few and fragmentary in other samples of archaic *H. sapiens*.

These Neanderthal features are largely unknown among the earliest known European hominids, such as Arago, Bilzingsleben and Petralona; only the suggestion of anterior dental enlargement anticipates the Neanderthals. By the time of the Steinheim and Swanscombe specimens, and certainly among the Early Neanderthals, many of these particular traits are in full evidence. Mid-facial prognathism is less accentuated than among the Neanderthals. But several occipital traits, such as the absence of an external occipital protuberance and the presence of a supra-iniac fossa, are clearly marked on Early Neanderthal occipitals, such as those from Biache, Salzgitter-Lebenstedt, Steinheim, Swanscombe and both sites at La Chaise.

These considerations suggest that in Europe during the late Middle and early Upper Pleistocene, certainly from the time of the Steinheim and Swanscombe specimens to the Neanderthals, there was a gradual evolution of European hominids. This is the period during which the Lower to Middle Palaeolithic transition took place, which suggests that there was no period of rapid biological evolution associated with the cultural transition.

The relative gradualness of the transition from the grade 1 sample to later hominids is difficult to assess. The Arago, Mauer, Petralona and Vértesszöllös specimens almost certainly antedate the Steinheim and Swanscombe fossils by a considerable period of time (Stringer 1980b). However, the Bilzingsleben fossils, which are among the more archaic, and certainly the most robust, of the European fossil hominids, may be relatively late. They are associated with a Holsteinian fauna and flora (Mania 1974), but "Holsteinian" deposits may span much of the Middle Pleistocene. However, recent thorium-uranium determinations on travertine at Bilzingsleben provide an age of about 228,000 BP (Harmon *et al.* 1980), suggesting a late Holsteinian or oxygen isotope stage 7 age. If these radiometric determinations are accurate, it may be difficult to maintain an interpretation of gradual hominid evolution in Europe during the Middle Pleistocene, even if one accepts a high degree of sexual dimorphism for these hominids. Yet, if there was a period of rapid morphological evolution in Europe at this time, it would predate the time of the Lower to Middle Palaeolithic industrial transition.

In the past several researchers (e.g. Vallois 1958) have argued for multiple hominid lineages within Europe during the Middle and Upper Pleistocene, an interpretation best known as the Pre-Sapiens Hypothesis. The available fossil data does not document more than one lineage in Europe (Hublin, in press; Pilbeam 1975; Stringer 1974; Trinkaus 1973). Furthermore, the coexistence of two genetically distinct lineages, adapting to the same cultural adaptive niche for a long period of time, would violate the competitive exclusion principle (Hardin 1960). Therefore, the Pre-Sapiens Hypothesis is currently untenable.

Asian Archaic *Homo sapiens*

The hominid fossil record for the remainder of the Old World during the late Middle and early Upper Pleistocene is far less complete than that of Europe. In some regions certain grades of archaic *H. sapiens* are represented by little more than fragments, and the chronological ordering of specimens is frequently tenuous. Yet, if one is willing to tolerate a certain degree of uncertainty, it is possible to see evolutionary patterns among these non-European archaic *H. sapiens*.

The Near East has yielded a number of fossils which are closely aligned with, but not identical to, the European Neanderthals in cranial and postcranial morphology. These include specimens from Amud, Shanidar and Tabun C (McCown and Keith 1939; Stewart 1977; Suzuki and Takai 1970). All of these individuals appear to have been contemporaneous with the European Neanderthals. They are followed, during the Middle Palaeolithic, by early representatives of anatomically modern humans at Qafzeh and Skhul (McCown and Keith 1939; Vandermeersch 1977).

There is only one diagnostic fossil from the Near East prior to the Neanderthals, an upper facial piece from Zuttiyeh (Keith 1927). A femoral shaft from Tabun E conforms to the morphological pattern of European archaic *H. sapiens* and *H. erectus* femora (McCown and Keith 1939; Trinkaus 1976b). The Zuttiyeh fossil appears to derive from deposits with an "Acheulian of Yabrudian facies" industry (Gisis and Bar-Yosef 1974), which have been dated to between 95,000 BP and 148,000 BP by thorium-uranium dating on travertine (Schwarcz 1980). It therefore predated or was contemporaneous with the Lower to Middle Palaeolithic transition in the Near East. Certain features of the Zuttiyeh specimen, especially its apparently modest mid-facial prognathism and absence of a retreating zygomatic profile (Stringer 1978; Vandermeersch 1981), suggest a non-Neanderthal or Early Neanderthal morphology. Yet, some of the older Near Eastern Neanderthals, such as Shanidar 2 and 4, exhibit less prognathism, minimally inflated maxillae and more robust zygomatics than the more recent Near Eastern Neanderthals (Stewart 1961b; Stringer and Trinkaus 1981). This implies that there was an increase in mid-facial prognathism during the early Upper Pleistocene of the Near East. Zuttiyeh would represent the beginning of the sequence and specimens like Shanidar 1 and 5 would represent the end of the evolutionary trend. Shanidar 2 and 4, and perhaps Amud 1, and Tabun C1, would represent the middle of the sequence. The fossil evidence from this region of the Old World therefore fits an interpretation of gradual evolution without significant evolutionary shifts.

A series of discoveries in China, particularly at Changyang, Dali, Hsuchiayao and Maba, have helped to fill out the east Asian archaic *H. sapiens* fossil record (Chia 1957; Chia *et al.* 1979; Woo and Peng 1959; Wu and You 1979). No one chronological ordering of these specimens can be confirmed with currently available data, but a sequence of Dali and Hsuchiayao to Maba to Changyang appears reasonable on faunal and stratigraphic grounds.

These fossils support Weidenreich's (1943) interpretation that there was an evolutionary continuum within eastern Asia. The Dali cranium shows similarities to the *H. erectus* remains from Zhoukoudian (Choukoutien), especially in the orbital region, although it is clearly more gracile and has a larger cranial capacity. The Hsuchiayao occipital and parietal fragments suggest a similar interpretation. The occipital has a transverse torus, but the vault bones are only moderately curved. The more recent Maba and Changyang fossils are more gracile versions of the same morphological pattern, of an evolutionary grade similar to the Neanderthals. Yet, none of these fossils, where sufficiently intact, indicate any particular affinity to the western Eurasian Neanderthals. Although the Chinese fossils are insufficient to establish a pattern of gradual evolution within eastern Asia, they are consistent with such an interpretation.

For peninsular southeast Asia fossils are lacking for most of the archaic *H. sapiens* grades. The only sample which has been referred to this group, the Ngandong fossils, probably derive from Middle Pleistocene deposits and represents either late *H. erectus* or very early archaic *H. sapiens* (Jacob 1978; Santa Luca 1980). There is an absence of more recent fossils in this region until the appearance of robust anatomically modern humans at Wadjak (Dubois 1921) and sites in Australia (Kirk and Thorne 1976). Although similarities exist between the Javanese *H. erectus* specimens and some of the Upper Pleistocene Australian fossils (Thorne and Wolpoff 1981), the tempo of that evolution cannot be ascertained.

African Archaic *Homo sapiens*

Archaic *H. sapiens* remains are considerably more numerous in Africa than in eastern Asia. However, similar problems of dating and fragmentation make their interpretation difficult (Rightmire 1980; Tobias 1968). These African fossils fall into two geographical groups, one in north Africa and the other in east and south Africa; these will be considered separately.

With the exception of the Irhoud 1 cranium, all of the north African archaic *H. sapiens* fossils are incomplete. Morphologically and chronologically they form two groups; an early one including Rabat and Sidi Abderrahman, and probably Salé and Témara (Arambourg and Biberson 1956; Jaeger 1975; Saban 1975; Vallois and Roche 1958), and a later one including Haua Fteah and Irhoud (Ennouchi 1962, 1968; Hublin and Tillier 1981; Tobias 1967). It is not possible to see a morphological continuum between the only reasonably complete crania, those from Salé and Irhoud, given the disparity in their ages. Yet, there appears to be a gradual morphological shift in the mandibular corpus from the Rabat and Sidi Abderrahman mandibles to that from Témara to the Irhoud 3 mandible (Hublin and Tillier 1981; Saban 1975). Through this sequence there is a retention of considerable robustness in the corpus, despite a gradual decrease in size. This is weak evidence on which to base an evolutionary interpretation, but the fossil evidence is consistent nontheless with a model of gradual change from the Acheulian associated Rabat, Sidi Abderrahman and Témara specimens to the Mousterian associated Irhoud remains.

It should be noted that although the Irhoud and Haua Fteah fossils have been called Neanderthals (Ennouchi 1962, 1968; McBurney *et al.* 1953), it is now apparent that they are distinct from the European and Near Eastern Neanderthals. In particular, they lack the mid-facial prognathism and the complex of occipito-mastoid traits which are characteristic of the Neanderthals (Hublin 1978; Hublin and Tillier 1981; Stringer 1978; Tobias 1967).

A series of late Middle and early Upper Pleistocene hominid fossils have been discovered in Sub-Saharan Africa, but it is still difficult to sort them in a conclusive manner. As with the north African sample, the fossils appear to separate into early and late archaic *H. sapiens* samples. The early, Middle Pleistocene sample consists of Bodo, Broken Hill and Saldanha, and perhaps Eyasi and Ndutu (Clark 1976; Conroy *et al.* 1978; Pycraft 1928; Singer 1954; Weinert 1939), whereas the Upper Pleistocene sample includes Florisbad, Laetoli 18 (Ngaloba), Omo-Kibish 2 and perhaps Singa (Day 1969; Day *et al.* 1980; Rightmire 1978; Stringer 1979). (Omo-Kibish 1 is anatomically modern and therefore distinct from these hominids (Day 1969).) The older sample resembles late *H. erectus* and European specimens such as Petralona, particularly in frontal and occipital morphology, and to a lesser extent in facial shape (Clark 1976; Conroy *et al.* 1978; Morant 1928; Stringer *et al.* 1979). The younger sample, which is highly heterogeneous, parallels the Neanderthals in evolutionary grade but is morphologically distinct from the European fossils. The relatively flat frontal associated with an angulated occipital on the Omo-Kibish 2 cranium recalls the earlier Eyasi and Saldanha crania, although the Omo-Kibish 2 cranium is considerably higher (Day 1969). The Laetoli 18 cranium shows resemblances to the Irhoud crania, especially in maxillary and occipital morphology. And Florisbad appears largely as a less archaic version of the older Broken Hill and Saldanha specimens (Rightmire 1978). Some of this heterogeneity may be due to the extensive geographical and possibly temporal spread of these fossils.

The critical problem in assessing the relative evolutionary patterns indicated by these Sub-Saharan archaic *H. sapiens* is chronological (Protsch 1975; Rightmire 1978). It is not possible to say whether the earlier sample is very distant temporally from the later group. Specimens which may be temporally intermediate, such as the Cave of Hearths mandible (Tobias 1971), are too incomplete to be very helpful. As with the east Asian fossil record, dating and, to a lesser extent, fragmentation of the fossils prevent a resolution of the problem of archaic *H. sapiens* continuity posed here.

Summary and Conclusion

This brief review of the fossil evidence for archaic *Homo sapiens* provides a background for evaluating the question originally posed: was there a period of rapid biological evolution associated with the Lower to Middle Palaeolithic transition? It is only within Europe and the Near East that this question can be answered with confidence. The fossil record for this phase of hominid evolution elsewhere in the Old World is too incomplete and/or poorly dated to permit firm conclusions.

In Europe there is a gradual evolution from the Early Neanderthals associated with Lower Palaeolithic industries to the Neanderthals found with Middle Palaeolithic assemblages. A variety of cranio-facial patterns, which are most prominent among the Neanderthals, appear among the Early Neanderthals and become increasingly accentuated during the penultimate glacial and last interglacial. In the Near East, early fossils are scarce, but the Zuttiyeh specimen may be considered representative of the hominids associated with the later phases of the Near Eastern Lower Palaeolithic. There appears to be a gradual shift from its morphology to that of the older Near Eastern Neanderthals, who in turn foreshadow the later Near Eastern Neanderthals. There is no evidence from elsewhere in the Old World to contradict this pattern, and the few specimens from China and north Africa strongly suggest that it was also representative of those regions.

It therefore appears reasonable to conclude that a gradualist model of hominid evolution is accurate for the period represented by archaic *Homo sapiens* in the fossil record and by the Lower to Middle Palaeolithic transition in the archaeological record.

Table 1

Examples of archaic *Homo sapiens* fossils for six regions of the Old World, arranged according to a grade system modified from the grade system of Stringer *et al.* (1979). The placement of fossils into grades is strictly morphological. "?" after a site name means that the placement of the fossil(s) from that site is subject to debate.

Grade	Europe	Near East	East Asia	Southeast Asia	North Africa	Sub-Saharan Africa
3a	Neanderthal La Chapelle Circeo, Spy La Ferrassie Krapina Régourdou	Amud Shanidar Tabun C	Changyang Maba	–	Haua Fteah Irhoud	Florisbad Laetoli 18 Omo-Kib. 2
2 or 3a	Krapina	Tabun E Zuttiyeh	–	–	–	Singa
2	Biache La Chaise Ehringsdorf Fontéchevade Saccopastore Salz.-Leben.	–	–	–	Témara	Cave of Hearths ?
1 or 2	Atapuerca Steinheim Swanscombe	–	Dali Hsuchiayao	–	Rabat Sidi Abder.	Eyasi
1	Arago, Mauer Bilzingsleben Petralona Vértesszöllös	–	–	Ngandong ?	Salé	Bodo Broken Hill Saldanha Ndutu ?

Acknowledgments

I would like to express my sincere appreciation to the many individuals who gave me access to original fossil material in their care. Drs. C.B. Stringer and W.W. Howells provided helpful criticism during the preparation of the paper. This research has been supported in part by NSF grants BNS76-14344 and BNS80-04578.

References

Aguirre, E. and M.-A. de Lumley 1977 Fossil Men from Atapuerca, Spain: Their Bearing on Human Evolution in the Middle Pleistocene. *Journal of Human Evolution* 6: 681-688.

Arambourg, C. and P. Biberson 1956 The fossil human remains from the Palaeolithic site of Sidi Abderrahman (Morocco). *American Journal of Physical Anthropology* 14: 467-490.

Billy, G. and H.V. Vallois 1977 La Mandibule Pré-Rissienne de Montmaurin. *L'Anthropologie* 81: 273-312, 411-458.

Butzer, K.W. and G.L. Isaac (eds.) 1975 *After the Australopithecines* Mouton Pub., The Hague.

Chia, L.P. 1957 Notes on the Human and some Mammalian Remains from Changyang, Hupei *Vertebrata PalAsiatica* 1: 247-257.

Chia, L.P., Q. Wei and C.R. Li 1979 Report on the Excavation of Hsuchiayao Man Site in 1976 *Vertebrata PalAsiatica* 17: 277-293.

Clark, R.J. 1976 New cranium of *Homo erectus* from Lake Ndutu, Tanzania *Nature* 262: 485-487.

Conroy, G.C., C.J. Jolly, D. Cramer and J.E. Kalb 1978 Newly discovered fossil hominid skull from the Afar depression, Ethiopia *Nature* 275: 67-70.

Day, M.H. 1969 Early *Homo sapiens* Remains from the Omo River Region of Southwest Ethiopia *Nature* 222: 1135-1138.

Day, M.H. 1971 Postcranial Remains of *Homo erectus* from Bed IV, Olduvai Gorge, Tanzania *Nature* 232: 383-387.

Day, M.H., M.D. Leakey and C. Magori 1980 A new hominid fossil skull (L.H. 18) from the Ngaloba Beds, Laetoli, northern Tanzania *Nature* 284: 55-56.

Dubois, E. 1921 The Proto-Australian Fossil Man from Wadjak, Java *Proceedings of the Koninklijke Nederlandse Akademie van Wetenschappen* 23: 1013-1051.

Ennouchi, E. 1962 Un Néandertalien: L'homme du Jebel Irhoud (Maroc) *L'Anthropologie* 66: 279-299.

Ennouchi, E. 1968 Le Deuxième Crâne de l'Homme d'Irhoud *Annales de Paléontologie (Vertébrés)* 54: 117-128.

Gisis, I. and O. Bar-Yosef 1974 New Excavation in Zuttiyeh Cave, Wadi Amud, Israel *Paléorient* 2: 175-180.

Gould, S.J. and N. Eldredge 1977 Punctuated equilibria: the tempo and mode of evolution reconsidered *Paleobiology* 3: 115-151.

Hardin, G. 1960 The Competitive Exclusion Principle *Science* 131: 1292-1297.

Harmon, R.S., J. Glazek and K. Nowak 1980 $^{230}Th/^{234}U$ dating of travertine from the Bilzingsleben archaeological site *Nature* 284: 132-135.

Heim, J.-L. 1978 Contribution du massif facial à la morphogenèse du crâne Néanderthalien. *In: Les Origines Humaines et les Époques de l'Intelligence* Masson & Cie., Paris. pp. 183-215.

Howells, W.W. 1978 Position phylétique de l'homme de Néanderthal. *In: Les Origines Humaines et les Époques de l'Intelligence* Masson & Cie., Paris. pp. 217-237.

Howells, W.W. 1980 *Homo erectus* – Who, When and Where: A Survey. *Yearbook of Physical Anthropology* 23: 1-23.

Hublin, J.-J. 1978 *Le Torus Occipital Transverse et les Structures Associées: Évolution dans le Genre Homo.* Thèse de 3e Cycle, Université de Paris VI.

Hublin, J.-J. in press Que reste-t-il des Présapiens européens? *L'Anthropologie.*

Hublin, J.-J. and A.-M. Tillier 1981 The Mousterian Juvenile Mandible from Irhoud (Morocco): A Phylogenetic Interpretation. *In:* C.B. Stringer (ed.) *Aspects of Human Evolution* Taylor & Francis, London. pp. 167-185.

Jacob, T. 1978 The Puzzle of Solo Man *Modern Quaternary Research in Southeast Asia* 4: 31-40.

Jaeger, J.J. 1975 The Mammalian Faunas and Hominid Fossils of the Middle Pleistocene of the Maghreb. *In:* K.W. Butzer and G.L. Isaac (eds.) *After the Australopithecines* Mouton Pub., The Hague. pp. 399-418.

Keith, A. 1927 A Report on the Galilee Skull *Researches in Prehistoric Galilee* British School of Archaeology in Jerusalem. pp. 53-106.

Kirk, R.L. and A.G. Thorne (eds.) 1976 *The Origin of the Australians* Australian Institute of Aboriginal Studies, Canberra.

Kukla, G.J. 1977 Pleistocene Land-Sea Correlations. I. Europe. *Earth-Science Reviews* 13: 307-374.

Kurtén, M. 1981 The Gestalt of Hominid Evolution. *In:* D. Ferembach (ed.) *Les processus de l'hominisation* C.N.R.S. Paris. pp. 61-65.

Lumley, H. de 1975 Cultural Evolution in France in its Palaeoecological Setting during the Middle Pleistocene. *In:* K.W. Butzer and G.L. Isaac (eds.) *After the Australopithecines* Mouton Pub., The Hague. pp. 745-808.

Lumley, H. de and M.-A. de Lumley 1979 L'homme de Tautavel *Dossiers de l'Archéologie* 36: 54-59.

Lumley, M.-A. de 1972 L'Os Iliaque Anténéandertalien de la Grotte du Prince (Grimaldi, Ligurie italienne) *Bulletin du Musée d'Anthropologie Préhistorique de Monaco* 18: 89-112.

Lumley, M.-A. de 1973 Anténéandertaliens et Néandertaliens du bassin mediterranéen occidental européen *Études Quaternaires* 2: 1-626.

Lumley, M.-A. de 1978 Les Anténéandertaliens. *In: Les Origines Humaines et les Époques de l'Intelligence* Masson et Cie., Paris. pp. 159-182.

Mania, D. 1974 Bilzingsleben, Kr. Artern. Eine altpaläolithische Travertinfundstelle im nördlichen Mitteleuropa (Vorbericht) *Zeitschrift für Archäologie* 8: 157-173.

McBurney, C.B.M., J.C. Trevor and L.H. Wells 1953 The Haua Fteah Fossil Jaw *Journal of the Royal Anthropological Institute* 83: 71-85.

McCown, T.D. and A. Keith 1939 *The Stone Age of Mount Carmel II: The Fossil Human Remains from the Levalloiso-Mousterian* Clarendon Press, Oxford.

Morant, G.M. 1930 Studies of Palaeolithic Man III. The Rhodesian Skull and its Relations to Neanderthaloid and Modern Types *Annals of Eugenics* 3: 337-360.

Musgrave, J.H. 1971 How Dextrous was Neanderthal Man? *Nature* 233: 538-541.

Musgrave, J.H. 1977 The Neanderthals from Krapina, northern Yugoslavia: An inventory of the hand bones *Zeitschrift für Morphologie und Anthropologie* 68: 150-171.

Patte, E. 1955 *Les Néanderthaliens* Masson et Cie., Paris.

Pilbeam, D.R. 1975 Middle Pleistocene Hominids. *In:* K.W. Butzer and G.L. Isaac (eds.) *After the Australopithecines* Mouton Pub., The Hague. pp. 809-856.

Piveteau, J. 1970 Les Grottes de La Chaise (Charente). Paléontologie Humaine 1, L'Homme de l'Abri Suard *Annales de Paléontologie (Vertébrés)* 56: 174-225.

Piveteau, J. 1976 Les Anté-Néandertaliens du Sud-Ouest. *In:* H. de Lumley (ed.) *La Préhistoire Française.* Éditions du C.N.R.S., Paris 1: 561-566.

Protsch, R. 1975 The absolute Dating of Upper Pleistocene Sub-Saharan Fossil Hominids and Their Place in Human Evolution *Journal of Human Evolution* 4: 297-322.

Pycraft, W.P. 1928 *Rhodesian Man and Associated Remains* British Museum (Natural History), London.

Rightmire, G.P. 1978 Florisbad and Human Population Succession in Southern Africa *American Journal of Physical Anthropology* 48: 475-486.

Rightmire, G.P. 1980 *Homo erectus* and Human Evolution in the African Middle Pleistocene. *In:* L.-K. Konigsson (ed.) *Current Argument on Early Man* Pergamon Press, N.Y. pp. 70-85.

Saban, R. 1975 Les Restes Humains de Rabat (Kébibat) *Annales de Paléontologie (Vertébrés)* 61: 153-207.

Santa Luca, A.P. 1978 A re-examination of presumed Neandertal-like fossils *Journal of Human Evolution* 7: 619-636.

Santa Luca, A.P. 1978 *The Ngandong Fossil Hominids: A Comparative Study of a Far Eastern Homo erectus Group* Yale University Publications in Anthropology 78: 1-175.

Schoetensack, O. 1908 *Der Unterfeifer des Homo heidelbergensis aus den Sanden von Mauer bei Heidelberg* Verlag von Wilhelm Engelmann, Leipzig.

Schwarcz, H.P. 1980 Absolute age determination of archaeological sites by uranium series dating of travertines *Archaeometry* 22: 3-24.

Shackleton, N.J. and N.D. Opdyke 1976 Oxygen-Isotope and Palaeomagnetic Stratigraphy of Pacific Core V28-239. Late Pliocene to Latest Pleistocene *Geological Society of America Memoir* 145: 449-464.

Singer, R. 1954 The Saldanha Skull from Hopefield, South Africa *American Journal of Physical Anthropology* 12: 345-362.

Smith, F.H. 1976 *The Neandertal Remains from Krapina* Department of Anthropology, University of Tennessee, Report of Investigations 15: 1-359.

Smith, F.H. and G.C. Ranyard 1980 Evolution of the Supraorbital Region in Upper Pleistocene Fossil Hominids from South-Central Europe *American Journal of Physical Anthropology* 53: 589-610.

Stanley, S.M. 1979 *Macroevolution* W.H. Freeman & Co., San Francisco.

Stewart, T.D. 1961a A neglected primitive feature of the Swanscombe skull. *In: Homenaje a Pablo Martínez del Río* Instituto Nacional de Antropología e Historia, Mexico. pp. 207-217.

Stewart, T.D. 1961b The Skull of Shanidar II *Sumer.* 17: 97-106.

Stewart, T.D. 1970 The Evolution of Man in Asia as Seen in the Lower Jaw *VIIIth International Congress of Anthropological and Ethnological Sciences* 1: 263-266.

Stewart, T.D. 1977 The Neanderthal skeletal remains from Shanidar Cave, Iraq: A summary of findings to date *Proceedings of the American Philosophical Society* 121: 121-165.

Stringer, C.B. 1974 Population relationships of later Pleistocene hominids: a multivariate study of available crania *Journal of Archaeological Science* 1: 317-342.

Stringer, C.B. 1978 Some problems in Middle and Upper Pleistocene Hominid relationships. *In:* D.J. Chivers and K.A. Joysey (eds.) *Recent Advances in Primatology* Academic Press, London 3: 395-418.

Stringer, C.B. 1979 A re-evaluation of the fossil human calvaria from Singa, Sudan *Bulletin of the British Museum (Natural History)* 32: 77-83.

Stringer, C.B. 1980 The Phylogenetic Position of the Petralona Cranium *Anthropos* 7: 81-95.

Stringer, C.B. 1981 The dating of European middle Pleistocene hominids and the existence of *Homo erectus* in Europe *Anthropologie (Brno)* 19: 3-14.

Stringer, C.B. and R. Burleigh 1981 The Neanderthal problem and the prospects for direct dating of Neanderthal remains *Bulletin of the British Museum (Natural History) (Geology)* 35(3): 225-241.

Stringer, C.B., F.C. Howell and J.K. Melentis 1979 The Significance of the Fossil Hominid Skull from Petralona, Greece *Journal of Archaeological Science* 6: 235-253.

Stringer, C.B. and E. Trinkaus 1981 The Shanidar Neanderthal crania. *In:* C.B. Stringer (ed.) *Aspects of Human Evolution* Taylor & Francis, London. pp. 129-165.

Suzuki, H. and F. Takai (eds.) 1970 *The Amud Man and His Cave Site* Academic Press of Japan, Tokyo.

Thoma, A. 1966 L'Occipital de l'Homme Mindélien de Vértesszöllös *L'Anthropologie* 70: 495-534.

Thorne, A.G. and M.H. Wolpoff 1981 Regional Continuity in Australasian Pleistocene Hominid Evolution *American Journal of Physical Anthropology* 55: 337-349.

Tobias, P.V. 1967 The hominid skeletal remains of Haua Fteah. *In:* C.B.M. McBurney (ed.) *The Haua Fteah* Cambridge University Press, Cambridge. pp. 338-352.

Tobias, P.V. 1968 Middle and Early Upper Pleistocene Members of the Genus *Homo* in Africa. *In:* G. Kurth (ed.) *Evolution and Hominisation* Gustav Fischer Verlag, Stuttgart. pp. 176-194.

Tobias, P.V. 1971 Human Skeletal Remains from the Cave of Hearths, Makapansgat, northern Transvaal *American Journal of Physical Anthropology* 34: 335-368.

Trinkaus, E. 1973 A Reconsideration of the Fontéchevade Fossils *American Journal of Physical Anthropology* 39: 25-36.

Trinkaus, E. 1976a The Morphology of European and Southwest Asian Neandertal pubic bones *American Journal of Physical Anthropology* 44: 95-104.

Trinkaus, E. 1976b The evolution of the hominid femoral diaphysis during the Upper Pleistocene in Europe and the Near East *Zeitschrift für Morphologie und Anthropologie* 67: 291-319.

Trinkaus, E. 1978 Functional implications of the Krapina Neandertal lower limb remains. *In:* M. Malez (ed.) *Krapinski Pračovjek i Evolucija Hominida* Jugoslavenska Akademija Znanosti i Umjetnosti, Zagreb. pp. 155-192.

Vallois, H.V. 1958 *La Grotte de Fontéchevade II: Anthropologie* Archives de l'Institut de Paléontologie Humaine 29: 1-164.

Vallois, H.V. and J. Roche 1958 La mandibule acheuléene de Témara, Maroc. *Comptes rendus de l'Académie des Sciences,* Paris. Série D 246:3113-3116.

Vandermeersch, B. 1965 Position stratigraphique et chronologique des restes humains du paléolithique moyen du sud-ouest de la France *Annales de Paléontologie (Vertébrés)* 51: 69-126.

Vandermeersch, B. 1977 *Les Hommes Fossiles de Qafzeh (Israël)* Thèse de Doctorat d'État, Université de Paris VI.

Vandermeersch, B. 1978 La Crâne Pré-Würmien de Biache-Saint-Vaast (Pas-de-Calais). *In: Les Origines Humaines et les Époques de l'Intelligence* Masson et Cie., Paris. pp. 153-157.

Vandermeersch, B. 1981 Les Premiers *Homo sapiens* au Proche Orient. *In:* D. Ferembach (ed.) *Les Processus de l'Hominisation* C.N.R.S., pp. 97-100.

Vlček, E. 1978 A New Discovery of *Homo erectus* in Central Europe *Journal of Human Evolution* 7: 239-251.

Weidenreich, F. 1941 The Extremity Bones of Sinanthropus Pekinensis *Palaeontologica Sinica.* Series D. 5.

Weidenreich, F. 1943 The Skull of Sinanthropus Pekinensis *Palaeontologica Sinica.* Series D. 10.

Weinert, H. 1936 Der Urmenschenschädel von Steinheim *Zeitschrift für Morphologie und Anthropologie* 35: 463-518.

Weinert, H. 1939 Africanthropus njarasensis *Zeitschrift für Morphologie und Anthropologie* 38: 252-308.

Wolpoff, M.H. 1975 Some aspects of human mandibular evolution. *In:* J.A. McNamara Jr. (ed.) *Determinants of Mandibular Form and Growth.* Center for Human Growth and Development, Ann Arbor. pp. 1-64.

Wolpoff, M.H. 1977 Some Notes on the Vértesszöllös Occipital *American Journal of Physical Anthropology* 47: 357-364.

Wolpoff, M.H. 1979 The Krapina dental remains *American Journal of Physical Anthropology* 50: 67-114.

Wolpoff, M.H. 1980a *Palaeoanthropology* Alfred A. Knopf, New York.

Wolpoff, M.H. 1980b Cranial Remains of Middle Pleistocene European Hominids *Journal of Human Evolution* 9: 339-358.

Woo, J.K. and R.C. Peng 1959 Fossil Human Skull of Early Palaeoanthropic Stage found at Mapa, Shaoquan, Kwangtung Province *Vertebrata PalAsiatica* 3: 176-182.

Wu, X.S. and Y.Z. You 1979 A Preliminary Observation of Dali Man Site (in Chinese) *Vertebrata PalAsiatica* 17: 294-303.

Discussion

E. Wreschner: Do you suggest a gradual evolutionary progress? I am speaking about your main point, the accelerating sequence in time as you approach the period of *Homo sapiens sapiens*. Was human evolution a unilineal development with some variations? I am rather skeptical about the unilineal development, since it would allow a very short time for the biological change from Neanderthal to *Homo sapiens.*

E. Trinkaus: The question, to be answered properly, requires a lengthy answer. I will try to be relatively brief. First of all, there is no one-to-one correlation between human biological form as we know it in the fossil record and the behavioral manifestations which we see in the archaeological record. I need only point to the association of two forms of hominids with Mousterian industries in the Near East, whether or not they were contemporaneous. There is no question that behavior during the early Upper Pleistocene became considerably more elaborate in a variety of different ways, and this is associated, at least in Europe where we have the best evidence, with an Early Neanderthal to Neanderthal lineage exhibiting increasing brain size and decreasing amounts of massiveness of the skeleton (although the Neanderthals were still very massive). One can infer a general correlation between the morphology and behavior, even though we do not know the functional significance of some of the morphological traits I pointed out in my talk. There has been speculation on the functional significance of these traits, but we do not have good biological explanations for some of them.

Your second question concerns how one can get rapid evolutionary change in a very short period of time. There is a considerable body of biological evidence, from living animals as well as the palaeontological record, that shows that significant anatomical evolution can occur in very short periods of time, within a matter of a small number of generations, perhaps a couple or a couple of dozen generations. I see no reason why we could not have had an evolutionary change from human populations similar to the Neanderthals, perhaps the Classic Neanderthals in Europe and the Near East or perhaps in another population of the same grade, to something that was basically modern. We simply do not know the genetic basis for the anatomical differences, and we will probably never know the precise genetic basis for these anatomical differences given the nature of the palaeontological record. Without knowing the nature of the genetic differences, there is no way that we can reasonably speculate on possible rates of evolution. If the genetic basis for the differences was relatively simple, for example in the form of regulator genes, and it is probable that many of these traits were genetically intercorrelated since they tend to occur together in most cases, then the genetic change required could have been relatively small. We could thus have major anatomical changes at this time period.

G. Bosinski: You did not take into consideration the finds from Eastern Europe, in the first place Staroselje, which was published as a *Homo sapiens* of Early Wurm.

E. Trinkaus: There are a number of good Neanderthal specimens from the Crimea, such as those from Kiik-Koba and Zaskalnaya VI. With reference to the Staroselje specimen, first of all, it is a very young infant, and secondly, its association with the Mousterian level is questionable. A fluorine test suggested that it might be intrusive. It is a very young infant, and young Neanderthal infants, approximately one year of age, are similar to infants of modern humans.

B. Vandermeersch: I would like to have the opinion of Ms. Tillier, who is studying Neanderthal children.

A.-M. Tillier: I don't think that the Staroselje infant is Neanderthal; it is probably a *Homo sapiens sapiens*.

E. Trinkaus: I would still question its association with the Mousterian level[1]. If the Staroselje 1 specimen was in fact contemporary with the Mousterian level of the site, it would have been associated with a Final Mousterian industry. Therefore, its apparent morphology would indicate that in the Crimea, as in the Levant, the Final Mousterian was made by anatomically modern humans rather than by Neanderthals.[2]

A. Belfer-Cohen: I can't see how you can judge that the Amud, Tabun C and Shanidar remains predate those of Skhul and Qafzeh.

E. Trinkaus: A good question. I am sure that Jelinek could probably answer your question better than I. Let me just say that it is open to debate whether the Tabun C skeleton comes from layer C or D. It may well come from layer D, and we know that at least Tabun layer D predates layer B at Skhul. The actual chronological position of the Qafzeh specimens is open to considerable debate. It depends upon how one interprets the fauna. The archaeological evidence and the palaeoclimatological evidence would permit a late date for the Qafzeh hominids. (I believe that Farrand would agree that there is considerable leeway in the chronological indications provided by the palaeoclimatological evidence.)

Notes added after the conference:

1 Fluorine analysis of Staroselje 1 produced no measurable fluorine, whereas supposedly associated animal bone provided 100F/P2O5 = 0.13 (Formozov 1958, *Mater. Issled. Arkheol. SSSR.* 71).

2 Alexeyev 1976 *J. Hum. Evol.* 5: 413.

A.J. Jelinek: I have noticed in the last two sessions in which I have participated that Anna* has a wonderful way of bringing up questions that hit a nerve. The essence of what we are going to be talking about here during the next few minutes is that at the base of all of these discussions and at the base of the resolution of the view of Trinkaus and Vandermeersch is the problem of chronology. Both of them have admitted that this is crucial, but we all agree that these chronological issues cannot be resolved with the evidence now in hand. If there was a clear resolution to these chronological problems we would not have to spend all of this time talking about them. What we have to consider in the chronological relations is that there is a little evidence derived from absolute dating techniques and that evidence doesn't help us very much because of problems in the nature of amino acid dating. At Qafzeh Masters has mentioned the problems posed by the addition of preservatives to the bone, and the possibility that the marked discrepancy between the dates on faunal and hominid material from the deep layers may be due to that cause. The other interpretation of amino acid racemization dating suggests that the hominids may be significantly younger than the fauna that overlies them; in the absence of evidence of stratigraphic intrusion this seems impossible.

There is faunal evidence from Tabun and Qafzeh that has been interpreted as showing that Qafzeh is older than Tabun. In the paper that I distributed I gave my reasons for taking exception to this. It is an open question and Tchernov himself has cautioned, as recently as the Lyon Colloquium last summer,** that one cannot date sites by the fauna. I think that the range of predatory birds at the two sites might well overlap, but it also extended into very different environments. This meant that the birds from Qafzeh could sample the fauna in the warmer Jordan Valley margins where more archaic forms might have survived after they disappeared from the Mount Carmel area. So this does not provide definitive evidence either.

Carbon-14 evidence simply does not exist for Qafzeh. This leaves us with only the technological evidence which, though it is also certainly not definitive, suggests a unidirectional sequence in man's manipulative abilities in the manufacture of stone tools. If this can be verified it supports a developmental sequence from Galilee to Tabun to Skhul to Qafzeh. So this evidence would tend to support the ideas of Trinkaus.

This, then, is the present state of the problem. I tend to disagree with Vandermeersch on two basic issues. The first is his view of the general relationships of the Neanderthals of Western Europe with those of the Near East. I think that there is cultural evidence for the isolation of Western Europe during the time of the Middle Palaeolithic industries in the later part of the Early Wurm. I think that the development of the Mousterian in Western Europe seems somewhat different than it does in Central and Eastern Europe, which suggests isolation. We do not see MTAb, for instance, in Central and Eastern Europe; it seems an exclusively Western European development.

At the Grotte du Renne there seems to be a progression or evolution from MTA to Chatelperronian that occurs *in situ*. The appearance of a Neanderthal at Saint-Cesaire with a Chatelperronian industry, then, may not be too surprising since they are also associated with earlier Mousterian industries (although perhaps not with MTA). However, when we see two very distinctive industries such as Chatelperronian and Aurignacian interstratified at a site like Roc de Combe, this suggests the presence in one area of two distinct and contemporary cultures, with the Aurignacian having probably originated in Central Europe. It is possible, I think, that there were two contemporary populations in

* A. Belfer-Cohen (ed.).

***Préhistoire du Lévant.* CNRS 1981 (ed.).

Western Europe for a short time; one a more evolved *Homo sapiens* from Central Europe associated with an Aurignacian industry and the other of Neanderthals associated with the Chatelperronian.

In the Near East, on the other hand, there is stronger evidence for cultural continuity without a break or an intrusion that tends to argue against Vandermeersch's model of an intrusive biological population of Neanderthals. There seems to be a single development from the Mugharan Tradition in its various phases leading into the Mousterian and its sequential development. This suggests that we have one population in the Levant in a continuous pattern of cultural and biological development. One assumes that if an intrusive biological population had come in, we would see a disruption of the cultural sequence.

This is, essentially, how I would see the most probable chronology. I would emphasize once again that there is not now any way to resolve these disagreements. What we can do is to see where the problems lie, where we need to go to gather more evidence, and what kinds of evidence we need.

As a footnote, let me say that I think that the assignment of the Tabun I burial to Layer D is still far from established. I have merely put this possibility forward as a suggestion on the basis of what we can now see in the stratigraphy. I think that it is not particularly important for these arguments whether it was in C or D because, technologically, Layer C is still well ahead of Skhul and Qafzeh in time if this kind of evidence is valid.

B. Vandermeersch: I have some different answers. First I think it is difficult to admit that evolution from Neanderthal to the typical *Homo sapiens sapiens* took place in the Near East. Even if the Neanderthals are older, a little older than the sapiens, the morphological differences between the two are too important. The Neanderthals are very specialized. The occipital bone is very particular, the temporal bone too. The face is very particular also, and I think we have never seen these particularities on other skulls than the Neanderthals. They are special Neanderthal characteristics. We have some of them on skulls in the Near East, and for this reason it is possible to call them Neanderthals. If you do not accept the migration from Europe to the Near East, how do you explain the origin of the Near Eastern Neanderthals? All the pre-Neanderthals are from Europe, and in Europe we have a very clear evolution from the end of the Mindel Glacial or the beginning of the Mindel-Riss Interglacial to the typical Neanderthals at the beginning of the Wurm.

In the other parts of the Old World we have no pre-Neanderthals. Thus it seems to me more simple to make the Near Eastern Neanderthals come from the European pre-Neanderthals. It is possible that the date of this migration is not clear. But I do have a question for the prehistorians: is the Yabrudian culture of local origin? In the Yabrudian there are many implements with the Quina retouch, and I would like to have your answer if it is of local origin.

A.J. Jelinek: Well, I'll answer the cultural part of your argument; the other part of your question is largely anatomical.

I'll reiterate the point that I tried to make in the paper here and in the paper at the Lyon conference*. What we have been calling the "Yabrudian culture" is simply a facies of a more comprehensive cultural tradition that I have labeled the Mugharan. The Mugharan shows gradual shifts in emphasis on scrapers relative to handaxes. In terms of

* *Préhistoire du Lévant.* CNRS 1981 (ed.).

Quina retouch, we checked the strongest samples of Acheulian facies against the strongest samples of Yabrudian facies in the Mugharan and found that the frequencies of Quina retouch were virtually identical. In other words we do not get any less Quina retouch when we get fewer scrapers and more handaxes than when there are no handaxes at all. Quina retouch is more frequent in the Mugharan than in the preceding Late Acheulian and may be another sign of the greater industrial specialization that seems to be represented in the Mugharan.

But the general cultural basis for the Mugharan tradition is present in the underlying Late Acheulian, where the full typological range of the Mugharan is already present. This suggests very strongly that the Mugharan developed from the Late Acheulian. There is a gap between the two that is represented, in part, by Garrod's Layer F.

Most of the tools in Layer F that I have seen are handaxes, and this dominance in the typology was stressed by Garrod. While I have looked at between 20 and 30 collections from Garrod's excavation I have not seen much Quina retouch in the Layer F material. We have only about 7% full Quina retouch in our Mugharan layers, so there is not very much of it in Layer E either. It has been noted by other workers because the similarity of the transverse scrapers to those of the Quina Mousterian of Western Europe is quite striking, but it is not present in any great quantity at Tabun.

So I see clear evidence for continuity at Tabun in this sequence, and I see no evidence for an intrusion by an outside population. Nor have I seen any evidence from the other collections from the Levant that I have examined to suggest an intrusion anywhere in the Middle to Lower Palaeolithic sequence.

B. Vandermeersch: But how was the Acheulian culture in Eastern Europe at the same time?

A.J. Jelinek: In some ways this was quite similar.

E. Trinkaus: I would like to express a rather strong dislike with the free use of the concept of migration, which seems to be dominant in palaeoanthropology/prehistory, as well as in human palaeontology. I think that migration on a significant level was primarily a phenomenon of historic and protohistoric time periods. I think that we should unilaterally reject the concept of migrations, except in those situations where we have evidence for the appearance of a cultural and, if possible, an associated biological group within a very short period of time over a relatively large geographical area. Within the Palaeolithic, and human palaeontology, I am familiar with only one well documented case, and that is the appearance of the Aurignacian in Europe, primarily in Western Europe.

Most cultural information and most biological information (in the form of genes) is transmitted either by diffusion in the cultural sense or gene flow in the biological sense. It is possible to have rather major levels of gene flow with people interbreeding only between neighboring populations. It is possible to have rather major levels of gene flow within the period we are talking about. Remember that one thousand years, which is a moment in time from our perspective, represents a minimum of fifty generations, at twenty years to the generation. I think that to talk of major human populations, major cultural traditions, packing their bags, getting on their feet and moving across continents is a concept that we should employ only when we have very good evidence to support it.

G. Bosinski: I would suggest not to mix stones and bones at this level of discussion. We cannot use the evolution of stones as an argument in human evolution. This should not be done.

P.M. Masters: Well, there is some possibility that the racemization dates and the carbon date for the Carmel hominoids are fairly correct. Assuming that they are knowledgeably correct, could that much morphological change take place in less than ten thousand years time?

E. Trinkaus: Yes, since that is five hundred generations.

I. Gilead: How do you view the Border Cave specimens?

E. Trinkaus: That is a question that we have not talked about, and I think that it is very relevant to the questions here. The Border Cave specimens are several anatomically modern human fossils from South Africa, which, it has been suggested, date to between 90,000 and 110,000 years ago. I question whether they are that old, but there is little question, in my mind, that in southern Africa there were anatomically modern humans at least fifty thousand years ago. There is a variety of indirect evidence to support this. If that is indeed the case, then that would provide a population from which genes could flow, so to speak, at least into the Near East. A number of people have suggested that if one were to determine the morphological affinities of the Qafzeh 9 facial skeleton to a living human population, it would probably be most closely aligned with living sub-Saharan Africans, rather than with any other group of modern people. In this I refer only to Qafzeh 9 and not necessarily to the other Qafzeh specimens.

B. Vandermeersch: How do you explain the relation between the Near Eastern Neanderthals and the European ones?

E. Trinkaus: I think they developed out of distantly related but generally similar populations from the earlier Upper Pleistocene.

B. Vandermeersch: Are they of the same origin?

E. Trinkaus: We are talking about a human species which populated most of the temperate Old World by the Middle Pleistocene. Once there was a species so widely distributed, there was inevitably regional variation. Yet there was always some genetic contact between populations.

B. Vandermeersch: With evolved *Homo erectus*?

E. Trinkaus: Not necessarily. I think that we have different concepts as to how evolution works, and this is influencing how we interpret the fossil record.

Y. Rak: I agree with Eric Trinkaus about migration, especially since the first thing to know are the dates. If we do not know the dates, there is no way to say that there was migration from one direction to another, because it is also from one point in time to another. But another question: It seems that to understand the philogeny of this complex is, to a large extent, to provide an explanation for specialization. Even relative dating may be provided by knowing the final product of specialization. Now everybody is talking about specialization, but what is the case of classical Neanderthal?

R. Ray: I should like to ask about the impressions left by the brain.

E. Trinkaus: We can tell very little about internal brain organization from the impressions on the inside of the cranial vault. About all that we can say is that brain size increased relative to body size, and that suggests that there was a general pattern of increasing mental abilities. However, I would point out that relative to body size there was no difference in brain size between the Neanderthals and early anatomically modern humans.

R. Ray: No information can be given by endocranial casts?

E. Trinkaus: Endocranial casts tell us very little about changes in the internal organization of the brain, on the scale of evolution that we are talking about. If we compare the endocranial casts of *Australopithecus* to those of modern humans, the differences on the surface will tell us something. But between archaic *Homo sapiens* and modern *Homo sapiens* they tell us very little, if anything.

A.-M. Tillier: We are sometimes talking about evolved *Homo erectus*, sometimes archaic *Homo sapiens*. I think that we have to choose not on chronological basis but on morphological characters.

E. Wreschner: I would like to ask Dr. Trinkaus a provocative question: does your concept imply that in order to become modern man, hominoids had to pass through the Neanderthal stage?

E. Trinkaus: If the Neanderthal stage is defined as a general phase of human evolution throughout the Old World, and not necessarily represented simply by the Western European form – the answer is yes.

A.J. Jelinek: I think that there is no question but that absolute and relative chronological evidence must underlie our understanding of the significance of morphological variability in man.

B. Vandermeersch: Concerning the relations between Near Eastern and European Neanderthals. In the Near East we have Neanderthals but in Europe we have Neanderthals and also pre-Neanderthals. The pre-Neanderthals are only in Europe, I think. The pre-Neanderthals in Europe have some typical characteristics of the Neanderthals – for instance, the maxillary. I think that the Petralona and Arago skulls have the maxillary like the Neanderthals.

E. Trinkaus: As Jelinek has just said, except for the Galilee specimen from Zuttiyeh we simply do not know what was going on before the Neanderthals in the Near East, and there is obviously disagreement as to the interpretation of the Zuttiyeh morphology. That aside, on a more general level of interpretation, I would argue that during the Middle Pleistocene, especially the late Middle Pleistocene, there is no reason to expect that one would find identical fossils in Europe and the Middle East. Any species with a relatively low population density is going to vary naturally over geographical distances. And we would expect a clinal distribution of morphology across those areas where there are continuous populations from one end to the other. Therefore one would not necessarily find typical Early Neanderthals in the late Middle Pleistocene of the Near East, because those populations were probably connected genetically not only with populations in Europe, but also with populations in Asia and North Africa. Populations move slowly over the length of time that we are talking about.

B. Vandermeersch: So there are no Neanderthals in the Near East. You are obliged to give them another name.

E. Trinkaus: Maybe we should drop the name Neanderthal and refer to them simply as variants of archaic *Homo sapiens*.

6

CONCLUDING DISCUSSION

CONCLUDING DISCUSSION

G. Bosinski: I hope that, in spite of the fact that we are meeting in the Holy Land, we are not necessarily shifting to compassion, but are trying to consider the facts.

My impression is that during Riss glaciation – called the "crucial point" by Derek Roe – there is a change in industry. Things which are, in general terms, considered "Mousterian" – I do not at all like this word, but you understand it – occur for the first time in the Riss. This means a big variability of flake tools. I will concentrate on the Riss material, which comprises at least three or three and a half different types of industries.

1) Assemblages with a few handaxes. Maybe there are assemblages with numerous handaxes as well, but I know of only a few. Besides these few handaxes there is a large development of flake tools, as simple and double side-scrapers. Often, mostly I should say, these flake tools are made with the aid of Levallois technique. So Levallois technique occurs in glacial series together with some late Acheulian handaxes. There are almost no transversal scrapers in these assemblages, a fact linked with the occurrence of Levallois technique; if you have Levallois flake it is not necessary, or even not easy, to make transversal scrapers.

The assemblage I describe is a kind of Acheulian, but if we remove the handaxes, we get a Mousterian, in the broad sense of the word. This material is surely linked with the Acheulian. It had its roots and developed from the Acheulian, and is characterized by flake-tools made by Levallois technique.

2) A second type of assemblages corresponds to the Quina type. The Quina type is known from Riss Glacial, first of all from Western Europe. I may mention, for example, High Lodge in England, and Les Tares in Southern France. This Quina type of "Mousterian" is almost indistinguishable from the Wurmian material. In any case, I am not able to see a difference between La Quina itself, which is of early Wurm and Les Tares, as far as I know it, or the High Lodge Material. This Quina type of Charentian is influenced by the absence of Levallois technique, an absence which is responsible for all the important characters of the Quina type, the equivalent of simple and transversal side-scrapers and the almost total absence of double side-scrapers. It is linked with the technique of flaking that the transversal scrapers are characteristic of the Quina material.

The shape of the Quina tools is clearly linked with the thickness of the flakes. If you have a flat Levallois flake, you will not succeed in obtaining Quina retouche. So the importance of Quina retouche should not be overestimated. It may occur anywhere when the edge of thick flakes is retouched.

The Quina type of early Middle Palaeolithic from Riss looks, at first view, very different from the late Acheulian with Levallois technique and flake tools. In Europe there are no handaxes in the Quina assemblages. But it is not impossible, perhaps it is even likely, that the Quina material derived from some Acheulian too, and it is not excluded that the Quina type had its roots in the Lower Palaeolithic.

The Yabrudian seems a variant of the Quina type. As far as we learned from Farrand, the Yabrudian does not go as far back as the European Quina type. It first occurs in the Interglacial. We learned from several contributions that here, in the Near East, in Tabun and Yabrud, there are levels with handaxes named "Acheulo-Yabrudian". This is a marked difference from the situation in Western Europe.

3) A third type of early Middle Palaeolithic assemblages occurring during Riss is the Ferrassie-type Charentian. Here, again, I am not able to distinguish these series from later ones such as La Ferrassie itself. I tried, with the Rheindahlen material, to work out some differences between the Riss and Wurm assemblages but I did not succeed.

The Ferrassie type of Charentian is characterized by the occurrence of Levallois technique, and all the differences between Quina and Ferrassie are due to the different techniques of flaking. Ferrassie is formed by Levallois technique, Quina by its absence. This explains why the Ferrassie type got only seldom transversal side-scrapers, but an abundance of simple and double side-scrapers.

To a certain degree one could say that the Ferrassie type of early Middle Palaeolithic is similar to the late Acheulian without handaxes. As I mentioned before, if you take away the handaxes from later Acheulian assemblages such as Lebenstedt or Markkleeberg you will get the Ferrassie type. This may point to the roots of the Ferrassie material, possibly from late Acheulian.

The last point – the transition. What may it mean? What is the character of the transition from Lower Palaeolithic to Middle Palaeolithic in the sense of Mousterian? First of all, there is no break at all between Acheulian and the "Mousterian" type of industries.

In this connection, we should perhaps discuss again the importance of Levallois technique, which starts with early Middle Palaeolithic. Beginning with Middle Palaeolithic there is a marked differentiation between the assemblages. In spite of the evolution of handaxe types, the Lower Palaeolithic looks much more uniform than the Middle Palaeolithic. Now there is a marked variation compared with the earlier stages. If we take things as they are known now, we have to mention the geographical component too. The Quina and the Ferrassie types of Riss age are only known from Western and Central Europe. Yabrudian material is later. So perhaps these types originated in Europe, and I come back to one point of my contribution which dealt with the fact that now first Ice Age hunters appear, in the sense of people living in the open Steppe. Perhaps the Quina and Ferrassie types are the answer to new conditions of environment.

If I am asked to compare, or to value, the importance of the transition between Lower and Middle Palaeolithic, I should say that this transition is of about the same importance as the transition from Middle to Upper Palaeolithic.

D.A. Roe: I think I would agree with the general points made by Prof. Bosinski. I do not think this is really the time to undertake any definition, or to attempt an analysis of the whole situation in the different regions, but it seems to me that one can make some general points by way of interpretation, or of summary of the conference. I think, for myself, I have learned both how much we know, and how little we know, at the same time, and on the very same subjects. There are still very many imponderables in the evidence, whether it be in the matter of stone tool industries themselves, or whether it be in their dating.

There is still much flexibility left available to use, and I hope that we will certainly not try to make any very sharp division between Lower and Middle Palaeolithic, believing it to be valid for the whole of the Old World, or, indeed, for any large part of the Old World. There are simply too many uncertainties in the chronology and in the archaeological evidence itself.

For example, we refer with general agreement to the "Riss" Glaciation as being the most important time when all this happened. So I am sure it is, but how many "Riss" Glaciations were there, or how many phases within the one glacial complex should we include under the general name of "Riss"? and how long a time is thereby involved in that phase, that simple short word – "Riss"?

The same is true if we consider the actual archaeological evidence before us. I do not think we yet understand the nature of the stone industries very well. What variations may there be, manufactured by the very same people, either as they move about their territory, or as different seasons of the year succeed each other?

Did they need handaxes for every task, or might those who made handaxes in one part of a region themselves go elsewhere, as it might be, up to the hills, or down to the plain, and make an industry that looked rather different? It would be very difficult for us to tell that such different industries were produced by the very same people.

I know that I keep returning to the importance of microwear analysis, but I think much more needs to be done to understand the different functions of the stone tools in a given tool kit. An example one might take, from the work of Larry Keeley, was at Hoxne in Suffolk, England, where the material is in beautifully fresh condition and could be studied in this way, and he was able to show that two handaxes were both used for meat-cutting tasks. It could be seen that they had cut through meat and had touched also skin and bone. We might, therefore, say that these were butchering implements. But we must not forget that Keeley was able to say exactly the same of some unretouched flakes in the very same industry.

Now this, perhaps, casts a curious light on typology, but it is in typology that we are heavily dependent. At the same time, I always think to myself, as I try to measure handaxes and compare industries or the broader spectra of industries: what is the relationship between the assemblage as it was made and used by people who occupied the site? Two things particularly: how many excavations are complete? What may be left behind under the baulk that the archaeologist leaves in his excavation? Did somebody sit a few meters further in one direction at High Lodge, or whatever site it might be, making handaxes by the dozen, and because it was in one area of the site, and this is unexcavated, we think that we have a site with no handaxes? This is only too possible. It is, I am afraid, very much an imponderable, but it is the kind of thing that we ought to keep in mind when we talk rather too precisely sometimes about particular industry types.

The other aspect of that is: what did the people themselves take away from the site when they left it? One finds many examples, to return again to handaxes, of those implements found completely in isolation.

The best example that I can think of, in that case, is those curious *bout coupé* handaxes of the British Mousterian of Acheulian Tradition, which I mentioned in my own paper. If you take the sites where those occur in any kind of context with the presence of other artifacts, in Britain one would have perhaps ten such sites, but I know that there are more than 100 of those highly distinctive artifacts themselves, occurring in the large majority as stray finds, not necessarily ploughed up on the modern surface, but sometimes by themselves in a late Pleistocene context. And do they not represent the movements of the people themselves over their territory from one site to another, or going back to the Continent, because it is getting too nasty to stay in England, or making their journeys, whatever kind they may be? So there is certainly the circumstance of the people themselves taking away material from the site, but we can only judge what is left for us to find.

These are all really rather negative points, but I am making them with one particular thing in view: to say that we still know very little about the actual mechanisms that form the archaeological sites, although every year we learn of more and more extremely interesting sites and occurrences themselves in different parts of the world. I myself have learned many that are quite new to me, here, this past week.

Technology, again, can be very difficult to understand. The Levallois technique has been mentioned many times. Do we always remember that the Levallois technique in its classic form is one of the most wasteful ways in which it is possible to work with flint? The big Levallois cores that are thrown away as waste products in certain British sites, and I think elsewhere too, may weigh several pounds or even several kilos, and the amount of that block of flint which was actually used to produce a tool is a very small percentage indeed.

What does this imply? It means that you cannot afford to use Levallois technique unless you have abundant supplies of raw material, and that is not true in all parts of the world, or in all parts of the same region. So here again is another aspect that we have to consider when we talk very precisely about technology.

And then a final point along that line, continuing my negative vein of thought, if that is indeed what it is; is it necessary to assume that two industries which, to us, look very similar must imply a similar position in time? Often I have heard, this week, people referring to one of the industries in this part of the world, and comparing it very specifically to Tabun C or B or D, but would Tabun B and C and D always occur in that particular order if we had other sites with long stratigraphies?

One remembers the view of Prof. Bordes in France that, of all those many variants of Mousterian that he distinguishes, almost any of them can occur in almost any order. I think he once said any could occur in any position, but I think he now believes that the Mousterian of Acheulian Tradition Type B always is younger than Type A. But, of the others, I think he still believes, though my French colleagues may wish to correct me on this, that they can occur in any order at all. Now, may this perhaps not also be true of some of the industries in the Levant, which we are using with great faith as chronological markers?

Well, all these things seem to me to indicate that a great degree of flexibility remains in the archaeological evidence and that it would be very unwise to try to make a general definition of where the transition may occur between Lower Palaeolithic, on the one hand, and Middle Palaeolithic, on the other. It is possible that we need some terms that we can use for convenience in discussion, and I see no reason why these should not continue to be used for the moment very much on a regional basis, and that we should not worry ourselves too much if there is no synchronism between regions.

Indeed, I think precise synchronisms between regions, given the different geographical factors to which I have previously referred, are highly unlikely. People must move from one area to another, or ideas must move, or industries must move in whatever way. I am crucially aware, for example, that there is a time gap between the Early Acheulian of East Africa and the Early Acheulian of Southern Britain, which is likely to be greater than one million years, for industries which morphologically and typologically appear to have a very great deal in common.

So, I suppose, to summarize this, my plea would not be to make any very firm decisions at the moment, but to allow local factors to continue to have importance, to continue the research with problems like these in view, and perhaps to remember that almost every sharp boundary line that has ever been proposed by a geologist or a palaeontologist or an archaeologist or a human biologist has been severely questioned within a very few years by his colleagues, and, generally speaking, found, in the end, to be a great nuisance.

A.J. Jelinek: It is clear from my discussion up to now that I believe that all we can do here is explore these problems. Most of us sensed this before we arrived here and hoped that our attempts at explanations would lead us in the direction of understanding and resolving the issues.

Now, as concerns the transition from Lower to the Middle Palaeolithic, I suspect that we all agree it is not a chronological problem. We cannot, as we would do in terms of some problems of chronostratigraphical relationships in geology, set a particular date and say that at 9:00 o'clock in the morning we are no longer in the Lower Palaeolithic. There are some instances where this is useful in relating materials and in understanding particular developments, but this is not that kind of problem, this is a cultural problem. The diversity of worldwide cultural development has led Roe to precisely anticipate the difficulty that we would have in even attempting to formulate sets of cultural criteria with which we and our colleagues would be satisfied for distinguishing between these two kinds of culture. I think that the views of Prof. Bosinski – to draw the line between Lower and Middle Palaeolithic in Europe at the time of the appearance of those three industries in the Riss – reflect our detailed knowledge of the sequence in Western Europe and the beginnings of that kind of knowledge in the Levant. And perhaps we should think more about these kinds of discussions, on the local level, because we will have to understand many local situations before we can understand the meaning of this transition in a global sense.

I would resist any proposal of universal typological-technological criteria for a transition on the basis of the evidence that we have now. We might draw up such a document here, but I think that our colleagues would quickly reject it. I think that, instead, we should look at the nature of this sort of transition in those areas where we understand it best, to see what kinds of things were going on culturally and biologically, so that we can begin to anticipate what we may find in other areas and what we can do with that new evidence. When we talk about these kinds of transitions, we are talking about changes from something that is qualitatively one thing, to something qualitatively different. Most of us in prehistory would agree that there was a dramatic transition, late in the Last Glaciation, or perhaps toward the middle of the Last Glaciation (depending on our terminology) from what we customarily refer to in Europe and the Near East as Middle Palaeolithic to something we recognize as Upper Palaeolithic. With this transition there is no question about the presence of modern man; with this transition we see all manner of qualitatively different forms of behavior such as recognition of self (ornaments), representational and abstract art, and utilization of different kinds of materials in different ways (bone and antler industries). These different patterns indicate an expanding awareness of self and of the world, and an expanding ability to manipulate elements of the surrounding world. With this development we see a virtually geometric progression in the kinds of industries that we can distinguish on the basis of typology as we go through time and across space. Here we see something really qualitatively different. I have tended, in several articles in the past few years, to distinguish between this kind of cultural behavior associated with modern *Homo sapiens* and what I call palaeocultural behavior. The latter is characterized by long periods of stability and very broad regional patterns, such as the similarities pointed out in the Acheulian, by Dr. Roe, from Africa to India. The uniformity of some of these industries is staggering, considering the time periods and geographical ranges involved. Even in those industries that we would all agree are Middle Palaeolithic, it seems to me that the level of innovation and the uniformity of patterning involved appears to be much more closely related to those earlier palaeocultural developments than to the Upper Palaeolithic cultural patterns. So I would say that it appears that the industries at the end of the palaeocultural sequence are antecedent to the industries of modern man, but when we look for transitions in that palaeocultural sequence we are on less firm ground. We can see that at one end of the sequence the level of cultural behavior in the Lower Pleistocene at such sites as Olduvai Gorge is quite distinct from the other end of the sequence, whether we place that in the Riss, Riss/Wurm, or in the early Wurm. As we follow this sequence through time, we see only a slow progression and no big steps reflected in the artifactual record. I think that whatever artifactual criteria are employed we will see a continuum and not a series of steps reflecting rapid change to qualitatively different levels. To divide this sequence of palaeocultural development into Lower and Middle Palaeolithic seems to me to require an arbitrary decision that would be difficult to defend objectively. The obvious transition is between palaeocultural and fully cultural behavior. I think that, because of this, our search for a clear-cut transition between Lower and Middle Palaeolithic may always be a vain one. Perhaps, instead, we should be more concerned with the difference between earlier and later palaeocultural behavior.

We prehistorians have an urge to compartmentalize; we like to think in terms of types. Because we do this we are frequently tempted to draw artificial boundaries, even when the types do not reinforce these boundaries. Our temptation to draw lines on charts may obscure the incredible continuity and stability in the palaeocultural record that reflects a mentality very unlike that of modern *Homo sapiens*. This is tied, in a crucial way, to the appearance of modern man. I think that one of the chief points of significance of the Qafzeh hominids and their association with a Mousterian industry, wherever they may lie in time, may well be the illustration of the necessity for the presence of all the biological qualities of modern man before all of the synthetic reactions could take place in the brain that produce the behavior that we see as fully cultural. I think that Qafzeh may be an example of the presence of modern man in a palaeocultural context, just prior to the development of fully cultural behavior. I think this is one of the most important finds in Levantine prehistory.

This really sums up what I have to say. I know that I have sidestepped a number of issues that, especially, my colleague Bosinski might want to pursue further, but I think that we are not yet ready to talk about these issues, except on a local level. I think that on a local level I would agree with him that in Europe we see some changes in the Riss industries that signal and are typologically antecedent to what is going to follow for the next 100 to 200 thousand years. But I do not think that they reflect as much of a qualitative "jump" as we see when we get to the very clear transition from palaeocultural to fully cultural behavior.

E. Trinkaus: I would just like to comment on one line of palaeontological evidence which was referred to only briefly during our discussions this morning, perhaps because it is rare in most of our Middle Pleistocene samples. This is the postcranial or limb skeleton of archaic *Homo sapiens*. What we find when we look at the postcranial evidence from the earliest fossils of Neanderthals or Neanderthal-grade hominids in the early Upper Pleistocene is a remarkable continuity of form. It is a form which is marked by extreme massiveness, it is a form which does decrease very slowly in massiveness during this time, and it is a form which can be distinguished for many parts of the limb skeleton from the form associated with early anatomically modern humans, including the samples from Skhul and Qafzeh and the early Upper Palaeolithic fossils of western Europe. I think that this is strong biological support for Jelinek's hypothesis that we have a remarkable continuum of behavioral patterns throughout this time period, since it is my belief that anatomy must reflect, in some way, the habitual behaviors of the individual, certainly over this long period of time.

A.J. Jelinek: I think that this might be an appropriate time to once again thank all of our colleagues who suffered through our peculiar language and made things so much easier for all of us who can't do as well with theirs. More important, beyond this, I think it is time that we all join in a unanimous resolution of thanks and appreciation to Prof. Ronen, to his colleagues, to his staff, to the University of Haifa, and the other sponsoring institutions, who brought us together for this remarkable exchange of ideas and I propose that we do so with a standing ovation.

A. Ronen: Thank you very much. It is a great pleasure to have with us Prof. Picard, the Doyen and father of quaternary geology in Israel.

L. Picard: I would not say the father, I would have said the grandfather. I was very pleased to hear your remarks. They reminded me that I came to geology through prehistory. I was born on the shore of the Lake of Constance and my first sampling of objects of nature was, naturally, the lake dwellers; and that went on until finally I turned to geology. But I have never neglected the importance of prehistory to the stratigraphy of the glacials and pluvials of the quaternary period. What I have heard, in the very short time I was here, reminded me indeed that the more we know – the less we know. But this is inspiring – this is not depressing. I could tell you one thing, just what I heard from Prof. Roe: what is still missing in our research of the Middle East, or of the south of the Mediterranean, is a little more stratigraphy. Stratigraphy, not in the sense only of tools or only of palaeontological remains, but a general synthesis of the stratigraphy. You speak of Wurm and Riss; what is Wurm and what is Riss here? Unfortunately, or fortunately, we have no glacial deposits except for a few signs in the high Lebanon. But we do have here, as you all know, what we call the Pluvial and the stratigraphy of the Pluvials has been completely neglected. In the last ten years, we heard constantly of Wurm, rarely of Riss or any other previous period. I would like to tell you: work together with geologists and bring your very important documents together with theirs to a real synthesis. I would like to invite those of you from abroad – come more to this country, it has enormous material still to be done. Many people have written, many people have published, but I am still missing this synthesis. I tried to do it when I first came to this country, in my first work*, but it is still missing and to see it is all I wish.

I congratulate you for what I have heard here. Indeed, it reminds me of the discussions I had as a very young geologist with Abbe Breuil, with Boule, with all those old grandfathers. I really thank you that I could take part here just to hear and see how lively this subject is, what an enormous life there is in the Quaternary research. All that you do, I wish, should remain.

* Picard, L. 1943 *Structure and Evolution of Palestine*. Jerusalem.